IUTAM Bookseries

Volume 36

The IUTAM Bookseries publishes the refereed proceedings of symposia organized by the International Union of Theoretical and Applied Mechanics (IUTAM).

Every two years the IUTAM General Assembly decides on the list of IUTAM Symposia. The Assembly calls upon the advice of the Symposia panels. Proposals for Symposia are made through the Assembly members, the Adhering Organizations, and the Affiliated Organizations, and are submitted online when a call is launched on the IUTAM website.

The IUTAM Symposia are reserved to invited participants. Those wishing to participate in an IUTAM Symposium are therefore advised to contact the Chairman of the Scientific Committee in due time in advance of the meeting. From 1996 to 2010, Kluwer Academic Publishers, now Springer, was the preferred publisher of the refereed proceedings of the IUTAM Symposia. Proceedings have also been published as special issues of appropriate journals. From 2018, this bookseries is again recommended by IUTAM for publication of Symposia proceedings.

Indexed in Ei Compendex and Scopus.

More information about this series at http://www.springer.com/series/7695

Jörg Fehr · Bernard Haasdonk

Editors

IUTAM Symposium on Model Order Reduction of Coupled Systems, Stuttgart, Germany, May 22–25, 2018

MORCOS 2018

 Springer

Editors
Jörg Fehr
Institute of Engineering and
Computational Mechanics
University of Stuttgart
Stuttgart, Baden-Württemberg, Germany

Bernard Haasdonk
Institute of Applied Analysis
and Numerical Simulation
University of Stuttgart
Stuttgart, Baden-Württemberg, Germany

ISSN 1875-3507   ISSN 1875-3493   (electronic)
IUTAM Bookseries
ISBN 978-3-030-21012-0   ISBN 978-3-030-21013-7   (eBook)
https://doi.org/10.1007/978-3-030-21013-7

IUTAM Symposium on "Model Order Reduction of Coupled System" MORCOS 2018

About the Symposium and the Proceedings

For the understanding and development of complex technical systems like automotive/aeronautic systems, bionic systems, mechatronic systems, the human body, civil structures, material modeling, or multiscale systems, an integrated, multiphysics, and multidisciplinary point of view is essential. The combination of different physical domains or different subcomponents can increase functionality, allow optimization, and reduce cost. Nowadays, many problems can be solved by simulation within one physical domain, e.g., by using the well-known finite element method, where the models can have more than 10^7 degrees of freedom, based on the meshing of 3D-data from design or CT-scans. However, for correct prediction, optimization, and control of nowadays complex systems the different simulation domains respectively substructures need to be connected with each other. Frequently, this combination is only possible by using advanced and modern reduced-order models where the large-scale system is approximated with a system of much smaller dimension. Here the most dominant features, input–output behavior, passivity, stability, etc. of the large-scale system are to be retained in the reduced system as much as possible.

The field of model order reduction (MOR) is interdisciplinary as rapid simulation requests are indispensable in all engineering application areas. It was, therefore, the aim of this IUTAM symposium to encourage the interdisciplinary work between researchers from Engineering, Mathematics, and Computer Science to identify, explore, and compare the potentials, challenges, and limitations of recent and new advances. The symposium took place at the University of Stuttgart, Campus Vaihingen, in Stuttgart, Germany, from May 22 to 25, 2018, and was hosted by the Institute of Engineering and Computational Mechanics, the Institute of Applied

Analysis and Numerical Simulation and the Cluster of Excellence in Simulation Technology (SimTech).

For many years, SimTech has been active in these fields and has established numerous cooperations and contacts with international scientists and international companies. Hence, the symposium perfectly fitted to the continuously running MOR-seminar and to the MORML 2016 Workshop on "Data-driven Model Order Reduction and Machine Learning". But also at the Institute of Engineering and Computational Mechanics, this IUTAM symposium nicely resumed the tradition of IUTAM symposia, as this was the sixth IUTAM symposium hosted by this institute after the following:

- the IUTAM symposium on "Nonlinear Dynamics in Engineering Systems" in 1989,
- the IUTAM symposium on "Optimization of Mechanical Systems" in 1995,
- the IUTAM symposium on "Multiscale Problems in Multibody System Contacts" in 2006,
- the IUTAM symposium on "Dynamical Analysis of Multibody Systems with Design Uncertainties" in 2014, and
- the IUTAM symposium on "Advances in Biomechanics of Hearing" in 2016.

The IUTAM symposium was supervised by the following international Scientific Committee: Francisco Chinesta (France), Jörg Fehr (Germany, Chairman), Bernard Haasdonk (Germany, Co-Chairman), Gianluigi Rozza (Italy), Anthony T. Patera (USA), Wil Schilders (Netherlands), Taichi Shiiba (Japan), and Peter Eberhard (Germany, IUTAM Representative). Four keynote presentations were given by the following:

- Kathrin Smetana (University of Twente, Netherlands),
- Olivier Brüls (University of Liege, Belgium),
- David Knezevic (Akselos, Cambridge, USA), and
- Tommaso Tamarozzi (KU Leuven/Siemens PLM, Belgium).

From the abstracts submitted for the symposium, 37 papers had been selected for oral presentation. Furthermore, 10 additional poster presentations where given including a poster flash. In these 47 contributions, many aspects related to model order reduction were discussed. The importance of linking different physical domains by using MOR techniques was discussed. The contributions clearly showed the increase of functionality, the advanced possibilities for optimization, and the cost reduction. The detailed scientific program of the symposium was as follows:

Scientific Program

Tuesday, May 22, 2018

Session 1—Chair: Prof. Dr. B. Haasdonk	
Advances in Reduced Order Methods for Computational Fluid Dynamics Problems in Applied Sciences and Engineering: Perspectives *Gianluigi Rozza**	09:30–10:00
POD-Based Economic Model Predictive Control of Heat Convection Phenomena *Luca Mechelli*, Stefan Volkwein*	10:00–10:30
Fully online ROMs based on LUPOD *Maria-Luisa Rapuń, Filippo Terragni, José M. Vega**	10:30–11:00
Randomized Generation of Localized Approximation Spaces for Parameterized Partial Differential Equations *Andreas Buhr*, Kathrin Smetana*	12:15–12:45
Session 2—Chair: Dr. C. Himpe	
Model Order Reduction of Coupled, Parametrized Elastic Bodies for Shape Optimization *Benjamin Fröhlich*, Florian Geiger, Jan Gade, Manfred Bischoff, Peter Eberhard*	11:30–12:00
Basis Selection for Non-Intrusive Modal Substructuring of Geometric Nonlinear Finite Element Models *Morteza K. Mahdiabadi*, Daniel J. Rixen*	12:00–12:30
Towards an Input-Aware System-Theoretic Model Order Reduction Approach for Nonlinear Systems *Björn Liljegren-Sailer*, Nicole Marheineke*	12:30–13:00
Keynote Presentation 1—Chair: Prof. W. Schilders	
Optimal Interface Reduction for Static Condensation or Substructuring *Kathrin Smetana**	14:15–15:00
Session 3—Chair: Prof. Gianluigi Rozza	
Hybrid Hyper-Reduced Modeling for Contact Problems in Elastostatics *Jules Fauque*, Isabelle Ramiére, David Ryckelynck*	15:00–15:30
A Novel Penalty Based Reduced Order Modelling Method for Dynamic Analysis of Jointed Structures with Localized Nonlinearities *Jie Yuan*, Loic Salles*	15:30–16:00
Session 4—Chair: Prof. K. Smetana	
Model Order Reduction for Drilling Automation *Harshit Bansal*, Laura Iapichino, Wil H.A. Schilders, Nathan van de Wouw*	16:30–17:00
Structured Cross-Covariance-Based Model Reduction Applied to Gas Network Models *Peter Benner, Sara Grundel, Christian Himpe**	17:00–17:30
Poster Flash and Poster Presentation—Chair: Prof. P. Eberhard	
Reduced-Order Modelling and Computational Homogenisation in Magnetomechanics *Benjamin Brands*, Julia Mergheim, Paul Steinmann*	18:30–18:32

(continued)

(continued)

Wednesday, May 23, 2018

(continued)

(continued)

Session 6—Chair: Dr. S. Rave	
Randomized Generation of Localized Approximation Spaces for Parameterized Partial Differential Equations *Andreas Buhr*, Kathrin Smetana*	12:15–12:45
Session 7—Chair: Prof. H. Matthies	
Smart Sparse Sampling *Rubén Ibáñez-Pinillo*, Emmanuelle Abisset-Chavanne, Elías Cueto, Francisco Chinesta*	14:00–14:30
A Reduced Model Approach for the Optimal Control of Dielectric Elastomer Actuated Systems *Tristan Schlögl, Sigrid Leyendecker**	14:30–15:00
Two-Stage Data-Assisted Mechanical Homogenization *Felix Fritzen, Oliver Kunc**	15:00–15:30
Session 8—Chair: Dr. D. Knezevic	
Reduced Order Models Using a Data-Driven and Equation-Free Method *Soledad Le Clainche*, José M. Vega*	16:00–16:30
Proper Orthogonal Decomposition (POD) Combined with Hierarchical Tensor Approximation (HTA) in the Context of Uncertain Parameters *Steffen Kastian*, Stefanie Reese, Dieter Moser, Lars Grasedyck*	16:30–17:00
Parameterised Reduced Order Models *Hermann G. Matthies*, Roger Ohayon*	17:00–17:30

Thursday, May 24, 2018

Keynote Presentation 3—Chair: Prof. T. Shiiba	
Component-Based Model Reduction for Industrial-Scale Problems *David J. Knezevic*	09:00–09:45
Session 9—Chair: Dr. F. Fritzen	
Online-Adaptive Localized Reduced Basis Approximation of Parameterized Parabolic Problems *Mario Ohlberger, Stephan Rave*, Felix Schindler*	09:45–10:15
Experimental Dynamic Substructuring on a 3 MW Wind Turbine *Andreas Schulze*, János Zierath, Roman Rachholz, Reik Bockhahn, Sven-Erik Rosenow, Johannes Luthe, Christoph Woernle*	10:15–10:45
Session 10—Chair: Prof. O. Brüls	
Recent Advances on Nonlinear Vibration Analysis Using Nonlinear Modes as Reduced Basis *Malte Krack*, Johann Groß, Maren Scheel*	11:15–11:45
An Open Source Description for (Semi-)Automatic Generation and Model Reduction of Machine Tool Network Models *Norman Lang, Andreas Naumann, Jens Saak*, Stefan Sauerzapf*	11:45–12:15

Friday, May 25, 2018

Keynote Presentation 4—Chair: Prof. W. Schiehlen	
Nonlinear Projection Methods for Mechanical Structures and Systems *Olivier Brüls**	09:00–09:45
Session 11—Chair: Dr. F. Naets	
Combined Frequency-Time Reduction Methods for Calculating Periodic Solutions of Unilaterally Constrained Systems *Frederic Schreyer, Remco I. Leine**	09:45–10:15
Model Order Reduction of Linear Switched Systems with Constrained Switching *Ion V. Gosea*, Igor P. Duff, Peter Benner, Athanasios C. Antoulas*	10:15–10:45
Session 12—Chair: Dr. T. Tamarozzi	
Index-Aware MOR for Gas Transport Networks *Nicodemus Banagaaya*, Sara Grundel, Peter Benner*	11:15–11:45
Efficient Analysis of Impact Between Reduced Flexible Bodies *Stephan Tschigg*, Pascal Ziegler, Robert Seifried*	11:45–12:15
Two-Stage Parametric Model Order Reduction for the Design Optimization of a Coupled Structural and Controller Model *Frank Naets*, Wim Desmet*	12:15–12:45
Session 13—Chair: Dr. J. Saak	
Order-Reduction for Magneto-Quasistatic Fields Including Magnetic Material Characteristics of Saturation Type *Daniel Klis, Rolf Baltes*, Romanus Dyczij-Edlinger*	14:00–14:30
A Model Order Reduction Method for Electro-Magnetic Vibration Analysis of Electric Motors *Akira Saito**	14:30–15:00
Magnetically Levitated Vehicles: Coupling Multibody and Mechatronic Systems with Elastic Structures Subject to Model Order Reduction *Werner Schiehlen**	15:00–15:30

The symposium had 56 participants from the following 11 countries: Austria, Belgium, France, Germany, Italy, Japan, the Netherlands, Nigeria, Spain, United Kingdom, and the United States. In the 47 presentations, given in the course of the symposium, many application examples from civil structures to automotive and aeronautic systems, or the efficient simulation of gas networks. Applications from industry included the improved control via MOR of laser machines from TRUMPF, one of the many companies around Stuttgart.

The scientific achievements and progress of this IUTAM symposium were significant and substantial. The symposium was successful in bringing together mathematicians, engineers, and computer scientist with practical and theoretical background and to encourage and strengthen their interdisciplinary work. Furthermore, it helped to create a link and a mutual understanding between the different research communities.

The scientific articles in this volume are reflecting this range of topics and shall provide an insight into the different issues addressed at the symposium. The order of the paper is as follows. We start with an article about a keynote presentation given at the symposium. Kathrin Smetana reports about port/interface reduction which is of uttermost importance for the coupling of systems. Then, the more methodically oriented papers are presented. The work of Matthies and Ohayon provides a unifying formulation for parametric models of coupled system in a functional analytic view. The next paper authored by Gosea et al. deals with model reduction of switched system. The work of Le Clainche and Vega is concerned with the discovery of the underlying physics in spatiotemperal data via an equation-free method. For nonlinear systems, the proper orthogonal decomposition (POD) of snapshots of offline simulations is one of the most common used MOR techniques. Kastian and Reese present in their paper an adaptive procedure of how to choose significant snapshots. The work of Rapun et al. handles the acceleration of time-dependent solvers for nonlinear systems by a fully online reduced-order model, using a POD on the fly and a collocation method called LUPOD. The final work in the method section is the work of Bhatt et al. which considers the approximation error introduced by MOR and provides two methods for estimating the error in the time domain.

After the methods section, various examples from the application domains are presented. The papers by Karatzas et al. and Jehle et al. address the efficient simulation of heat transport with the help of MOR techniques. In the work of Karatzas et al., the Shifted Boundary Method is applied for a heat exchange problem, whereas in Jehle et al. different manners to compute the POD snapshots are compared to solve an optimal boundary control problem of a heat equation with convection. The paper by Toth and Kaltenbacher discusses the coupling of incompressible free-surface flow, acoustic fluids, and flexible structures via a modal basis. Parametrized model order reduction is considered in the paper by Fröhlich et al. for shape optimization of static components. Another example from mechanical engineering/aerospace engineering is the paper of Yuan et al. where jointed structures with localized nonlinearities are analyzed with the help of a combination of vibration modes, static modes, and trial vector derivatives. In the paper by Emamy et al., the nonlinear POD-DEIM technique is applied to a 0D/1D model used to simulate the propagation of action potentials through the myocardium or along skeletal muscle fibers. The paper by Banagaaya et al. uses index-aware MOR techniques for the efficient simulation of gas networks. A reduced-order finite element approach is used to speed up the stability analysis of a milling process in the paper by Ozoegwu.

To conclude this preface we want to gratefully acknowledge the financial support of this IUTAM symposium by a IUTAM grant as well as additional funding by the German Research Foundation (DFG, grant FE1583/2-1), SimTech, the University of Stuttgart, and the Robert Bosch GmbH.

Stuttgart, Germany   Jörg Fehr
January 2019   joerg.fehr@itm.uni-stuttgart.de

                                                  Bernard Haasdonk
                            haasdonk@mathematik.uni-stuttgart.de
                                                  Symposium Chairs

Contents

Static Condensation Optimal Port/Interface Reduction and Error Estimation for Structural Health Monitoring

Kathrin Smetana

Abstract Having the application in structural health monitoring in mind, we propose reduced port spaces that exhibit an exponential convergence for static condensation procedures on structures with changing geometries for instance induced by newly detected defects. Those reduced port spaces generalize the port spaces introduced in [K. Smetana and A.T. Patera, SIAM J. Sci. Comput., 2016] to geometry changes and are optimal in the sense that they minimize the approximation error among all port spaces of the same dimension. Moreover, we show numerically that we can reuse port spaces that are constructed on a certain geometry also for the static condensation approximation on a significantly different geometry, making the optimal port spaces well suited for use in structural health monitoring.

Keywords Interface reduction · Model order reduction · (component-based) static condensation · Substructuring · Component mode synthesis

1 Introduction

Manual or automated inspection of large structures such as offshore platforms is carried out on a regular basis; the effects of any detected defects must be assessed rapidly in order to avoid further damage or even catastrophic failure. This can be facilitated by relying on numerical simulations. One step towards a fast numerical simulation response for such large structures is to exploit their natural decomposition into components and apply static condensation to obtain a (Schur complement) system of the size of the degrees of freedom (DOFs) on all interfaces or ports in the system. However, as the size of this Schur complement system may still be very large

K. Smetana (✉)
Department of Applied Mathematics, University of Twente, P.O. Box 217,
7500 AE Enschede, The Netherlands
e-mail: k.smetana@utwente.nl

Department of Mechanical Engineering, Massachusetts Institute of Technology,
Cambridge, MA 02139, USA

© Springer Nature Switzerland AG 2020
J. Fehr and B. Haasdonk (eds.), *IUTAM Symposium on Model Order Reduction of Coupled Systems, Stuttgart, Germany, May 22–25, 2018*, IUTAM Bookseries 36, https://doi.org/10.1007/978-3-030-21013-7_1

it is vital to reduce the number of DOFs on the interfaces or ports and thus consider reduced interface or port spaces.

In the popular component mode synthesis (CMS) approach [3, 4, 10, 12, 15] this reduced space is spanned via certain eigenmodes. In [13] generalized Legendre polynomials are used and in [9] deformation patterns from an analysis of the assembled structure are employed. Moreover, local reduced models are generated from parametrized Lagrange or Fourier modes and coupled via FE basis functions in [14]. Finally, empirical modes generated from local solutions of the PDE are suggested in [5, 7, 20].

Recently, port spaces that are optimal in the sense of Kolmogorov and thus minimize the approximation error among all port spaces of the same dimension have been introduced in [24]. The approach in [24] generalizes the idea of separation of variables by connecting two components at the port for which we wish to construct the port space and consider the space of all local solutions of the partial differential equation (PDE) with arbitrary Dirichlet boundary conditions on the ports that lie on the boundary of the two-component system. From separation of variables we anticipate an exponential decay (of the higher modes) of the Dirichlet boundary conditions to the interior of the system. To quantify which information of the Dirichlet boundary conditions reaches the shared port of the system, a (compact) transfer operator that acts on the space of local solutions of the PDE is introduced. Solving the transfer eigenproblem for the composition of the transfer operator and its adjoint yields the optimal space. For related work in the context of the generalized finite element method we refer to [1, 2].

In [6] it has been shown that by employing methods from randomized numerical linear algebra an extremely accurate approximation of those optimal port spaces can be computed in close to optimal computational complexity. To account for variations in a material or geometric parameter in [24] a *parameter-independent* port space is generated from the optimal parameter-dependent port spaces via a spectral greedy algorithm. It is further numerically demonstrated in [24] that the optimal port spaces often outperform other approaches such as Legendre polynomials [13] or empirical modes [7]; also an exponential convergence of the approximation can be observed. Finally, those optimal port modes have been used in structural integrity management of offshore structures in [17] and optimal local approximation have been exploited in the context of data assimilation in [26].

In this article we want to investigate the applicability of optimal port spaces for structural health monitoring and more specifically extend the concept of [24] to geometry changes. First, we show how to construct one port space for several different geometries such as a beam and a beam with a crack or hole via a spectral greedy algorithm. Moreover, if during an inspection a defect is detected, unfortunately, often the precise geometry of the newly detected defect is not amongst the component geometries the reduced model has been trained for. Therefore, we demonstrate numerically that for realistic error tolerances the optimal port spaces constructed on one geometry can often be reused on another. In order to assess whether the resulting reduced model is accurate enough we suggest to employ the error estimator for port reduction introduced in [23] as this error estimator is both an upper and lower bound of the error and based on local error indicators associated with the ports; also the latter are a lower bound of the local error on the component pair that shares the respective port. Error estimation for port or interface reduction has also been considered in [5, 7, 15]. Finally, we note that also in [5] local reduced order models for geometry changes are suggested. However, the authors of [5] neither reuse existing reduced models nor build one reduced model for different geometries.

The remainder of this paper is organized as follows. In Sect. 2 we introduce the problem setting and recall the algebraic (port reduced) static condensation procedure. Subsequently, we recall the optimal port spaces introduced in [24] in Sect. 3. In Sect. 4 we propose quasi-optimal port spaces for parametrized problems including geometry changes such as from a beam to a beam with a crack. Subsequently, we discuss in Sect. 5 how to deal with systems with many components and recall in Sect. 6 the a posteriori error estimator from [23]. Finally, we present numerical experiments in Sect. 7 and draw some conclusions in Sect. 8.

2 Preliminaries

2.1 Problem Setting

Let $\Omega_{gl} \subset \mathbb{R}^d$, $d = 2, 3$, be a large, bounded domain with Lipschitz boundary and assume that $\partial \Omega_{gl} = \Sigma_D \cup \Sigma_N$, where Σ_D denotes the Dirichlet and Σ_N the Neumann boundary, respectively. We consider a linear, elliptic PDE on Ω_{gl} with solution u_{gl}, where u_{gl} equals g_D on Σ_D and satisfies homogeneous Neumann boundary conditions on Σ_N noting that the extension to non-homogenous Neumann boundary conditions is straightforward.

To compute an approximation of u_{gl} we decompose the large domain Ω_{gl} into (many) non-overlapping subdomains. To simplify the presentation we consider henceforth two subdomains Ω_1, $\Omega_2 \subset \Omega_{gl}$ and their union Ω with $\bar{\Omega} = \bar{\Omega}_1 \cup \bar{\Omega}_2$ as illustrated in Fig. 1; the approximation of the whole system associated with Ω_{gl} will be discussed in Sect. 5. Moreover, we introduce the shared interface $\Gamma_{in} := \bar{\Omega}_1 \cap \bar{\Omega}_2$ and $\Gamma_{out} := \partial \Omega \setminus \partial \Omega_{gl}$.

We consider the following problem on Ω: For given f find u such that

$$\mathscr{A} u = f \text{ in } \Omega, \quad \text{and} \quad u = u_{gl} \text{ on } \Gamma_{out}, \tag{1}$$

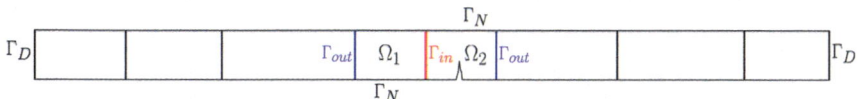

Fig. 1 Illustration of Ω_{gl}, Ω, defined as $\bar{\Omega} = \bar{\Omega}_1 \cup \bar{\Omega}_2$, and the ports Γ_{in} and Γ_{out}

where $\mathscr{A}$ is a linear, elliptic, and continuous differential operator. We may then introduce a conforming Finite Element (FE) discretization and a FE approximation u whose FE coefficients $\underline{u} \in \mathbb{R}^N$ solve the following linear system of equations

$$\underline{A}\underline{u} = \underline{f}. \tag{2}$$

Here, $\underline{A} \in \mathbb{R}^{N \times N}$ discretizes the (weak form of the) differential operator $\mathscr{A}$ and $\underline{f} \in \mathbb{R}^N$ accounts for the discretization both of f and enforcing the non-homogeneous Dirichlet boundary conditions $u_{gl}|_{\Gamma_{out}}$; we assume that in the rows associated with the Dirichlet DOFs the non-diagonal entries are zero and the diagonal entries equal one.

2.2 Static Condensation

To obtain a linear system of equations of the size of the number of DOFs N_{in} on the interface Γ_{in} we perform static condensation. To that end, we first sort the DOFs in DOFs associated with Ω_1, Ω_2, and Γ_{in} to rewrite (2) as follows:

$$\begin{bmatrix} \underline{A}_{\Gamma_{in}} & \underline{A}_{\Gamma_{in},\Omega_1}^T & \underline{A}_{\Gamma_{in},\Omega_2}^T \\ \underline{A}_{\Gamma_{in},\Omega_1} & \underline{A}_{\Omega_1} & 0 \\ \underline{A}_{\Gamma_{in},\Omega_2} & 0 & \underline{A}_{\Omega_2} \end{bmatrix} \begin{bmatrix} \underline{u}_{\Gamma_{in}} \\ \underline{u}_{\Omega_1} \\ \underline{u}_{\Omega_2} \end{bmatrix} = \begin{bmatrix} \underline{f}_{\Gamma_{in}} \\ \underline{f}_{\Omega_1} \\ \underline{f}_{\Omega_2} \end{bmatrix}. \tag{3}$$

We may then apply static condensation to remove the DOFs corresponding to the interior of Ω_1 and Ω_2: We define the Schur complement matrix and the Schur complement right-hand side as

$$\underline{A}_{SC} = \underline{A}_{\Gamma_{in}} - \underline{A}_{\Gamma_{in},\Omega_1}^T \underline{A}_{\Omega_1}^{-1} \underline{A}_{\Gamma_{in},\Omega_1} - \underline{A}_{\Gamma_{in},\Omega_2}^T \underline{A}_{\Omega_2}^{-1} \underline{A}_{\Gamma_{in},\Omega_2} \quad \in \mathbb{R}^{N_{in} \times N_{in}} \tag{4}$$

$$\underline{f}_{SC} = \underline{f}_{\Gamma_{in}} - \underline{A}_{\Gamma_{in},\Omega_1}^T \underline{A}_{\Omega_1}^{-1} \underline{f}_{\Omega_1} - \underline{A}_{\Gamma_{in},\Omega_2}^T \underline{A}_{\Omega_2}^{-1} \underline{f}_{\Omega_2} \quad \in \mathbb{R}^{N_{in}} \tag{5}$$

such that the vector of interface coefficients solves the *Schur complement system*

$$\underline{A}_{SC}\underline{u}_{\Gamma_{in}} = \underline{f}_{SC} \quad \text{of size } N_{in} \times N_{in}. \tag{6}$$

We note that computing $\underline{A}_{\Omega_i}^{-1} \underline{A}_{\Gamma_{in},\Omega_i}$ corresponds to solving the PDE on Ω_i, $i = 1, 2$, N_{in} times with homogeneous Dirichlet boundary conditions on Γ_{in} and right-hand sides that occur from lifting the respective N_{in} FE basis functions on Γ_{in}.

2.3   Port or Interface Reduction

As indicated in the introduction, unfortunately, for many real-world applications the size of (6) is still too large, such that a further reduction in size is desirable. We assume that we have a reduced basis $\underline{\phi}_1, \ldots, \underline{\phi}_n \in \mathbb{R}^{N_{in}}$, $n \ll N_{in}$, at our disposal, which we store in the columns of a matrix $\underline{\Phi}_n \in \mathbb{R}^{N_{in} \times n}$. We may then introduce a *port reduced static condensation approximation* [7] u^n with FE coefficients $\underline{u}^n \in \mathbb{R}^N$, where the coefficients on the interface $\underline{u}_{\Gamma_{in}}^n$ satisfy the reduced Schur complement system

$$\underline{\Phi}_n^T \underline{A}_{SC} \underline{\Phi}_n \underline{u}_{\Gamma_{in}}^n = \underline{\Phi}_n^T \underline{f}_{SC} \qquad \text{of size } n \times n \tag{7}$$

and the DOFs of $\underline{u}^n$ in the interior of $\Omega_i, i = 1, 2$ can be obtained in a standard manner via the definition of $\underline{A}_{SC}$ and $\underline{f}_{SC}$. The question of how to construct a reduced basis $\underline{\phi}_1, \ldots, \underline{\phi}_n \in \mathbb{R}^{N_{in}}$ which yields a rapidly convergent approximation and is even in some sense optimal will be addressed in the next section.

3   Optimal Port Spaces

Rather than assuming a priori knowledge about the shape of the global system associated with Ω_{gl}, we wish to enable maximum flexibility in terms of system assembly on the user's side. In other words, we wish to supply the user with many components (or subdomains), each equipped with (local) reduced models, which the user can then use to build the desired system and thus implicitly define Ω_{gl}. As a consequence, due to the a priori unknown geometry of Ω_{gl}, we assume that the trace of the global solution u_{gl} on Γ_{out} is *unknown* to us when constructing the reduced basis $\underline{\phi}_1, \ldots, \underline{\phi}_n \in \mathbb{R}^{N_{in}}$. We thus aim at approximating all local solutions of (1) with *arbitrary Dirichlet boundary conditions* on Γ_{out}. Before presenting the construction of the reduced basis in Sect. 3.1 we illustrate in a motivating example taken from [24, Remark 3.3] why we may hope to be able to find a low-dimensional port space that approximates the set of all local solutions well.

Remark 1 We consider two components $\Omega_i \subset \mathbb{R}^2$, $i = 1, 2$ each of height H in x_2 and length L in x_1, such that Γ_{out} is at $x_1 = -L$ and $x_1 = L$ and Γ_{in} is at $x_1 = 0$. We consider the Laplacian and impose homogeneous Neumann conditions on $x_2 = 0$ and $x_2 = H$ in both subdomains. Proceeding with separation of variables, we can infer that all local solutions of the PDE for this problem are of the form

$$u(x_1, x_2) = a_0 + b_0 x_1 + \sum_{n=1}^{\infty} \cos(n\pi \frac{x_2}{H}) \left[a_n \cosh(n\pi \frac{x_1}{H}) + b_n \sinh(n\pi \frac{x_1}{H}) \right], \tag{8}$$

where the coefficients $a_n, b_n \in \mathbb{R}, n = 0, \ldots, \infty$ are determined by the Dirichlet data on Γ_{out}. Thanks to the cosh function we can observe a very rapid and exponential decay of the local solutions (8) in the interior of Ω. Therefore, most of the local solutions (8) have negligibly small values on Γ_{in}, which is why we expect a low-dimensional port space on Γ_{in} to be able to provide a very good approximation of all local solutions (8). The construction procedure described below generalizes the separation of variables ansatz.

3.1 Construction of Optimal Port Spaces via a Transfer Operator

First, we address the case $f = 0$; the general case will be dealt with at the end of this subsection. Motivated by the separation of variables procedure, and the fact that the global solution u_{gl} on Ω_{gl} satisfies the PDE locally on Ω, we consider the space of all local solutions of the PDE

$$\mathcal{H} := \{w : \mathcal{A}w = 0 \text{ in } \Omega, \ w = 0 \text{ on } \Sigma_D \cap \partial\Omega\}. \tag{9}$$

As in [2, 6, 24] we may then introduce a *transfer operator* $\mathcal{T} : \mathcal{S} \to \mathcal{R}$ that takes arbitrary data ζ on Γ_{out} as an input, solves the PDE $\mathcal{A}u = 0$ on Ω with that data ζ as Dirichlet boundary conditions on Γ_{out}, and finally restricts the local solution to Γ_{in}. Introducing the source and range spaces $\mathcal{S} := \{w|_{\Gamma_{out}} : w \in \mathcal{H}\}$ and $\mathcal{R} := \{(w - P_{\ker(\mathcal{A})}(w))|_{\Gamma_{in}} : w \in \mathcal{H}\}$ the transfer operator is thus defined as

$$\mathcal{T}(w|_{\Gamma_{out}}) = \left(w - P_{\ker(\mathcal{A})}(w)\right)\big|_{\Gamma_{in}} \qquad \text{for } w \in \mathcal{H}. \tag{10}$$

Here, $P_{\ker(\mathcal{A})}(w)$ denotes the orthogonal projection of w on the kernel of the differential operator. Note that for instance for the Laplacian $\ker(\mathcal{A})$ equals the constant functions and in the case of linear elasticity $\ker(\mathcal{A})$ is the space of the rigid body motions. Following up Remark 1 note that the transfer operator allows us to assess how much of the data on Γ_{out} reaches the inner interface Γ_{in}. It can then be shown that thanks to the Caccioppoli inequality $\mathcal{T}$ is compact and that certain eigenfunctions of $\mathcal{T}^*\mathcal{T}$ span the optimal port space, where $\mathcal{T}^* : \mathcal{R} \to \mathcal{S}$ denotes the adjoint operator (see [2, 21, 24] for details). Here, we use the concept of optimality in the sense of Kolmogorov [19]: A subspace $\mathcal{R}^n \subset \mathcal{R}$ of dimension at most n for which holds

$$d_n(\mathcal{T}(\mathcal{S}); \mathcal{R}) = \sup_{\psi \in \mathcal{S}} \inf_{\zeta \in \mathcal{R}^n} \frac{\|\mathcal{T}\psi - \zeta\|_{\mathcal{R}}}{\|\psi\|_{\mathcal{S}}}$$

is called an optimal subspace for $d_n(\mathcal{T}(\mathcal{S}); \mathcal{R})$, where the Kolmogorov n-width $d_n(\mathcal{T}(\mathcal{S}); \mathcal{R})$ is defined as

$$d_n(\mathcal{T}(\mathcal{S}); \mathcal{R}) := \inf_{\substack{\mathcal{R}^n \subset \mathcal{R} \\ \dim(\mathcal{R}^n)=n}} \sup_{\psi \in \mathcal{S}} \inf_{\zeta \in \mathcal{R}^n} \frac{\|\mathcal{T}\psi - \zeta\|_{\mathcal{R}}}{\|\psi\|_{\mathcal{S}}}.$$

We summarize the findings about the optimal port spaces in the following proposition.

Proposition 1 (Optimal port spaces) *[24]) The optimal port space is given by*

$$\mathcal{R}^n := \mathrm{span}\{\phi_1^{sp}, \ldots, \phi_n^{sp}\}, \qquad \text{where } \phi_j^{sp} = \mathcal{T}\varphi_j, \quad j = 1, \ldots, n, \qquad (11)$$

and λ_j are the largest n eigenvalues and φ_j the corresponding eigenfunctions that satisfy the transfer eigenvalue problem: *Find $(\varphi_j, \lambda_j) \in (\mathcal{S}, \mathbb{R}^+)$ such that*

$$(\mathcal{T}\varphi_j, \mathcal{T}w)_{\mathcal{R}} = \lambda_j(\varphi_j, w)_{\mathcal{S}} \quad \forall w \in \mathcal{S}. \qquad (12)$$

Moreover, the following holds:

$$d_n(\mathcal{T}(\mathcal{S}); \mathcal{R}) = \sup_{\xi \in \mathcal{S}} \inf_{\zeta \in \mathcal{R}^n} \frac{\|\mathcal{T}\xi - \zeta\|_{\mathcal{R}}}{\|\xi\|_{\mathcal{S}}} = \sqrt{\lambda_{n+1}}. \qquad (13)$$

Remark 2 We note that, as can be seen from (12), the optimal modes are those that maximize the energy on the inner interface Γ_{in} relative to the energy they have on Γ_{out}. The optimal port space is thus spanned by the modes that relatively still contain the most information on Γ_{in}. For our motivating example discussed in Remark 1 we obtain $\mathcal{R}^n := \mathrm{span}\{\cos(\pi\frac{x_2}{H}), \cos(2\pi\frac{x_2}{H}), \ldots, \cos(n\pi\frac{x_2}{H})\}$. Moreover, we can exploit the separation of variables solution to solve (12) in closed form: $\lambda_j = (\cosh(L\sigma_{j-1}))^{-2}$, $j = 1, 2, 3, \ldots$, where the eigenproblem in x_2 in the separation of variables procedure yields separation constants $\sigma_j = (j\pi)/H$, $j = 0, 1, 2, \ldots$. This simple model problem also foreshadows the potentially very good performance of the associated optimal space (11) in light of (13) and Proposition 2: we obtain exponential convergence.

For $f \neq 0$ we solve the problem: Find u^f such that

$$\mathcal{A}u^f = f \quad \text{in } \Omega \qquad \text{and} \qquad u = 0 \quad \text{on } \Gamma_{out}$$

and augment the space $\mathcal{R}^n$ with $u^f|_{\Gamma_{in}}$ to arrive at

$$\mathcal{R}^n_{data,ker} := \mathrm{span}\{\phi_1^{sp}, \ldots, \phi_n^{sp}, u^f|_{\Gamma_{in}}, \eta_1|_{\Gamma_{in}}, \ldots, \eta_{\dim(\ker(\mathcal{A}))}|_{\Gamma_{in}}\}, \qquad (14)$$

where $\{\eta_1, \ldots, \eta_{\dim(\ker(\mathcal{A}))}\}$ denotes a basis for $\ker(\mathcal{A})$.

Using the optimal port space $\mathcal{R}^n_{data,ker}$ within the static condensation procedure allows proving the following *a priori* error bound for the static condensation approximation. We note that Proposition 2 gives a bound for the continuous analog on u^n of u^n, the latter being defined in Sect. 2.3. To simplify the notation we do not give a precise definition of u^n and refer to that end to [24]. Note however, that the convergence behavior of u^n towards u is very similar to the continuous setting, differing only due to the FE approximation.

Proposition 2 (A priori error bound) *[24]) Let u be the (exact) solution of* (1) *and* u^n *the continuous static condensation approximation employing the optimal port space* $\mathscr{R}^n_{data,ker}$. *Moreover, denote with* $\|\cdot\|_{\mathscr{E}}$ *the norm induced by the bilinear form associated with the differential operator* $\mathscr{A}$. *Then we have the following a priori error bound:*

$$\frac{\|u - u^n\|_{\mathscr{E}}}{\|u\|_{\mathscr{E}}} \leq C_1(\Omega)\sqrt{\lambda_{n+1}}, \tag{15}$$

where λ_{n+1} *is the* $n + 1$*th eigenvalue of* (12) *and* $C_1(\Omega)$ *is a constant which depends neither on u nor on* u^n.

3.2 Approximation of the Optimal Spaces

In this subsection we show how an approximation of the continuous optimal local spaces $\mathscr{R}^n_{data,ker}$ can be computed with the FE method. First, in order to define a matrix form of the transfer operator we introduce DOF mappings $\underline{D}_{\Gamma_{out}\to\Omega} \in \mathbb{R}^{N\times N_{out}}$ and $\underline{D}_{\Omega\to\Gamma_{in}} \in \mathbb{R}^{N_{in}\times N}$ that map the DOFs on Γ_{out} to the DOFs of Ω and the DOFs of Ω to the DOFs of Γ_{in}, respectively; N_{out} denotes the number of DOFs on Γ_{out}. By denoting with $\underline{\zeta} \in \mathbb{R}^{N_{out}}$ the coefficients of a FE function ζ on Γ_{out} and denoting by $\underline{K}_\Omega$ the matrix of the orthogonal projection $P_{ker(\mathscr{A}),\Omega}$ on $ker(\mathscr{A})$ on Ω we obtain the following matrix representation $\underline{T} \in \mathbb{R}^{N_{in}\times N_{out}}$ of the transfer operator:

$$\underline{T}\,\underline{\zeta} = \underline{D}_{\Omega\to\Gamma_{in}}\left(1 - \underline{K}_\Omega\right)\underline{A}^{-1}\underline{D}_{\Gamma_{out}\to\Omega}\,\underline{\zeta}. \tag{16}$$

Finally, we denote by $\underline{M}_S$ the inner product matrix of the FE source space S and by $\underline{M}_R$ the inner product matrix of the FE range space R. Possible inner products for S and R are the L^2-inner product and a lifting inner product. To obtain the latter we solve for instance for a function ξ defined on Γ_{in} the PDE on Ω_1 and Ω_2 numerically with Dirichlet data ξ on Γ_{in} and homogeneous Dirichlet boundary conditions on Γ_{out}—for further details we refer to [24] and for the FE implementation to the Supplementary Materials of [24]. The FE approximation of the transfer eigenvalue problem then reads as follows: Find the eigenvectors $\underline{\zeta}_j \in \mathbb{R}^{N_S}$ and the eigenvalues $\lambda_j \in \mathbb{R}^+_0$ such that

$$\underline{T}^t\underline{M}_R\underline{T}\,\underline{\zeta}_j = \lambda_j\,\underline{M}_S\,\underline{\zeta}_j. \tag{17}$$

Note that in actual practice we would not assemble $\underline{T}$ but instead solve successively the linear system of equations

$$\underline{A}\underline{u}_i = D_{\Gamma_{out}\to\Omega}\underline{e}_i \quad \text{with the standard unit vectors } \underline{e}_i \tag{18}$$

and assemble $(\underline{T}^T\underline{M}_R\underline{T})_{i,j} = (\underline{D}_{\Omega\to\Gamma_{in}}\underline{u}_j, \underline{D}_{\Omega\to\Gamma_{in}}\underline{u}_i)_R$. The coefficients of the FE approximation of the basis functions $\{\phi_1^{sp}, ..., \phi_n^{sp}\}$ of the optimal local approximation space $R^n := \operatorname{span}\{\phi_1^{sp}, \ldots, \phi_n^{sp}\}$ are then given by $\underline{\phi}_j^{sp} = \underline{T}\,\underline{\zeta}_j$, $j = 1, \ldots, n$. Adding the representation of the right-hand side and a basis of $\ker(\mathscr{A})$ yields the optimal space $R_{data,\ker}^n$.

Remark 3 We may also define the discrete transfer operator implicitly via (16) and pass it together with its implicitly defined adjoint to a Lanczos method. This is in general much more favorable from a computational viewpoint compared to solving (18) N_{out} times. However, it turns out that employing techniques from randomized linear algebra can be even more computationally beneficial than a Lanczos method as it requires only about n local solutions of the PDE with random boundary conditions while yielding an approximation of the eigenvectors $\underline{\zeta}_j$ of (17) at any required accuracy [6].

4 Extension to Parameter-Dependent Problems and Problems with Geometric Changes

Many applications require a rapid simulation response for many different material parameters such as Young's modulus or a real-time simulation response for a different geometry such as a beam with a newly detected crack. Therefore, it is desirable to have a port-reduced static condensation procedure that is able to deal efficiently with parameter-dependent PDEs and geometric changes. Recall however that the optimal port space as presented in Sect. 3 is based on the space of functions that solve the (now parametrized) PDE on a specific domain Ω and therefore also depends on the parameter and the geometry of Ω. As constructing a new optimal port space "from scratch" for each new parameter value is in general not feasible, the goal of this section is to show how to construct a low-dimensional and quasi-optimal port space that is independent of the parameter and the geometry but yields an accurate approximation for the full parameter set and all geometries of interest. To that end, we present in Sect. 4.3 a spectral greedy approach which constructs a reduced basis to approximate the n eigenspaces associated with the n largest eigenvalues of the parameter and geometry dependent generalized (transfer) eigenvalue problem. Here, we slightly extend the spectral greedy algorithm introduced in [24] to the case of varying geometries. At the beginning we state the parametrized PDE of interest in Sect. 4.1 and recall the port-reduced static condensation procedure for parameter-dependent PDEs in Sect. 4.2.

4.1 Parametrized Partial Differential Equations with Geometric Changes

We consider a setting where $\Omega(\mu)$ accounts for different geometries such as a beam, a beam with a crack or a beam with a hole. Note that in contrast to "standard" model order reduction approaches we do accommodate here geometries that cannot be transformed into one another by a C^1-map. Moreover, we allow different discretizations in the interior of $\Omega_i(\boldsymbol{\mu})$, $i = 1, 2$. However, we have to insist that the geometry of $\Gamma_{in}(\boldsymbol{\mu})$ is parameter-independent and that the meshes associated with all considered components coincide on $\Gamma_{in}(\boldsymbol{\mu})$; translation is of course possible resulting in the parameter dependency of $\Gamma_{in}(\boldsymbol{\mu})$.

In detail, we consider a discrete geometry parameter set $\mathscr{P}_{Geo}$ being the union of the considered geometries. Geometric changes via smooth maps can additionally be accounted for via the parameter-dependent operator $\mathscr{A}(\boldsymbol{\mu})$, where $\boldsymbol{\mu}$ belongs to the compact parameter set $\mathscr{P}_{PDE} \subset \mathbb{R}^p$. Again we assume that the port is not geometrically deformed. Then, we consider the following problem on $\Omega(\boldsymbol{\mu})$: For any $\boldsymbol{\mu} \in \mathscr{P} := \mathscr{P}_{Geo} \times \mathscr{P}_{PDE}$ and given $f(\boldsymbol{\mu}) \in L^2(\Omega(\boldsymbol{\mu}))$ find $u(\boldsymbol{\mu})$ such that

$$\mathscr{A}(\boldsymbol{\mu})u(\boldsymbol{\mu}) = f(\boldsymbol{\mu}) \text{ in } \Omega(\boldsymbol{\mu}), \quad \text{and} \quad u(\boldsymbol{\mu}) = u_{gl}(\boldsymbol{\mu}) \text{ on } \Gamma_{out}(\boldsymbol{\mu}). \tag{19}$$

Again, we introduce a conforming FE discretization to arrive at the linear system of equations $\underline{A}(\boldsymbol{\mu})\underline{u}(\boldsymbol{\mu}) = \underline{f}(\boldsymbol{\mu})$ of size $N(\boldsymbol{\mu}) \times N(\boldsymbol{\mu})$ and FE approximation $u(\boldsymbol{\mu})$.

4.2 Port Reduced Static Condensation for Parametrized Equations

We assume that we have given a *parameter-independent* reduced port basis $\underline{\phi}_1, \ldots,$ $\underline{\phi}_m \in \mathbb{R}^{N_{in}}$, $m \ll N_{in}$ that we store in the columns of the matrix $\underline{\Phi}_m \in \mathbb{R}^{N_{in} \times m}$. Proceeding as above we can then define a *parameter-dependent port reduced static condensation approximation* [7] $u^m(\boldsymbol{\mu})$ with FE coefficients $\underline{u}^m(\boldsymbol{\mu}) \in \mathbb{R}^N$, where the coefficients on the interface $\underline{u}^m_{\Gamma_{in}}(\boldsymbol{\mu})$ satisfy the parametrized and reduced Schur complement system

$$\underline{\Phi}_m^T \underline{A}_{SC}(\boldsymbol{\mu}) \underline{\Phi}_m \underline{u}^m_{\Gamma_{in}}(\boldsymbol{\mu}) = \underline{\Phi}_m^T \underline{f}_{SC}(\boldsymbol{\mu}) \qquad \text{of size } m \times m \tag{20}$$

and the Schur complement matrix $\underline{A}_{SC}(\boldsymbol{\mu}) \in \mathbb{R}^{N_{in} \times N_{in}}$ and the Schur complement right-hand $\underline{f}_{SC}(\boldsymbol{\mu}) \in \mathbb{R}^{N_{in}}$ are defined as follows

$$\underline{A}_{SC}(\boldsymbol{\mu}) = \underline{A}_{\Gamma_{in}}(\boldsymbol{\mu}) - \underline{A}_{\Gamma_{in},\Omega_1}^T(\boldsymbol{\mu})\underline{A}_{\Omega_1}^{-1}(\boldsymbol{\mu})\underline{A}_{\Gamma_{in},\Omega_1}(\boldsymbol{\mu}) - \underline{A}_{\Gamma_{in},\Omega_2}^T(\boldsymbol{\mu})\underline{A}_{\Omega_2}^{-1}(\boldsymbol{\mu})\underline{A}_{\Gamma_{in},\Omega_2}(\boldsymbol{\mu}),$$
$$\tag{21}$$
$$\underline{f}_{SC}(\boldsymbol{\mu}) = \underline{f}_{\Gamma_{in}}(\boldsymbol{\mu}) - \underline{A}_{\Gamma_{in},\Omega_1}^T(\boldsymbol{\mu})\underline{A}_{\Omega_1}^{-1}(\boldsymbol{\mu})\underline{f}_{\Omega_1}(\boldsymbol{\mu}) - \underline{A}_{\Gamma_{in},\Omega_2}^T(\boldsymbol{\mu})\underline{A}_{\Omega_2}^{-1}(\boldsymbol{\mu})\underline{f}_{\Omega_2}(\boldsymbol{\mu}).$$

We note that in order to facilitate a simulation response at low marginal cost one would in actual practice also use model order reduction techniques to approximate $\underline{A}_{\Omega_i}^{-1}(\boldsymbol{\mu})\underline{A}_{\Gamma_{in},\Omega_i}(\boldsymbol{\mu})$, $i = 1, 2$. This is however not the topic of this paper and we refer for details to [13]. We only note that if one wishes to perform many simulations for different parameters or a real-time simulation after a (non-smooth) geometry change a new intra-element reduced space has to be generated. Depending on the smoothness of the parameter-to-solution map the construction of the intra-element reduced space can be more expensive than the generation of the interface space. However, the total online computational time for the construction of all reduced spaces is reduced if we reuse the interface space after a geometry change. Moreover, in particular in cases where no intra-element reduced space is required reusing the reduced interface space can be appealing from a computational perspective.

4.3   Spectral Greedy Algorithm

The process defined in Sect. 3 yields for every $\mu \in \mathscr{P}$ the (optimal) port space $R^n_{data,ker}(\mu)$ for this specific parameter $\mu \in \mathscr{P} = \mathscr{P}_{Geo} \times \mathscr{P}_{PDE}$. The spectral greedy algorithm as introduced in [24] and which we extend here to the case of geometry changes constructs *one* quasi-optimal parameter-independent port space R^m which approximates those parameter-dependent spaces $R^n_{data,ker}(\mu)$ with a given accuracy on a finite dimensional training set $\Xi = \mathscr{P}_{Geo} \times \Xi_{PDE}$ with $\Xi_{PDE} \subset \mathscr{P}_{PDE}$. In the spectral greedy algorithm we exploit the fact that, although the solutions on the component pair may vary significantly with the parameter $\mu \in \mathscr{P}_{PDE}$ and the geometry, we expect that the port spaces $R^n_{data,ker}(\mu)$, and in particular the spectral modes that correspond to the largest eigenvalues, are much less affected by a variation in the parameter and changes in the geometry thanks to the expected very rapid decay of the higher eigenfunctions in the interior of $\Omega(\mu)$.

The spectral greedy as described in Algorithm 4.1 then proceeds as follows. After the initialization we compute for all $\mu \in \Xi$ the parameter-dependent optimal port spaces $R^n_{data,ker}(\mu)$.[1] Also in the parameter-dependent setting we can prove an *a priori* error bound [24] for the error between $u(\mu)$ and the continuous port-reduced static condensation approximation $u^n(\mu)$ corresponding to the *parameter-depedent* optimal port space $\mathscr{R}^n_{data,ker}(\mu)$:

$$\frac{\|u(\mu) - u^n(\mu)\|_{\mathscr{E}(\mu)}}{\|u(\mu)\|_{\mathscr{E}(\mu)}} \le c_1(\mu)c_2(\mu)C_1(\Omega(\mu), \mu)\sqrt{\lambda_{n+1}(\mu)}. \tag{22}$$

Here, the norm $\|\cdot\|_{\mathscr{E}(\mu)}$ is the norm induced by the parameter-dependent bilinear form associated with $\mathscr{A}(\mu)$ and $c_1(\mu)$ and $c_2(\mu)$ are chosen such that we have $c_1(\mu)\|\cdot\|_{\mathscr{E}(\bar{\mu})} \le \|\cdot\|_{\mathscr{E}(\mu)} \le c_2(\mu)\|\cdot\|_{\mathscr{E}(\bar{\mu})}$ for all $\mu \in \mathscr{P}_{PDE}$ and a fixed reference parameter $\bar{\mu} \in \mathscr{P}_{PDE}$.[2] To ensure that for every parameter $\mu \in \Xi$ we include all necessary information that we need to obtain a good approximation for this spe-

---

[1] Note, that this may restrict the applicability of the spectral greedy to training sets Ξ_{PDE} of moderate cardinality.

[2] We note that in order to prove (22) it is necessary to define the lifting inner product on the ports for one reference parameter $\bar{\mu} \in \mathscr{P}_{Geo}$ and use the equivalence of the norm induced by the lifting inner product and the $H^{1/2}$-norm on the ports. Exploiting that the latter is the same for all considered geometries allows switching between the geometries in the proof.

Algorithm 4.1: spectral greedy [24]

---

input : train sample $\varXi \subset \mathscr{P}$, tolerance ε
output: set of chosen parameters P_m, port space R^m

1 **Initialize**

$P_{\dim(\ker(\mathscr{A}))} \leftarrow \emptyset$, $R^{\dim(\ker(\mathscr{A}))} \leftarrow \operatorname{span}\{\eta_1|_{\varGamma_{in}}, \dots, \eta_{\dim(\ker(\mathscr{A}))}|_{\varGamma_{in}}\}$, $m \leftarrow \dim(\ker(\mathscr{A}))$

2 **foreach** $\mu \in \varXi$ **do**

3 $\quad\bigg|\quad$ Compute $R^n_{data,\ker}(\mu)$ such that $c_1(\mu)c_2(\mu)C_1(\varOmega(\mu), \mu)\sqrt{\lambda_{n+1}(\mu)} \leq \frac{\varepsilon}{2}$.

4 **end**

5 **while** *true* **do**

6 $\quad\bigg|\quad$ **if** $\max_{\mu \in \varXi} E(S(R^n_{data,\ker}(\mu)), R^m) \leq \varepsilon/(\varepsilon + 2C_2(\varOmega(\mu), \mu)c_1(\mu)c_2(\mu))$ **then**

7 $\quad\bigg|\quad\bigg|\quad$ **return**

8 $\quad\bigg|\quad$ **end**

9 $\quad\bigg|\quad$ $\mu^* \leftarrow \arg\max_{\mu \in \varXi} E(S(R^n_{data,\ker}(\mu)), R^m)$

10 $\quad\bigg|\quad$ $P_{m+1} \leftarrow P_m \cup \mu^*$

11 $\quad\bigg|\quad$ $\kappa \leftarrow \arg\sup_{\rho \in S(R^n_{data,\ker}(\mu^*))} \inf_{\zeta \in R^m} \|\rho - \zeta\|_R$

12 $\quad\bigg|\quad$ $R^{m+1} \leftarrow R^m + \operatorname{span}\{\kappa\}$

13 $\quad\bigg|\quad$ $m \leftarrow m + 1$

14 **end**

---

cific parameter μ we choose the dimension of $R^n_{data,\ker}(\mu)$ for each $\mu \in \varXi$ such that $c_1(\mu)c_2(\mu)C_1(\varOmega(\mu), \mu)\sqrt{\lambda_{n+1}(\mu)} \leq \frac{\varepsilon}{2}$ for a given tolerance ε. Although precise estimates for $C_1(\varOmega(\mu), \mu)$ can be obtained, setting $C_1(\varOmega(\mu), \mu) = 1$ yields in general good results as another value would just result in rescaling ε; for further details see [24]. After having collected all vectors on $\varGamma_{in}$ that are essential to obtain a good approximation for all vectors $\underline{D}_{\varOmega(\mu) \to \varGamma_{in}} \underline{u}(\mu)$, $\mu \in \varXi$, we must select a suitable basis from those vectors. This is realized in an iterative manner in Lines 5-14.

In each iteration we first identify in Line 9 the port space $R^n_{data,\ker}(\mu^*)$ that maximizes the deviation

$$E(S(R^n_{data,\ker}(\mu)), R^m) := \sup_{\xi \in S(R^n_{data,\ker}(\mu))} \inf_{\zeta \in R^m} \|\xi - \zeta\|_R, \quad \mu \in \varXi,$$

where possible choices of $S(R^n_{data,\ker}(\mu)) \subset R^n_{data,\ker}(\mu)$ will be discussed below. Subsequently, we determine in Line 11 the function $\kappa \in S(R^n_{data,\ker}(\mu^*))$ that is worst approximated by the space R^m and enhance R^m with the span of κ. The spectral greedy algorithm terminates if for all $\mu \in \varXi$ we have

$$\max_{\mu \in \varXi} E(S(R^n_{data,\ker}(\mu)), R^m) \leq \varepsilon/(\varepsilon + 2C_2(\varOmega(\mu), \mu)c_1(\mu)c_2(\mu)) \qquad (23)$$

for a constant $C_2(\varOmega(\mu), \mu)$, which can in general be chosen equal to one. A slight modification of the stopping criterion (23) and a different scaling of ε in the threshold for the a priori error bound in Line 3 allows to prove that after termination of the spectral greedy we have [24]

$$\|u(\boldsymbol{\mu}) - u^m(\boldsymbol{\mu})\|_{\mathscr{E}(\boldsymbol{\mu})} / \|u(\boldsymbol{\mu})\|_{\mathscr{E}(\boldsymbol{\mu})} \leq \varepsilon, \tag{24}$$

where $u^m(\boldsymbol{\mu})$ is the continuous port-reduced static condensation approximation corresponding to $\mathscr{R}^m$; $\mathscr{R}^m$ being the continuous outcome of the spectral greedy.

Choice of the Subset $S(R^n_{data,\mathrm{ker}}(\boldsymbol{\mu}))$

First, we emphasize that in contrast to the standard greedy as introduced in [27] we have an ordering of the basis functions in $R^n_{data,\mathrm{ker}}(\boldsymbol{\mu})$ in terms of their approximation properties thanks to the transfer eigenvalue problem. To obtain a parameter-independent port space that yields a (very) good static condensation approximation already for moderate m it is therefore desirable that the spectral greedy algorithm selects the more important basis functions sooner rather than later during the while-loop. The sorting of the basis functions in terms of their approximation properties is implicitly saved in their norms as $\|\phi_j(\boldsymbol{\mu})\|^2_R = \lambda_j(\boldsymbol{\mu})$, $j = 1, \ldots, n$ where $\phi_j(\boldsymbol{\mu})$ denotes the spectral basis of $R^n_{data,\mathrm{ker}}(\boldsymbol{\mu})$. As suggested in [24] we thus propose to consider

$$S(R^n_{data,\mathrm{ker}}(\boldsymbol{\mu})) := \{\zeta(\boldsymbol{\mu}) \in R^n_{data,\mathrm{ker}}(\boldsymbol{\mu}) : \sum_{i=1}^{\dim(R^n_{data,\mathrm{ker}}(\boldsymbol{\mu}))} (\underline{\zeta}(\boldsymbol{\mu})_i)^2 \leq 1\} \tag{25}$$

with $\zeta(\boldsymbol{\mu}) = \sum_{i=1}^{\dim(R^n_{data,\mathrm{ker}}(\boldsymbol{\mu}))} \underline{\zeta}(\boldsymbol{\mu})_i \phi_i(\boldsymbol{\mu})$. The deviation $E(S(R^n_{data,\mathrm{ker}}(\boldsymbol{\mu})), R^m)$ can then be computed by solving the eigenvalue problem: Find $(\underline{\psi}_j(\boldsymbol{\mu}), \sigma_j(\boldsymbol{\mu})) \in (\mathbb{R}^{\dim(R^n_{data,\mathrm{ker}}(\boldsymbol{\mu}))}, \mathbb{R}^+)$ such that

$$\underline{Z}(\boldsymbol{\mu})\underline{\psi}_j(\boldsymbol{\mu}) = \sigma_j(\boldsymbol{\mu})\underline{\psi}_j(\boldsymbol{\mu}), \tag{26}$$

where $\underline{Z}_{i,l}(\boldsymbol{\mu}) := (\phi_l(\boldsymbol{\mu}) - \sum_{k=1}^m (\phi_l(\boldsymbol{\mu}), \phi_k)_R \phi_k, \phi_i(\boldsymbol{\mu}) - \sum_{k=1}^m (\phi_i(\boldsymbol{\mu}), \phi_k)_R \phi_k)_R,$

$$\tag{27}$$

where ϕ_k denotes the basis of R^m and the underscore denotes the coefficients of a vector in $R^n_{data,\mathrm{ker}}(\boldsymbol{\mu})$ expressed in the spectral basis $\phi_l(\boldsymbol{\mu})$. We thus obtain $E(S(R^n_{data,\mathrm{ker}}(\boldsymbol{\mu})), R^m) = \sqrt{\sigma_1(\boldsymbol{\mu})}$, for all $\boldsymbol{\mu} \in \varXi$, and $\kappa = \psi_1(\boldsymbol{\mu}^*)$ at each iteration. To further motivate this choice of $S(R^n_{data,\mathrm{ker}}(\boldsymbol{\mu}))$ let us assume that all spectral modes in $R^n_{data,\mathrm{ker}}(\boldsymbol{\mu})$ are orthogonal to the space R^m for all $\boldsymbol{\mu} \in \varXi$, which is the case for instance for $m = \dim(R^n_{data,\mathrm{ker}}(\boldsymbol{\mu}))$ but also often for higher m. In this case the matrices $\underline{Z}(\boldsymbol{\mu})$ reduce to diagonal matrices with diagonal entries $\underline{Z}_{i,i}(\boldsymbol{\mu}) = \|\phi_i(\boldsymbol{\mu})\|^2_R$, $i = 1, \ldots, \dim(R^n_{data,\mathrm{ker}}(\boldsymbol{\mu}))$, $\boldsymbol{\mu} \in \varXi$. A spectral greedy based on $E(S(R^n_{data,\mathrm{ker}}(\boldsymbol{\mu})), R^m)$ would therefore select the parameter $\boldsymbol{\mu}^*$ such that the associated function $\psi_1(\boldsymbol{\mu}^*)$ has maximal energy with respect to the $(\cdot, \cdot)_R$-inner product. Note that this is consistent with our aim to include the weighting induced by the transfer eigenvalue problem into the basis selection process by the spectral greedy.

Remark 4 Note that were we to consider the norm $\| \cdot \|_R$ in (25) the sorting of the spectral basis $\phi_i(\mu)$ of $R^n_{data,\mathrm{ker}}(\mu)$ in terms of approximation properties is neglected in the `while` loop of Algorithm 4.1; for further explanations see [24]. As a consequence it may and often would happen in actual practice, also due to numerical inaccuracies, that a spectral greedy algorithm based on the $\| \cdot \|_R$ norm in (25) selects first functions that have been marked by the transfer eigenvalue problem as *less* important. Therefore, we would observe an approximation behaviour of the static condensation approximation based on the so constructed port space that is not satisfactory for moderate m.

5 Approximating the Whole System Associated with Ω_{gl}

To allow a maximal topological flexibility during assembly of the system associated with Ω_{gl}, we assume that we neither know the size, the composition, nor the shape of the system when generating the reduced model. Therefore, we perform the spectral greedy algorithm for all interfaces that may appear in the large structure on the component pairs that share the interface. Multiplying the left-hand side of the inequality in Line 3 in Algorithm 4.1 and $2C_2(\Omega(\mu), \mu)c_1(\mu)c_2(\mu)$ in Line 6 in Algorithm 4.1 by an estimate for the number of times we expect the interface to appear in the large system ensures that the relative approximation error on the whole domain Ω_{gl} associated with the system will lie below ε (see [24] for the proof).[3] We note that in actual practice numerical experiments show a very weak scaling in the number of ports such that the scaling might not be necessary [24].

6 A Posteriori Error Estimation

In order to assess after the detection of a new defect in the assembled system whether the quality of the reduced port space is still sufficient we wish to have an *a posteriori* error estimator for the error between the port reduced solution $u^m(\mu)$ and the FE solution $u(\mu)$ available. To that end, we employ the error estimator derived in [23]. We exploit that the FE solution satisfies a *weak flux continuity* at the interface Γ_{in}

$$\underline{f}_{SC}(\mu) - \underline{A}_{SC}(\mu)\underline{u}_{\Gamma_{in}}(\mu) = 0. \tag{28}$$

Regarding the term "weak flux continuity" we recall first that (28) is the discrete version of a Steklov-Poincaré interface equation. The latter is the weak counterpart

---

[3] As indicated above it is necessary to slightly modify the spectral greedy algorithm to prove convergence.

of the Neumann condition $\frac{\partial u|_{\Omega_1}}{\partial n} = \frac{\partial u|_{\Omega_2}}{\partial n}$ on Γ_{in} for the outer normal n, requiring continuity of the flux across the interface. For further details we refer to [22].

Also the reduced solution $\underline{u}^m(\mu)$ satisfies a weak flux continuity with respect to the reduced test space:

$$\underline{\Phi}_m^T \underline{f}_{SC}(\mu) - \underline{\Phi}_m^T \underline{A}_{SC}(\mu) \underline{\Phi}_m \underline{u}_{\Gamma_{in}}^m(\mu) = 0.$$

However, the reduced solution $\underline{u}^m(\mu)$ does *not* satisfy a weak flux continuity with respect to the full test space:

$$\underline{\Phi}_{N_{in}}^T \underline{f}_{SC}(\mu) - \underline{\Phi}_{N_{in}}^T \underline{A}_{SC}(\mu) \Phi_m \underline{u}_{\Gamma_{in}}^m(\mu) \neq 0. \tag{29}$$

Here, the first m columns of $\underline{\Phi}_{N_{in}} \in \mathbb{R}^{N_{in} \times N_{in}}$ contain the basis $\underline{\phi}_1, \ldots, \underline{\phi}_m$ generated by the spectral greedy and the remainder spans the orthogonal complement of R^m. Note that the left-hand side in (29) can also be interpreted as a *residual* on Γ_{in}. We use the violation of the weak flux continuity in (29) to assess how much the reduced solution differs from the FE solution at the interface Γ_{in}. To utilize this information for a posteriori error estimation in [23] the concept of conservative fluxes defined according to Hughes et al. [11] is adapted to the setting of port reduction. In a slight generalization of [23] we define the *jump of the conservative flux* $\underline{\zeta}^m(\mu)$ as the solution of

$$\underline{\Phi}_{N_{in}}^T \underline{M}_R \underline{\Phi}_{N_{in}} \underline{\zeta}^m(\mu) = \underline{\Phi}_{N_{in}}^T \underline{f}_{SC}(\mu) - \underline{\Phi}_{N_{in}}^T \underline{A}_{SC}(\mu) \Phi_m \underline{u}_{\Gamma_{in}}^m(\mu). \tag{30}$$

If $\phi_1, \ldots, \phi_m$ are orthonormal with respect to the inner product in R the linear system of equations (30) simplifies to

$$\underline{\zeta}^m(\mu) = \underline{\Phi}_{N_{in}}^T \underline{f}_{SC}(\mu) - \underline{\Phi}_{N_{in}}^T \underline{A}_{SC}(\mu) \Phi_m \underline{u}_{\Gamma_{in}}^m(\mu). \tag{31}$$

The computation of the jump of the conservative flux thus reduces to assembling the residual. Therefore, the computational costs scale linearly in $(N_{in} - m)$ and m.

Proposition 3 (A posteriori error estimator for port reduction [23]) *Equip R with the L^2-norm and define*

$$\Delta^m(\mu) := \frac{(\max_{i=1,2} c_{t^*,i})\sqrt{1 + c_p^2}}{\alpha_{app}(\mu)} \|\zeta^m\|_{L^2(\Gamma_{in})}, \tag{32}$$

where $c_{t^*,i}$ is the discrete trace constant in $\|v\|_{L^2(\Gamma_{in})} \leq c_{t^*,i} \|v\|_{H^1(\Omega_i)}$, c_p is the constant in the Poincaré-Friedrichs-inequality, and $\alpha_{app}(\mu)$ an approximation of the FE coercivity constant $\alpha_h(\mu)$ of the bilinear form associated with $\mathscr{A}(\mu)$. If $\alpha_{app}(\mu) \leq \alpha_h(\mu)$, there holds

$$\frac{1}{\gamma_h(\mu) c_h h^{-1/2} c_a} \Delta^m(\mu) \leq \|\nabla u(\mu) - \nabla \mathsf{u}(\mu)\|_{L^2(\Omega_\mu)} \leq \Delta^m(\mu). \tag{33}$$

where c_a is the continuity constant of the discrete extension operator, c_h is the constant in the inverse inequality $\|v\|_{H^{1/2}(\Gamma_{in})} \leq c_h h^{-1/2} \|v\|_{L^2(\Gamma_{in})}$, and $\gamma_h(\mu)$ the FE continuity constant of the bilinear form associated with $\mathscr{A}(\mu)$.

Remark 5 (Error estimation on Ω_{gl}) Let us assume that the system associated with Ω_{gl} has P^{Γ} ports, which are denoted by Γ_p, $p = 1, \ldots, P^{\Gamma}$. Moreover, denote by ζ_p^m the jump of the conservative flux at port Γ_p. Then we can define an error estimator on Ω_{gl} as follows [23]:

$$\Delta^m(\mu) := \frac{c_{t^*}\sqrt{1 + c_p^2}}{\alpha_{app}(\mu)} \left(\sum_{p=1}^{P^{\Gamma}} \|\zeta_p^m\|_{L^2(\Gamma_p)}^2 \right)^{1/2}. \tag{34}$$

Here, c_{t^*} denotes the maximum over the discrete trace constants in all components, where we estimate the L^2-norm on all ports of that component against the H^1-norm on that component. c_p is the constant in the Poincaré-Friedrichs-inequality with respect to Ω_{gl}. Again, one can show that the effectivity of the error estimator (34) is bounded [23]. Moreover, we have that the effectivity of all local error indicators defined as in (32) is bounded. Those local error indicators associated with one port in the system can thus be used within an adaptive scheme to decide where to enrich the port space first.

We note that due to coercivity constant and the constant c_p the effectivities of $\Delta^m(\mu)$ are in general rather high. However, in [25] an error estimator is presented, which is solely based on local constants and in consequence provides a very sharp bound for the error. We finally note that the *a posteriori* error estimator introduced in [23] also assess the error due to an RB approximation of $\underline{A}_{\Omega_i}^{-1}(\mu)\underline{A}_{\Gamma_{in},\Omega_i}(\mu), i = 1, 2$ in (20).

7 Numerical Experiments

In this section we investigate the performance of the optimal port space $R_{data,\ker}^n(\mu)$ for changing geometries as occurring in structural health monitoring. We demonstrate in Sect. 7.2 that we can use a port space generated on a component pair of two undefecive (I-)beams also for a component pair with a defect such as a crack, obtaining a relative approximation error of less than 10^{-3}. Subsequently, we investigate the performance of the spectral greedy algorithm for geometry changes in Sect. 7.3. We begin in Sect. 7.1 with the description of our benchmark problem: isotropic, homogeneous linear elasticity.

For the implementation we used the finite element library libMesh [16] including rbOOmit [18]. The eigenvalue problems in the transfer eigenvalue problem and the computation of the deviation have been computed with the Eigen library [8].

7.1 Benchmark Problem: Isotropic, Homogeneous Linear Elasticity

We assume that $\Omega(\boldsymbol{\mu}) \in \mathbb{R}^d$, $d = 2, 3$, $\bar{\Omega}(\boldsymbol{\mu}) = \bar{\Omega}_1 \cup \bar{\Omega}_2(\boldsymbol{\mu})$ is filled with an isotropic, homogeneous material and consider defects in the sense that $\Omega_2(\boldsymbol{\mu})$ may have say a hole or a crack with a boundary $\Gamma_{defect}(\boldsymbol{\mu}) \subset \Gamma_N(\boldsymbol{\mu})$. We consider the following linear elastic boundary value problem: Find the displacement vector $u(\boldsymbol{\mu})$ and the Cauchy stress tensor $\sigma(u(\boldsymbol{\mu}))$ such that

$$
\begin{aligned}
-\nabla \cdot \sigma(u(\boldsymbol{\mu})) &= 0 & &\text{in } \Omega(\boldsymbol{\mu}), \\
\sigma(u(\boldsymbol{\mu})) \cdot n(\boldsymbol{\mu}) &= 0 & &\text{on } \Gamma_N(\boldsymbol{\mu}), \\
u(\boldsymbol{\mu}) &= g(\boldsymbol{\mu}) & &\text{on } \Gamma_D,
\end{aligned}
\tag{35}
$$

where g is a given Dirichlet boundary condition on the displacement.

Thanks to Hooke's law we can express for a linear elastic material the Cauchy stress tensor as $\sigma(u(\boldsymbol{\mu})) = C : \varepsilon(u(\boldsymbol{\mu}))$, where C is the stiffness tensor, $\varepsilon(u(\boldsymbol{\mu})) = 0.5(\nabla u(\boldsymbol{\mu}) + (\nabla u(\boldsymbol{\mu}))^T)$ is the infinitesimal strain tensor, and the colon operator : is defined as $C : \varepsilon(u(\boldsymbol{\mu})) = \sum_{i,j=1}^{d} C_{ij} \varepsilon_{ij}(u(\boldsymbol{\mu}))$. We assume in two spatial dimensions, i.e. for $d = 2$, that the considered isotropic, homogeneous material is under plane stress. Therefore, the stiffness tensor can be written as

$$
C_{ijkl} = \begin{cases} \frac{\nu}{(1-\nu)^2}\delta_{ij}\delta_{kl} + \frac{1}{2(\nu+1)}(\delta_{ik}\delta_{jl} + \delta_{il}\delta_{jk}), \ 1 \le i,j,k,l \le 2, & \text{if } d = 2, \\ \frac{\nu}{(1+\nu)(1-2\nu)}\delta_{ij}\delta_{kl} + \frac{1}{2(1+\nu)}(\delta_{ik}\delta_{jl} + \delta_{il}\delta_{jk}), \ 1 \le i,j,k,l \le 3, & \text{if } d = 3, \end{cases}
$$

where δ_{ij} denotes the Kronecker delta and we choose Poisson's ratio $\nu = 0.3$. We only consider parameters due to geometry changes such as a replacement of a beam with a cracked beam and no material parameters; therefore we have $\mathcal{P}_{PDE} = \emptyset$. As indicated in Sect. 4.1 we discretize the weak form of (35) by a conforming FE discretization.

The kernel of $\mathcal{A}$ for the present example equals the three-dimensional space of rigid body motions for $d = 2$ and the six-dimensional space of rigid body motions for $d = 3$. To construct a port space $R_{data,ker}^n(\boldsymbol{\mu})$ on $\Gamma_{in}(\boldsymbol{\mu})$ for each parameter we follow the procedure described in Sect. 3, where we use a lifting inner product (for further details on the latter see [24]). As we do not consider a load here, we obtain $\dim(R_{data,ker}^n) = n + 3$ for $d = 2$ and $\dim(R_{data,ker}^n) = n + 6$ for $d = 3$. In order to construct one joint port space on $\Gamma_{in}(\boldsymbol{\mu})$ we use the spectral greedy algorithm described in Sect. 4.3 using the L^2-inner product on Γ_{in} and $C_1(\Omega(\boldsymbol{\mu}), \boldsymbol{\mu}) = C_2(\Omega(\boldsymbol{\mu}), \boldsymbol{\mu}) = 1$; note that we have $c_1(\boldsymbol{\mu}) = c_2(\boldsymbol{\mu})$ thanks to $\mathcal{P}_{PDE} = \emptyset$.

7.2   Reusing the Port Space for a Component with Different Geometry

To provide a simulation response at low marginal cost it would be desirable if we could reuse the port space generated for certain geometries also for other geometries. Therefore, we investigate here the effect on the relative approximation error if we construct a port space on a component pair of two un-defective beams and use that port space for the approximation on a component pair consisting of one un-defective beam on Ω_1 and various defective beams on $\Omega_2(\mu)$.

To this end, we first connect components associated with the subdomains $\Omega_1 = (-7.5, -2.5) \times (-0.5, 0.5)$ and $\Omega_2(\mu) = (-2.5, 2.5) \times (-0.5, 0.5)$, i.e. two un-defective beams, and construct the associated port space. Both the FE spaces corresponding to Ω_1 and $\Omega_2(\mu)$ have the dimension of 1314 and the dimension of the FE port space is $N_{in} = 22$. In the online stage we then prescribe random Dirichlet boundary conditions[4] drawn uniformly from the interval $[-5, 5]$ on the non-shared ports such that the random Dirichlet values are mutually independent. We verify that the average relative error $\|u(\mu) - u^{n+3}(\mu)\|_{\mathscr{E}(\mu)}/\|u(\mu)\|_{\mathscr{E}(\mu)}$ over 20 realizations exhibits nearly the same convergence behavior as $\sqrt{\lambda_{n+1}(\mu)}$ (see Fig. 2). Subsequently, we replace the component associated with $\Omega_2(\mu)$ by defective components: a cracked beam, a beam where the crack is shifted towards the shared port, and a beam with a hole; the corresponding component meshes are depicted in Fig. 3. The corresponding FE spaces on $\Omega_2(\mu)$ have the dimensions 2426, 2580, and 1898, respectively, where N_{in} still equals 22 as all components share the same port mesh. Again we prescribe random Dirichlet boundary conditions on the non-shared ports and analyze the behavior of the average relative error for an increasing number of spectral modes that have been constructed by connecting two un-defective components. As anticipated the convergence behavior of the static condensation approximation for the defective components is (much) worse as that of the un-defective component (see Fig. 4). Analyzing the convergence behavior of the static condensation approximation for the cracked beam using a port space that has been constructed by connecting a beam with a cracked beam (see Fig. 2) demonstrates that this worse convergence behavior is solely due to the fact that for the results in Fig. 4 we have employed the port space for the un-defective components. However, we emphasize that already for six spectral modes (including the three rigid body modes) we obtain for the defective components a relative error of about 10^{-5} (see Fig. 4). Moreover, we observe that the error increases only slightly when we shift the crack towards the shared port.

Therefore, we conclude that in two space dimensions reusing the port space of the un-defective component yields a sufficiently accurate static condensation approximation. However, it should be noted that the port space contains only six port modes and is therefore rather small.

---

[4]Note that the values of the random Dirichlet boundary conditions do not belong to the parameter set.

Fig. 2 Eigenvalues $\sqrt{\lambda_{n+1}(\boldsymbol{\mu})}$ for the beam (b.) and the cracked beam (c. b.) and the average relative error $\|\mathsf{u}(\boldsymbol{\mu}) - \mathsf{u}^{n+3}(\boldsymbol{\mu})\|_{\mathscr{E}(\boldsymbol{\mu})} / \|\mathsf{u}(\boldsymbol{\mu})\|_{\mathscr{E}(\boldsymbol{\mu})}$ if the respective spectral modes are employed in the static condensation approximation

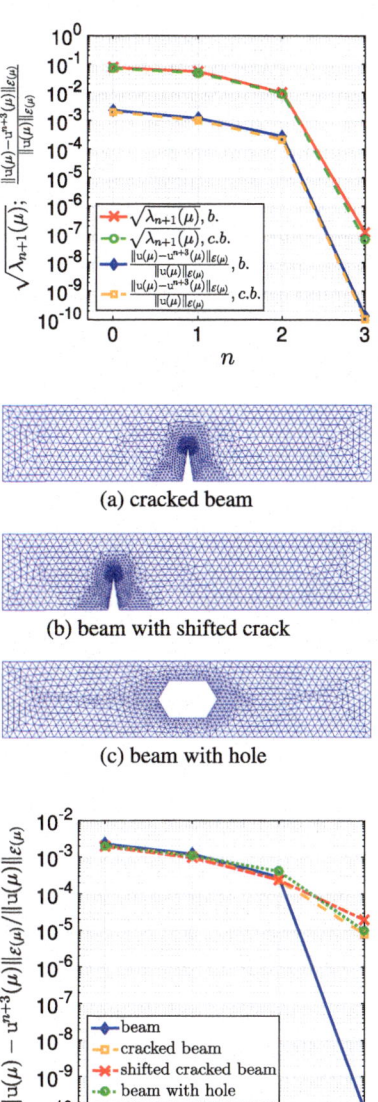

Fig. 3 Different component meshes

(a) cracked beam

(b) beam with shifted crack

(c) beam with hole

Fig. 4 Average relative error $\|\mathsf{u}(\boldsymbol{\mu}) - \mathsf{u}^{n+3}(\boldsymbol{\mu})\|_{\mathscr{E}(\boldsymbol{\mu})} / \|\mathsf{u}(\boldsymbol{\mu})\|_{\mathscr{E}(\boldsymbol{\mu})}$ for various defective components

Thus, we consider next an I-Beam and a cracked I-Beam, whose corresponding component meshes are depicted in Fig. 5a, b and the joint port meshes can be seen in Fig. 5c. The FE space associated with the I-Beam component has a dimension of 11781 and the FE space corresponding to the cracked I-Beam component has a dimension of 21705. Finally, the FE space associated with the joint port mesh is of dimension $N_{in} = 150$. We generate the reduced port space $R_{data,ker}^{n}(\boldsymbol{\mu})$ by connecting two un-defective I-Beams. Then, we prescribe homogeneous Dirichlet

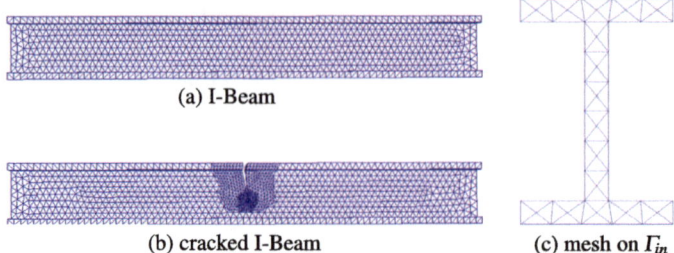

(a) I-Beam

(b) cracked I-Beam   (c) mesh on Γ_{in}

Fig. 5 Component mesh of i-beam (**a**) and i-beam with crack (**b**) and mesh of shared port (**c**)

Fig. 6 Relative error
$\|u(\mu) - u^{n+6}(\mu)\|_{\mathscr{E}(\mu)}/\|u(\mu)\|_{\mathscr{E}(\mu)}$
for I-Beam with and without
crack in $\Omega_2(\mu)$; Dirichlet
boundary conditions on outer
ports of Ω_1 and $\Omega_2(\mu)$ are
$(0, 0, 0)^T$ and $(1, 1, 1)^T$,
respectively

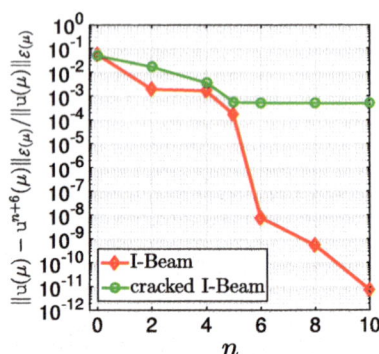

boundary conditions at the outer port of Ω_1 and $g = (1, 1, 1)^T$ at the outer port of $\Omega_2(\mu)$ and assess the relative error between the FE solution $u(\mu)$ and the port reduced static condensation approximation $u^{n+6}(\mu)$ based on that port space in Fig. 6; here $+6$ accounts for the six rigid body motions included in $R^n_{data, \ker}(\mu)$. We observe a stagnation of the relative error if we connect an I-Beam with a cracked I-Beam and use the spectral modes generated by connecting two un-defective I-Beams. However, again, we stress that we obtain a relative error of less than 10^{-3}. We also highlight the extremely fast convergence of the reduced static condensation approximation for the I-Beam and thus the convincing approximation capacities of the optimal port spaces for this test case.

7.3 Spectral Greedy Algorithm for Geometry Changes

If we perform the spectral greedy algorithm to generate a joint port space both for the defective and un-defective I-Beam we obtain a port space of size 23 for a tolerance of $2 \cdot 10^{-6}$; see also Fig. 7. Taking into account that for this tolerance the eigenspaces for the transfer eigenvalue problems for each geometry have a dimension of 16 and 15 (see Fig. 7) including the six rigid body modes, we observe that at least for this tight tolerance neither of the two eigenspaces is well suited to approximate the other.

Fig. 7 Eigenvalues $\sqrt{\lambda_{n+1}(\boldsymbol{\mu})}$ for the I-Beam with and without crack in $\Omega_2(\boldsymbol{\mu})$ and deviation $E(S(R_{data,\mathrm{ker}}^n(\boldsymbol{\mu})), R^{m+6})$ during the spectral greedy algorithm

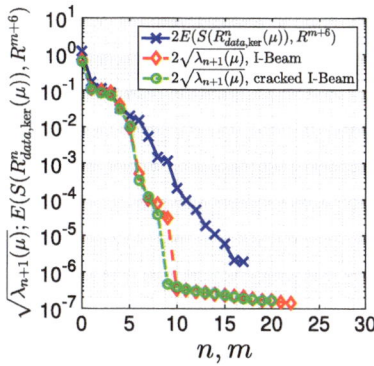

Note, that this is consistent with our observation of the stagnation of the relative error in Fig. 6. However, based on the latter we would expect that for higher tolerances the size of the port space generated by the spectral greedy is significantly smaller than the size of space obtained by uniting the two eigenspaces obtained by the transfer eigenvalue problem. This can indeed be observed in Fig. 7. If we prescribe for instance a tolerance of 10^{-2} the dimension of the port space obtained by the spectral greedy would be 14 while the dimension of the union of the two eigenspaces is 17. Increasing the tolerance further rises the gain we obtain by employing the spectral greedy rather than uniting the two eigenspaces as can be observed in Fig. 7, where we compare $2\sqrt{\lambda_{n+1}(\boldsymbol{\mu})}$ and the scaled deviation $2E(S(R_{data,\mathrm{ker}}^n(\boldsymbol{\mu})), R^{m+6})$. Note that the factor 2 comes from our chosen division of the tolerance in the spectral greedy, namely $\varepsilon/2$. Furthermore, we increase m by 6 to ease comparison with the eigenvalues; the first value of the deviation corresponds to the reduced space R^6 comprising the 6 rigid body modes.

Similar results are obtained in two space dimensions. We connect the 2d beam as introduced in the beginning of Sect. 7.2 subsequently with the 2d beam, the cracked beam depicted in Fig. 3a, and the beam with a hole (see Fig. 3c). For a tolerance of $2 \cdot 10^{-7}$ the spectral greedy yields a port space of dimension 13. As the three eigenspaces have the sizes 6 (un-defective beam and beam with hole) and 7, including the three rigid body modes, we observe that in this case the dimension of the port space generated by the spectral greedy equals the dimension of the union of the three eigenspaces. However, we emphasize that for larger tolerances as 10^{-3} or 10^{-2} we observe, again, that the spectral greedy is able to produce a very small port space which is able to yield accurate approximations for geometries which are rather different (see Fig. 8).

Finally, we analyze the convergence behavior of the relative error $\|u(\boldsymbol{\mu}) - u^m(\boldsymbol{\mu})\|_{\mathscr{E}(\boldsymbol{\mu})}/\|u(\boldsymbol{\mu})\|_{\mathscr{E}(\boldsymbol{\mu})}$ if we connect either an un-defective beam, a cracked beam, or a beam with a hole with an un-defective beam and consider random Dirichlet boundary conditions as above. Here, we use the port space constructed by the spectral greedy algorithm. Again, we observe that already for very few modes, in this case 5, we obtain for all geometries a relative error below 10^{-3}. However, if we insist

Fig. 8 Eigenvalues
$\sqrt{\lambda_{n+1}(\boldsymbol{\mu})}$ for the
un-defective beam, the
cracked beam, and the beam
with a hole, and the deviation
$E(S(R_{data,ker}^n(\boldsymbol{\mu})), R^{m+3})$
during the spectral greedy
algorithm

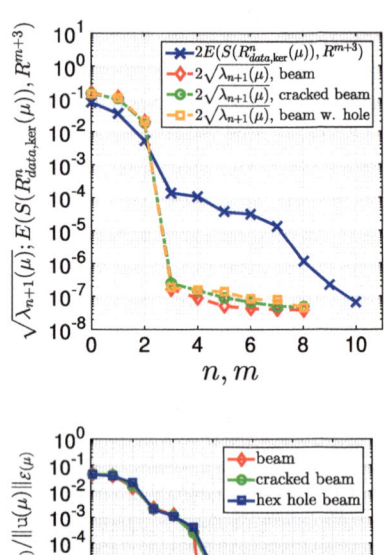

Fig. 9 Average relative error
$\|\mathsf{u}(\boldsymbol{\mu}) - \mathsf{u}^m(\boldsymbol{\mu})\|_{\mathscr{E}(\boldsymbol{\mu})}/\|\mathsf{u}(\boldsymbol{\mu})\|_{\mathscr{E}(\boldsymbol{\mu})}$ for
the un-defective beam, the
cracked beam, and the beam
with a hole

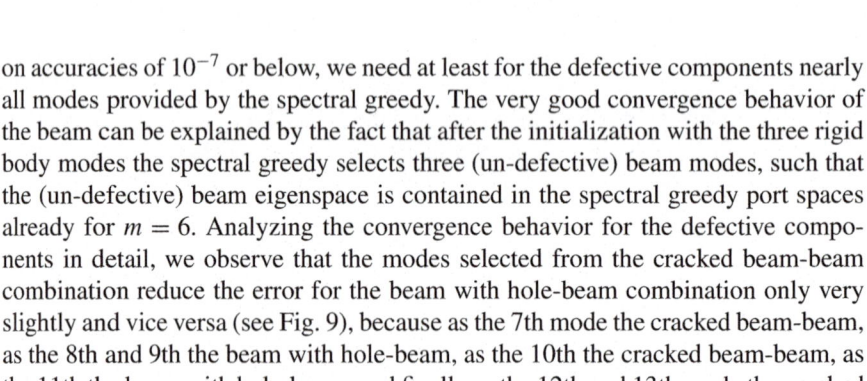

on accuracies of 10^{-7} or below, we need at least for the defective components nearly all modes provided by the spectral greedy. The very good convergence behavior of the beam can be explained by the fact that after the initialization with the three rigid body modes the spectral greedy selects three (un-defective) beam modes, such that the (un-defective) beam eigenspace is contained in the spectral greedy port spaces already for $m = 6$. Analyzing the convergence behavior for the defective components in detail, we observe that the modes selected from the cracked beam-beam combination reduce the error for the beam with hole-beam combination only very slightly and vice versa (see Fig. 9), because as the 7th mode the cracked beam-beam, as the 8th and 9th the beam with hole-beam, as the 10th the cracked beam-beam, as the 11th the beam with hole-beam, and finally as the 12th and 13th mode the cracked beam-beam combination has been selected during the spectral greedy.

8   Conclusions

Having the application in structural health monitoring in mind we have proposed quasi-optimal port spaces for parametrized PDEs on different geometries. To that end, we employed the optimal port spaces generated by a transfer eigenvalue problem as introduced in [24] and slightly generalized the there suggested spectral greedy algorithm to geometry changes. In the numerical experiments we showed that for tolerances of about 10^{-4}–10^{-2} the spectral greedy algorithm is able to construct a small port space that already yields an accurate approximation. Moreover, we demonstrated that using the optimal port space generated on a component pair of two un-defective beam yields a rather small relative approximation error on a component pair of an un-defective beam and a beam with a defect. We therefore expect that if one constructs port spaces for a library of defects, and then detects a new defect, very often reusing the constructed port space will result in a small relative approximation error.

Acknowledgements I would like to thank Prof. Dr. Anthony Patera for many fruitful discussions and comments on the content of this paper. This work was supported by OSD/AFOSR/MURI Grant FA9550-09-1-0613 and ONR Grant N00014-11-1-0713.

References

1. Babuška, I., Huang, X., Lipton, R.: Machine computation using the exponentially convergent multiscale spectral generalized finite element method. ESAIM Math. Model. Numer. Anal. **48**(2), 493–515 (2014)
2. Babuška, I., Lipton, R.: Optimal local approximation spaces for generalized finite element methods with application to multiscale problems. Multiscale Model. Simul. **9**(1), 373–406 (2011)
3. Bampton, M., Craig, R.: Coupling of substructures for dynamic analyses. AIAA J. **6**(7), 1313–1319 (1968)
4. Bourquin, F.: Component mode synthesis and eigenvalues of second order operators: discretization and algorithm. RAIRO Modél. Math. Anal. Numér. **26**(3), 385–423 (1992)
5. Buhr, A., Engwer, C., Ohlberger, M., Rave, S.: ArbiLoMod, a simulation technique designed for arbitrary local modifications. SIAM J. Sci. Comput. **39**(4), A1435–A1465 (2017)
6. Buhr, A., Smetana, K.: Randomized local model order reduction. SIAM J. Sci. Comput. **40**(4), A2120–A2151 (2018)
7. Eftang, J.L., Patera, A.T.: Port reduction in parametrized component static condensation: approximation and a posteriori error estimation. Int. J. Numer. Meth. Eng. **96**(5), 269–302 (2013)
8. Eigen: a C++ linear algebra library: http://eigen.tuxfamily.org/
9. Fehr, J., Holzwarth, P., Eberhard, P.: Interface and model reduction for efficient explicit simulations—a case study with nonlinear vehicle crash models. Math. Comput. Model. Dyn. Syst. **22**(4), 380–396 (2016)
10. Hetmaniuk, U., Lehoucq, R.B.: A special finite element method based on component mode synthesis. ESAIM Math. Model. Numer. Anal. **44**(3), 401–420 (2010)
11. Hughes, T.J.R., Engel, G., Mazzei, L., Larson, M.G.: The continuous Galerkin method is locally conservative. J. Comput. Phys. **163**(2), 467–488 (2000)

12. Hurty, W.C.: Dynamic analysis of structural systems using component modes. AIAA J. **3**(4), 678–685 (1965)
13. Huynh, D.B.P., Knezevic, D.J., Patera, A.T.: A static condensation reduced basis element method: approximation and a posteriori error estimation. ESAIM Math. Model. Numer. Anal. **47**(1), 213–251 (2013)
14. Iapichino, L., Quarteroni, A., Rozza, G.: Reduced basis method and domain decomposition for elliptic problems in networks and complex parametrized geometries. Comput. Math. Appl. **71**(1), 408–430 (2016)
15. Jakobsson, H., Bengzon, F., Larson, M.G.: Adaptive component mode synthesis in linear elasticity. Int. J. Numer. Meth. Eng. **86**(7), 829–844 (2011)
16. Kirk, B.S., Peterson, J.W., Stogner, R.H., Carey, G.F.: libMesh: a C++ library for parallel adaptive mesh refinement/coarsening simulations. Eng. Comput. **22**(3–4), 237–254 (2006)
17. Knezevic, D.J., Kang, H., Sharma, P., Malinowski, G., Nguyen, T.T., et al.: Structural integrity management of offshore structures via RB-FEA and fast full load mapping based digital twins. In: The 28th International Ocean and Polar Engineering Conference. International Society of Offshore and Polar Engineers (2018)
18. Knezevic, D.J., Peterson, J.W.: A high-performance parallel implementation of the certified reduced basis method. Comput. Meth. Appl. Mech. Eng. **200**(13–16), 1455–1466 (2011)
19. Kolmogoroff, A.: Über die beste Annäherung von Funktionen einer gegebenen Funktionenklasse. Ann Math. (2) **37**(1), 107–110 (1936)
20. Martini, I., Rozza, G., Haasdonk, B.: Reduced basis approximation and a-posteriori error estimation for the coupled Stokes-Darcy system. Adv. Comput. Math. **41**(5), 1131–1157 (2015)
21. Pinkus, A.: n-widths in Approximation Theory, vol. 7. Springer-Verlag, Berlin (1985)
22. Quarteroni, A., Valli, A.: Domain decomposition methods for partial differential equations. In: Numerical Mathematics and Scientific Computation. The Clarendon Press, Oxford University Press, New York, reprint (2005)
23. Smetana, K.: A new certification framework for the port reduced static condensation reduced basis element method. Comput. Meth. Appl. Mech. Eng. **283**, 352–383 (2015)
24. Smetana, K., Patera, A.T.: Optimal local approximation spaces for component-based static condensation procedures. SIAM J. Sci. Comput. **38**(5), A3318–A3356 (2016)
25. Smetana, K., Patera, A.T.: Fully localized a posteriori error estimation for the port reduced static condensation reduced basis element method (2018+). In preparation
26. Taddei, T., Patera, A.T.: A localization strategy for data assimilation; application to state estimation and parameter estimation. SIAM J. Sci. Comput. **40**(2), B611–B636 (2018)
27. Veroy, K., Prud'homme, C., Rovas, D.V., Patera, A.T.: A posteriori error bounds for reduced-basis approximation of parametrized noncoercive and nonlinear elliptic partial differential equations. In: Proceedings of the 16th AIAA Computational Fluid Dynamics Conference, vol. 3847 (2003)

Analysis of Parametric Models for Coupled Systems

Hermann G. Matthies and Roger Ohayon

Abstract In many instances one has to deal with parametric models. Such models in vector spaces are connected to a linear map. The reproducing kernel Hilbert space and affine-/linear- representations in terms of tensor products are directly related to this linear operator. This linear map leads to a generalised correlation operator, in fact it provides a factorisation of the correlation operator and of the reproducing kernel. The spectral decomposition of the correlation and kernel, as well as the associated Karhunen-Loève- or proper orthogonal decomposition are a direct consequence. This formulation thus unifies many such constructions under a functional analytic view. Recursively applying factorisations in higher order tensor representations leads to hierarchical tensor decompositions. This format also allows refinements for cases when the parametric model has more structure. Examples are shown for vector- and tensor-fields with certain required properties. Another kind of structure is the parametric model of a coupled system. It is shown that this can also be reflected in the theoretical framework.

Keywords Parametric models · Reproducing kernel Hilbert space · Correlation · Spectral decomposition · Coupled systems

MSC Classification 35B30 · 37M99 · 41A05 · 41A45 · 41A63 · 60G20 · 60G60 · 65J99 · 93A30

H. G. Matthies (✉)
Institute of Scientific Computing, TU Braunschweig, Brunswick, Germany
e-mail: wire@tu-bs.de

R. Ohayon
Laboratoire de Mécanique des Structures et des Systèmes Couplés, CNAM,
Paris, France
e-mail: roger.ohayon@lecnam.net

© Springer Nature Switzerland AG 2020
J. Fehr and B. Haasdonk (eds.), *IUTAM Symposium on Model Order Reduction of Coupled Systems, Stuttgart, Germany, May 22–25, 2018*, IUTAM Bookseries 36, https://doi.org/10.1007/978-3-030-21013-7_2

1 Introduction

Parametric models are used in many areas of science, engineering, and economics to describe variations or changes of some system. They can have many different uses such as evaluating the *design* of some system, or to *control* its behaviour, or to *optimise* the performance in some way. Another important case is when some of the parameters may be uncertain and are modelled by random variables (RVs), and one wants to perform *uncertainty quantification* or *identify* some of the parameters in a mathematical model. One important consideration is the preservation of structure which one knows to be present in the system. One such structure is the consideration of *coupled systems*, and it will be shown how this can be dealt with. In fact, the coupling conditions can be one of the possible parameters.

The representations of such parametric models leads directly to *reduced order models* which are used to lessen the possibly high computational demand in some of the tasks described above. Such reduced models hence become parametrised. The survey [1] and the recent collection [2], as well as the references therein, provide a good account of parametric reduced order models and some of the areas where they appear. The interested reader may find there further information on parametrised reduced order models and how to generate them.

This present work is a continuation of [16, 17], where the theoretical background of such parametrised models was treated in a functional analysis setting. For many of the theoretical details we thus refer to these publications, and especially to [17] for a more thorough account of the theory.

As an example, assume that some physical system is investigated, which is modelled by an evolution equation for its state $u(t) \in V$ at time $t \in [0, T]$, where V is assumed to be a Hilbert space for the sake of simplicity: $\dot{u}(t) = A(\mu; u(t)) + f(\mu; t)$; $u(0) = u_0$, where the superimposed dot signifies the time derivative, A is an operator modelling the physics of the system, and f is some external excitation. The model depends on some quantity $\mu \in \mathcal{M}$, where $\mathcal{M}$ denotes the set of possible parameters, and we assume that for all μ of interest the system is well-posed. Other than that, it is not assumed that the set $\mathcal{M}$ has any additional structure. In this way the system state becomes a function of the parameters, and can thus be written as $u(\mu; t)$. Later mainly the dependence on μ will be interesting, so that the other arguments may be dropped.

To fix ideas, take as a simple example a very simple fluid-structure interaction problem used in [14] to explain the basics of coupling algorithms: it is a mass-spring system coupled with a gas-filled piston. The governing equations are in the simplest case considered there (with slight change of notation):

$$m\,\ddot{w}(t) + k\,w(t) = S(p(t) - p_0); \qquad p(t) = p_0 \left(1 + \frac{\gamma - 1}{2} \frac{\dot{w}(t)}{c_0}\right)^{\frac{2\gamma}{\gamma - 1}}.$$

The mass m and spring constant k are properties of the mass-spring system with displacement $w(t)$ coupled over the surface S with the gas-filled piston with pressure

$p(t)$. The equilibrium pressure is p_0, and c_0 is the speed of sound in the gas with specific heat ratios γ. Introducing the velocity $v(t) = \dot{w}(t)$, the state of the system is given by $u = [w, v]^\mathsf{T}$, the system depends on the parameters $\mu = (m, k, S, c_0, \gamma - 1)$, the state Hilbert space is $V = \mathbb{R}^2$, the set $\mathcal{M}$ can be taken as $\mathcal{M} = \{(\mu_j)_{j=1..5} \mid \mu_j > 0\} \subset \mathbb{R}^5$, and the problem $\dot{u}(t) = A(\mu; u(t)) + f(\mu; t)$ becomes:

$$\dot{u}(t) = \begin{bmatrix} \dot{w}(t) \\ \dot{v}(t) \end{bmatrix} = \begin{bmatrix} v(t) \\ -\frac{k}{m} w(t) + \frac{Sp_0}{m} \left(1 + \frac{\gamma-1}{2} \frac{v(t)}{c_0}\right)^{\frac{2\gamma}{\gamma-1}} \end{bmatrix} + \begin{bmatrix} 0 \\ -\frac{Sp_0}{m} \end{bmatrix}. \quad (1)$$

A little twist can be given to this by assuming that some, or all, of the parameters are *random variables*; say for example the spring stiffness k. Formally, this is a measurable function from some probability space into the real numbers: $k : \Omega \to \mathbb{R}$. This makes also the displacement w and velocity v into random variables $w, v : \Omega \to \mathbb{R}$. If we assume that all involved random variables have finite variance, i.e. $k, w, v \in L_2(\Omega)$, then the parameter set could be taken as $\mathcal{M} = \mathbb{R}_+ \times L_2(\Omega) \times \mathbb{R}_+^3$, and the state space would be the Hilbert space $V = L_2(\Omega)^2$. Such probabilistic examples have prompted much of the theory and terminology [13], and such probabilistic problems are treated specifically in the present framework in [15]; but here we do not want to digress and keep the focus on coupled systems.

A bit more involved is the following example from [18], it is a kind of generic example of fluid-structure interaction. The fluid is described by the incompressible Navier–Stokes equation in arbitrary Lagrangean–Eulerian (ALE) formulation in a domain Ω_f:

$$\rho_f(\dot{v} + ((v - \dot{\chi}) \cdot \nabla)v) - \operatorname{div} \sigma + \nabla p = r_f, \quad (2)$$

$$2\sigma = \nu_f(\nabla v + (\nabla v)^\mathsf{T}), \quad \operatorname{div} v = 0, \quad (3)$$

where ρ_f and ν_f are the fluid mass-density and viscosity, $v(x, t)$ is the fluid velocity-field, $\sigma(x, t)$ is the viscous stress in the fluid, $r_f(x, t)$ are the volume forces in the fluid, and the pressure $p(x, t)$ is the Lagrange multiplier for the incompressibility constraint, whereas $\chi(x, t)$ is the movement of the ALE background reference system. On part of the boundary $\Gamma_c \subset \partial\Omega_f$ the fluid is coupled to an elastic solid, described in the solid domain Ω_s (with also $\Gamma_c \subset \partial\Omega_s$) in a Lagrangean or material frame by

$$\rho_s \ddot{w} - \operatorname{div}(FS) = r_s, \quad F = \nabla w, \quad (4)$$

$$S = \lambda_s(\operatorname{tr} E)I + 2\mu_s E, \quad 2E = C - I, \quad C = F^\mathsf{T} F, \quad (5)$$

where ρ_s is the solid mass-density and λ_s, μ_s are the elastic Lamé moduli, $w(x, t)$ is the solid displacement-field and F its gradient, $S(x, t)$ is the elastic 2nd Piola-Kirchhoff stress in the solid, $r_s(x, t)$ are the solid volume forces, and the Lagrange-Green strain E is stated in terms of the Cauchy-Green strain tensor C. The position of a solid particle which was at position x at time $t = 0$ is $w(x, t) + x$. Hence at the

each point $\chi(x, t) = w(x, t) + x$ on the coupling boundary Γ_c with normal $\boldsymbol{n}(x, t)$ one has the condition that the velocities of fluid and solid have to match

$$v(\chi(x, t), t) = \dot{w}(x, t). \tag{6}$$

The solution (v, p) to the fluid part Eqs. (2) and (3) lives in a Hilbert space $H(\mathrm{div})(\Omega_f) \times L_2(\Omega_f)$, whereas the displacement w of the solid part can be envisioned in $H^1(\Omega_s)^3$ with a velocity $\dot{w} \in H^1(\Omega_s)^3$. The state of the system $(v, p, w, \dot{w})$ is thus in the Hilbert space $\mathcal{V} = H(\mathrm{div})(\Omega_f) \times L_2(\Omega_f) \times H^1(\Omega_s)^6$. The parameters can for example be the material constants for fluid and solid $\mu = (\rho_f, \nu_f, \rho_s, \lambda_s, \mu_s)$ such that the set $\mathcal{M}$ can be taken again as $\mathcal{M} = \{(\mu_j)_{j=1..5} \mid \mu_j > 0\} \subset \mathbb{R}^5$, or it can be the smooth initial shape $\chi_0(x)$ of the coupling boundary Γ_c, such that $\mu = \chi_0$ and $\mathcal{M} \subset C^1(\Omega_c)$ with the boundary description $\Omega_c \subset \mathbb{R}^2$; or a combination of these two cases.

With these two examples in mind, turning again to the general description, one is interested in how the system changes when these *parameters* μ change. As we have seen in the above examples, these can be something specific describing the operator, or the initial condition, or specifying the excitation, etc. [1]. One may be interested in the state of the system $u(\mu; t)$, or some functional of it, say $\Psi(\mu)$. In the first example this could for example be the maximum acceleration $\Psi(\mu) = \max_t \ddot{w}(\mu; t)$. While evaluating $A(\mu; \hat{u}(t))$—for some *trial* state $\hat{u}(t)$—or $f(\mu; t)$ for a certain μ may be straightforward, there are situations where evaluating $u(\mu; t)$ or $\Psi(\mu)$ may be very costly.

In this situation one is interested in representations of $u(\mu; t)$ or $\Psi(\mu)$ which allow a cheaper evaluation, these are called *proxy-* or *surrogate*-models, among others. Any such parametric object can be analysed by linear maps which are associated with such representations. This association of parametric models and linear mappings has probably been known for a long time, see [13] for an exposition in the context of stochastic models.

It should be pointed out that the exposition here is a general technique which can be used to analyse any parametric model and its approximations, where the parameters can be of whatever nature, e.g. just some numbers, or functions, or random variables, etc., and not so much something to be directly implemented. It is shown though that in the framework of orthogonal bases it has a direct connection with proper orthogonal decomposition (POD) and Karhunen-Loève expansions. As will be seen, it also connects with representations in tensor product spaces, which allows numerically to use low-rank tensor approximations [10, 16]. It is furthermore also connected with non-orthogonal decompositions which are easier to compute, like the *proper generalised decomposition* (PGD) [4]. Here we shall only consider orthogonal bases for the sake of conciseness of exposition.

Whereas the parametric map may be quite complicated, the association with a linear map translates the whole problem into one of linear functional analysis, and into linear algebra upon approximation and actual numerical computation.

2 Parametric Models and Reproducing Kernel

This is a short recap of the developments in [17], where the interested reader may find more detail. Let $r : \mathcal{M} \to \mathcal{U}$ be one of the objects alluded to in the introduction, where $\mathcal{M}$ is some set without any further assumed structure, and $\mathcal{U}$ is assumed for the sake of simplicity as a separable Hilbert space with inner product $\langle \cdot | \cdot \rangle_{\mathcal{U}}$.

Assume without significant loss of generality that span $r(\mathcal{M}) = \text{span im } r \subseteq \mathcal{U}$, the subspace of $\mathcal{U}$ which is spanned by all the vectors $\{r(\mu) \mid \mu \in \mathcal{M}\}$, is dense in $\mathcal{U}$—otherwise we just restrict ourselves to the closure of span $r(\mathcal{M}) = \text{span im } r$. Then to each such function r one may associate a linear map $\tilde{R} : \mathcal{U} \ni u \mapsto \langle r(\cdot) | u \rangle_{\mathcal{U}} \in \mathbb{R}^{\mathcal{M}}$, the real-valued functions on the set $\mathcal{M}$. As a motivation, one may think of such functions as providing a 'co-ordinate system' on the otherwise unstructured set $\mathcal{M}$. By construction, $\tilde{R}$ restricted to span im $r = \text{span } r(\mathcal{M})$ is injective, and has an inverse on its restricted range range $\tilde{\mathcal{R}} := \tilde{R}(\text{span im } r) \subseteq \mathbb{R}^{\mathcal{M}}$. This may be used to define an inner product on $\tilde{\mathcal{R}}$ as

$$\forall \phi, \psi \in \tilde{\mathcal{R}} \quad \langle \phi | \psi \rangle_{\mathcal{R}} := \langle \tilde{R}^{-1} \phi | \tilde{R}^{-1} \psi \rangle_{\mathcal{U}}, \tag{7}$$

and to denote the completion of $\tilde{\mathcal{R}}$ with this inner product by $\mathcal{R}$. One immediately obtains that $\tilde{R}^{-1}$ is a bijective isometry between span im r and $\tilde{\mathcal{R}}$, hence extends to a *unitary* map between $\mathcal{U}$ and $\mathcal{R}$, as does $\tilde{R}$.

Given the maps $r : \mathcal{M} \to \mathcal{U}$ and $\tilde{R} : \mathcal{U} \to \mathcal{R}$, one may define the *reproducing kernel* [3, 11] given by $\varkappa(\mu_1, \mu_2) := \langle r(\mu_1) | r(\mu_2) \rangle_{\mathcal{U}}$. It is straightforward to verify that $\varkappa(\mu, \cdot) \in \tilde{\mathcal{R}} \subseteq \mathcal{R}$, and span$\{\varkappa(\mu, \cdot) \mid \mu \in \mathcal{M}\} = \tilde{\mathcal{R}}$, as well as the reproducing property $\phi(\mu) = \langle \varkappa(\mu, \cdot) | \phi \rangle_{\mathcal{R}}$ for all $\phi \in \tilde{\mathcal{R}}$.

On this *reproducing kernel Hilbert space* (RKHS) $\mathcal{R}$ one can build a first representation. As $\mathcal{U}$ is separable, so is $\mathcal{R}$, and one may choose a complete orthonormal system (CONS) $\{\varphi_m\}_{m \in \mathbb{N}}$ in $\mathcal{R}$. Then with the CONS $\{y_m \mid y_m = \tilde{R}^{-1}\varphi_m = \tilde{R}^*\varphi_m\}_{m \in \mathbb{N}}$ in $\mathcal{U}$, the unitary operator $\tilde{R}$, and its adjoint or inverse $\tilde{R}^* = \tilde{R}^{-1}$, and the parametric element $r(\mu)$ become [17]

$$\tilde{R} = \sum_m \varphi_m \otimes y_m; \quad \text{i.e.} \quad \tilde{R}(u)(\cdot) = \sum_m \langle y_m | u \rangle_{\mathcal{U}} \varphi_m(\cdot), \quad \tilde{R}^* = \tilde{R}^{-1} = \sum_m y_m \otimes \varphi_m;$$

$$r(\mu) = \sum_m \varphi_m(\mu) y_m = \sum_m \varphi_m(\mu) \tilde{R}^* \varphi_m. \tag{8}$$

Observe that the relations Eq. (8) exhibit the tensorial nature of the representation mapping. One sees that *model reductions* may be achieved by choosing only subspaces of $\mathcal{R}$, i.e. a—typically finite—subset of $\{\varphi_m\}_m$. Furthermore, the representation of $r(\mu)$ in Eq. (8) is *linear* in the new 'parameters' φ_m.

3 Correlation and Kernel Space

The RKHS construction $\mathcal{R}$ of Sect. 2 just mirrors or reproduces the inner product structure on the original space $\mathcal{U}$. There is presently no way of telling what is important in the parameter set $\mathcal{M}$. For this one needs additional information. As a way of indicating what is important on the set $\mathcal{M}$, assume that there is another inner product $\langle\cdot|\cdot\rangle_\mathcal{Q}$ for scalar functions $\phi \in \mathbb{R}^\mathcal{M}$, and denote the Hilbert space of functions with that inner product by $\mathcal{Q}$. We define a linear map in the same way as the map $\tilde{R}$ in Sect. 2, but as it now has a different range or image space—$\mathcal{Q}$ instead of $\mathcal{R}$—and denote the map by $R : \mathcal{U} \ni u \mapsto \langle r(\cdot)|u\rangle_\mathcal{U} \in \mathcal{Q}$. As the inner product on the image space has changed, the map R, unlike the map $\tilde{R}$ in Sect. 2, will in general not be unitary any more. Also assume that the subspace dom $R = \{u \in \mathcal{U} \mid \|Ru\|_\mathcal{Q} < \infty\}$ is, if not the whole space $\mathcal{U}$, at least dense in $\mathcal{U}$, and that the densely defined operator R is closed. These are essentially requirements that the topologies on $\mathcal{U}$ and $\mathcal{Q}$ fit in some way with the map $r : \mathcal{M} \to \mathcal{U}$. But to make things even simpler, assume here that R is defined on the whole space and hence continuous, and still injective, unless explicitly otherwise stated.

With this, one may define [13, 17] a densely defined map C in $\mathcal{U}$ through the bilinear form

$$\forall u, v \in \mathcal{U} : \quad \langle Cu|v\rangle_\mathcal{U} := \langle Ru|Rv\rangle_\mathcal{Q}. \tag{9}$$

The map C, which may also be written as $C = R^*R$, may be called the *'correlation'* operator. By construction it is self-adjoint and positive, and if R is continuous so is C. In case the inner product $\langle\cdot|\cdot\rangle_\mathcal{Q}$ comes from a measure ϖ on $\mathcal{M}$, so that for two functions ϕ and ψ on $\mathcal{M}$ one has

$$\langle\phi|\psi\rangle_\mathcal{Q} := \int_\mathcal{M} \phi(\mu)\psi(\mu)\, \varpi(\mathrm{d}\mu), \quad \text{such that from Eq. (9)}$$

$$\langle Cu|v\rangle_\mathcal{U} = \int_\mathcal{M} \langle r(\mu)|u\rangle_\mathcal{U}\langle r(\mu)|v\rangle_\mathcal{U}\, \varpi(\mathrm{d}\mu), \text{ i.e. } C = R^*R = \int_\mathcal{M} r(\mu) \otimes r(\mu)\, \varpi(\mathrm{d}\mu).$$

The space $\mathcal{Q}$ may then be taken as $\mathcal{Q} := L_2(\mathcal{M}, \varpi)$. A special case is when ϖ is a probability measure, $\varpi(\mathcal{M}) = 1$, this inspired the term 'correlation' [13].

In Sect. 2 it was the factorisation of $C = R^*R$ which allowed the RKHS representation in Eq. (8). For other representations, one needs other factorisations. Most common is to use the spectral decomposition (e.g. [8]) of C to achieve such a factorisation.

On infinite dimensional Hilbert spaces self-adjoint operators may have a continuous spectrum, e.g. [8]. To make everything as simple as possible to explain the main underlying idea, we shall from now on assume that C is a non-singular trace class or nuclear operator. This means that it is compact, the spectrum $\sigma(C)$ is a point spectrum, has a CONS $\{v_m\}_m$ consisting of eigenvectors, with each eigenvalue $\lambda_m \geq \lambda_{m+1} \cdots \geq 0$ positive and counted decreasingly according to their finite multiplicity, and has finite trace $\operatorname{tr} C = \sum_m \lambda_m < \infty$. Then a version of the spectral decomposition of C is

$$C = \sum_m \lambda_m (v_m \otimes v_m). \tag{10}$$

Define a new CONS $\{s_m\}_m$ in Q: $\lambda_m^{1/2} s_m := R v_m$, to obtain the corresponding *singular value decomposition* (SVD) of R and R^*:

$$R = \sum_m \lambda_m^{\frac{1}{2}} (s_m \otimes v_m); \quad \text{i.e.} \quad R(u)(\cdot) = \sum_m \langle v_m | u \rangle_{\mathcal{U}} s_m(\cdot), \quad R^* = \sum_m \lambda_m^{\frac{1}{2}} (v_m \otimes s_m);$$

$$r(\mu) = \sum_m \lambda_m^{\frac{1}{2}} s_m(\mu) v_m = \sum_m s_m(\mu) R^* s_m, \quad \text{as} \quad R^* s_m = \lambda_m^{\frac{1}{2}} v_m. \tag{11}$$

The set $\varsigma(R) = \{\lambda_m^{1/2}\}_m = \sqrt{\sigma(C)} \subset \mathbb{R}_+$ are the *singular values* of R and R^*. The last relation is the so-called *Karhunen-Loève expansion* or *proper orthogonal decomposition* (POD). The finite trace condition of C translates into the fact that r is in $\mathcal{U} \otimes Q$. If in that relation the sum is *truncated* at $n \in \mathbb{N}$, i.e.

$$r(\mu) \approx r_n(\mu) = \sum_{m=1}^{n} \lambda_m^{\frac{1}{2}} s_m(\mu) v_m = \sum_{m=1}^{n} s_m(\mu) R^* s_m, \tag{12}$$

we obtain the *best n-term approximation* to $r(\mu)$ in the norm of $\mathcal{U}$.

Observe that, similarly to Eq. (8), r is linear in the s_m. This means that by choosing the 'co-ordinate transformation' $\mathcal{M} \ni \mu \mapsto (s_1(\mu), \ldots, s_m(\mu), \ldots) \in \mathbb{R}^{\mathbb{N}}$ one obtains a *linear / affine* representation where the first co-ordinates are the most important ones.

A formulation of the spectral decomposition different from Eq. (10) does not require C to be nuclear [8], nor does C or R have to be continuous. The self-adjoint and positive operator $C : \mathcal{U} \to \mathcal{U}$ is unitarily equivalent with a multiplication operator M_γ,

$$C = V M_\gamma V^*, \tag{13}$$

where $V : L_2(\mathcal{T}) \to \mathcal{U}$ is unitary between some $L_2(\mathcal{T})$ on a measure space $\mathcal{T}$ and the Hilbert space $\mathcal{U}$, and M_γ is a multiplication operator, multiplying any $\psi \in L_2(\mathcal{T})$ with a real-valued function γ. In case C is bounded, so is $\gamma \in L_\infty(\mathcal{T})$. As C is positive, $\gamma(t) \geq 0$ for $t \in \mathcal{T}$, the essential range of γ is the spectrum of C, and M_γ is self-adjoint and positive. Its square root is $M_\gamma^{1/2} := M_{\sqrt{\gamma}}$, from which one obtains the square-root of $C^{1/2} = V M_{\sqrt{\gamma}} V^*$. The factorisation corresponding to $C = R^* R$ with the square-root is $C = (V M_{\sqrt{\gamma}})(V M_{\sqrt{\gamma}})^* =: G^* G$. Another possibility is $C = (C^{1/2})^* C^{1/2} = C^{1/2} C^{1/2}$. From this follows another formulation of the singular value decomposition (SVD) of R and R^* with a unitary $U : L_2(\mathcal{T}) \to Q$:

$$R = U M_{\sqrt{\mu}} V^*, \quad R^* = V M_{\sqrt{\mu}} U^*. \tag{14}$$

These are all examples of a general factorisation $C = B^*B$, where $B : \mathcal{U} \to \mathcal{H}$ is a map to a Hilbert space $\mathcal{H}$ with all the properties demanded from R—see the beginning of this section. It can be shown [17] that any two such factorisations $B_1 : \mathcal{U} \to \mathcal{H}_1$ and $B_2 : \mathcal{U} \to \mathcal{H}_2$ with $C = B_1^*B_1 = B_2^*B_2$ are *unitarily equivalent* in that there is a unitary map $X_{21} : \mathcal{H}_1 \to \mathcal{H}_2$ such that $B_2 = X_{21}B_1$. Equivalently, each such factorisation is unitarily equivalent to R, i.e. there is a unitary $X : \mathcal{H} \to \mathcal{Q}$ such that $R = XB$. For finite dimensional spaces, a favourite choice is the Cholesky factorisation $C = LL^\mathsf{T}$, where $B = L^\mathsf{T}$ and $B^* = L$.

In the situation where C has a purely discrete spectrum and a CONS of eigenvectors $\{v_m\}_m$ in $\mathcal{U}$, the map B from the decomposition $C = B^*B$ can be used to define a CONS $\{h_m\}_m$ in $\mathcal{H}$: $h_m := BC^{-1/2}v_m$, which is an eigenvector CONS of the operator $C_\mathcal{H} := BB^* : \mathcal{H} \to \mathcal{H}$, with $C_\mathcal{H}h_m := \lambda_m h_m$, see [17]. From this follows a SVD of B and B^* analogous to Eq. (11). Taking the special case $\mathcal{H} = \mathcal{Q}$ with $C_\mathcal{Q} = RR^*$, we see that $C_\mathcal{Q}s_m = \lambda_m s_m$, and $s_m = UV^*v_m$, as well as $C_\mathcal{Q} = UV^*CVU^*$.

From the factorisation and Karhunen-Loève expansion in Eq. (11) one has $r(\mu) = \sum_m s_m(\mu) R^*s_m$. With other equivalent factorisations $C = B^*B$ one obtains new representations in an analogous manner. The main result is [17] that in the case of a nuclear C every factorisation leads to a separated representation in terms of a series, and vice versa. The associated 'correlations' $C_\mathcal{Q} = RR^*$ on $\mathcal{Q}$ resp. $C_\mathcal{H} = BB^*$ on $\mathcal{H}$ have the same spectrum as C, and factorisations of $C_\mathcal{Q}$ resp. $C_\mathcal{H}$ induce new factorisations of C.

The abstract equation $C_\mathcal{Q} = UV^*CVU^* = UM_\gamma U^* = \sum_m \lambda_m s_m \otimes s_m$ can be spelt out in more analytical detail for the special case when the inner product on $\mathcal{Q}$ is given by a measure ϖ on $\mathcal{P}$. It then becomes for all $\varphi, \psi \in \mathcal{Q}$:

$$\langle C_\mathcal{Q}\varphi|\psi\rangle_\mathcal{Q} = \sum_m \lambda_m \langle\varphi|s_m\rangle_\mathcal{Q}\langle s_m|\psi\rangle_\mathcal{Q} = \langle R^*\varphi|R^*\psi\rangle_\mathcal{U} =$$

$$\iint_{M \times M} \varphi(\mu_1)\varkappa(\mu_1, \mu_2)\psi(\mu_2)\, \varpi(\mathrm{d}\mu_1)\varpi(\mathrm{d}\mu_2) =$$

$$\iint_{M \times M} \varphi(\mu_1)\left(\sum_m \lambda_m s_m(\mu_1)s_m(\mu_2)\right)\psi(\mu_2)\, \varpi(\mathrm{d}\mu_1)\varpi(\mathrm{d}\mu_2) =$$

$$\sum_m \lambda_m \left(\int_M \varphi(\mu)s_m(\mu)\, \varpi(\mathrm{d}\mu)\right)\left(\int_M s_m(\mu)\psi(\mu)\varpi(\mathrm{d}\mu)\right),$$

i.e. $C_\mathcal{Q}$ is a Fredholm integral operator and its spectral decomposition is nothing but the familiar theorem of Mercer [7] for the kernel $\varkappa(\mu_1, \mu_2) = \sum_m \lambda_m s_m(\mu_1)s_m(\mu_2)$. Factorisations of $C_\mathcal{Q}$ are then usually factorisations of the kernel $\varkappa(\mu_1, \mu_2)$.

An example is if on some measure space $(\mathcal{X}, \nu)$ it holds that $\varkappa(\mu_1, \mu_2) = \int_\mathcal{X} g(\mu_1, x)g(\mu_2, x)\, \nu(\mathrm{d}x)$, then the integral transform with kernel g will play the role of a factor as before did the mappings R or B, leading to a new Karhunen-Loève-like representation of r. The abstract setting outlined in this section can hence be applied to the analysis of a great number of different situations, of any kind of representation of $r(\mu)$, see [17] for more detail, and [15] for the case where essentially μ is a random variable.

Let us remark how this framework here may be extended when the assumptions at the beginning of this section are not satisfied. In case the parametric map $r : \mathcal{M} \to \mathcal{U}$ maps not into a Hilbert space, but only into a locally convex topological vector space (LCTVS), it is still possible to define a corresponding linear map and a correlation operator. Denote the continuous dual space of $\mathcal{U}$ by $\mathcal{U}^*$ and define a linear map by $S : \mathcal{U}^* \ni u^* \mapsto S(u^*) := \langle u^*, r(\cdot) \rangle_* \in \mathcal{Q}$, where $\langle \cdot, \cdot \rangle_*$ is the duality pairing on $\mathcal{U}^* \times \mathcal{U}$. The Hilbert space $\mathcal{Q}$ is still identified with its dual, although one could also make generalisations here. Then the dual map $S^* : \mathcal{Q} \to \mathcal{U}$—w.r.t. the weak* topology on $\mathcal{U}^*$—is defined as usual by $\langle u^*, S^*\phi \rangle_* := \langle Su^* | \phi \rangle_{\mathcal{Q}}$. This allows us to define a 'correlation' $C_S : \mathcal{U}^* \to \mathcal{U}$ by $C_S = S^*S$, i.e. $\langle u^*, C_S v^* \rangle_* := \langle Su^* | Sv^* \rangle_{\mathcal{Q}}$. The dual map $S^* : \mathcal{Q} \to \mathcal{U}$ can still provide a representation. Other factorisations of C_S such as $C_S = B^*B$, where $B : \mathcal{U}^* \to \mathcal{H}$—a Hilbert space identified with its dual—can provide alternative representations via the map $B^* : \mathcal{H} \to \mathcal{U}$. But spectral theory can not be used easily as we are not in a Hilbert space and domain and range or image space are not the same for C_S.

A frequent situation where some further development is possible is as follows: there is an injective continuous map $T : \mathcal{Z} \to \mathcal{U}$ from a Hilbert space $\mathcal{Z}$, which will be a pivot space identified with its dual into the space $\mathcal{U}$, such that the subspace $T(\mathcal{Z}) \subseteq \mathcal{U}$ is dense. Then the dual map $T^* : \mathcal{U}^* \to \mathcal{Z}$ is also injective and continuous, $\mathcal{T} := T^*(\mathcal{U}^*) \subseteq \mathcal{Z}$ is dense in $\mathcal{Z}$, and we have a Gel'fand-like triplet of spaces

$$ \mathcal{U}^* \xrightarrow{T^*} \mathcal{Z} \xrightarrow{T} \mathcal{U}. $$

On the dense subspace $\mathcal{T} \subseteq \mathcal{Z}$ the map T^* is invertible, $J := T^{-*}_{|\mathcal{T}} : \mathcal{T} \to \mathcal{U}^*$. This allows one to define a mapping R and a 'correlation' C densely defined on $\mathcal{T} \subseteq \mathcal{Z}$:

$$ R := S \circ J : \mathcal{T} \xrightarrow{J} \mathcal{U}^* \xrightarrow{S} \mathcal{Q} \quad \text{and} \quad C := R^*R = J^*S^*SJ = J^*C_S J : \mathcal{T} \to \mathcal{Z}. $$

We view C now as a self-adjoint positive operator densely defined in the Hilbert space $\mathcal{Z}$, and we are in the previously described Hilbert space setting.

A typical occurrence of such a situation is the case when $\mathcal{Z} \hookrightarrow \mathcal{U}$ is a *continuously embedded* Hilbert space, the map T is then just the identity. A concrete example of this is $r : \mathcal{M} \to \mathscr{S}'(\Omega)$, a parametric map into the Schwartz space of tempered distributions. As the dual of $\mathcal{U} := \mathscr{S}'(\Omega)$ in the weak* topology is $\mathcal{U}^* = \mathscr{S}(\Omega)$, the test space of rapidly decaying smooth functions, we have the continuous embeddings $\mathcal{U}^* = \mathscr{S}(\Omega) \hookrightarrow \mathcal{Z} := L_2(\Omega) \hookrightarrow \mathscr{S}'(\Omega) = \mathcal{U}$. Hence one may define the densely defined maps $R : \mathcal{T} := \mathscr{S}(\Omega) \to \mathcal{Q}$ and $C : \mathcal{Z} := L_2(\Omega) \to \mathcal{Z}$, given for $f, g \in \mathscr{S}(\Omega)$ by the bilinear form $\langle Cf|g \rangle_{L_2} = \langle Rf|Rg \rangle_{\mathcal{Q}} = \langle \langle r(\cdot), f \rangle_{\mathscr{S}} | \langle r(\cdot), g \rangle_{\mathscr{S}} \rangle_{\mathcal{Q}}$, where $\langle \cdot, \cdot \rangle_{\mathscr{S}}$ is the duality pairing on $\mathscr{S}'(\Omega) \times \mathscr{S}(\Omega)$.

4   Structure Preservation and Coupled Systems

The main feature up to now was the mapping $r : \mathcal{M} \to \mathcal{U}$ and the associated linear map $R : \mathcal{U} \to \mathcal{Q} \subseteq \mathbb{R}^{\mathcal{M}}$, and the resulting tensor representation of $r \in \mathcal{U} \otimes \mathcal{Q}$. Here we mention some possible refinements and extensions which try to make some known structure in the parametric model explicitly visible.

One frequent situation is that the parameter space is a product space, say $\mathcal{M} = \mathcal{M}_1 \times \mathcal{M}_2$ and a corresponding factorisation of the space $\mathcal{Q} = \mathcal{Q}_1 \otimes \mathcal{Q}_2$, where $\mathcal{Q}_1 \subseteq \mathbb{R}^{\mathcal{M}_1}$ and $\mathcal{Q}_2 \subseteq \mathbb{R}^{\mathcal{M}_2}$. Then the functions $\phi(\mu) \in \mathcal{Q}$ are linear combinations of products $\varphi(\mu_1)\psi(\mu_2)$, with $\varphi(\mu_1) \in \mathcal{Q}_1$ and $\psi(\mu_2) \in \mathcal{Q}_2$. Now such a product may be seen as a parametric mapping

$$\varrho : \mathcal{M}_2 \ni \mu_2 \mapsto \varphi(\cdot)\psi(\mu_2) \in \mathcal{Q}_1;$$

and by setting

$$\mathcal{Q} = \mathcal{U}_* \otimes \mathcal{Q}_* = \mathcal{Q}_1 \otimes \mathcal{Q}_2, \tag{15}$$

the theory of the preceding sections may now be applied to the parametric map ϱ and its tensor product representation in $\mathcal{U}_* \otimes \mathcal{Q}_*$ to get a refined representation of the complete model. This shows how the product structure of the parameter set $\mathcal{M} = \mathcal{M}_1 \times \mathcal{M}_2$ expresses itself in the representation. In some way the full parameter set is a *coupled* object, and some of this is reflected in the factorisation. But for a real coupled system we will demand a bit more, as will be explained later.

It is now not difficult to see that in case $\mathcal{M} = \prod_j \mathcal{M}_j$ with corresponding $\mathcal{Q} = \bigotimes_j \mathcal{Q}_j$ and $j \in \mathcal{J} \subset \mathbb{N}$, this may be further factorised by different associations depending on a partition of the parameter set $\mathcal{J} = \mathcal{J}_1 \cup \mathcal{J}_2$ into two disjoint sets $\mathcal{J}_1 \cap \mathcal{J}_2 = \emptyset$:

$$\mathcal{Q} = \mathcal{U}_* \otimes \mathcal{Q}_* = \left(\bigotimes_{k \in \mathcal{J}_1} \mathcal{Q}_k \right) \otimes \left(\bigotimes_{k \in \mathcal{J}_2} \mathcal{Q}_k \right). \tag{16}$$

Each of the factors can then be recursively factorised further, and this leads to hierarchical tensor approximations, e.g. [10, 16].

Of course it is possible to split the tensor product in different ways, and the grouping of indices can be viewed as a tree. The well-known *canonical polyadic* (CP) decomposition uses the flat tensor product in Eq. (16). It has also been published as so-called *proper generalised decomposition* (PGD) as a computational method to solve high-dimensional problems, see the review [5] and the monograph [4]. But recursive splittings of Eq. (16) yield *deep* or *hierarchical* tensor approximations. Particular formats are the *tensor train* (TT) and more generally the *hierarchical Tucker* (HT) decompositions, see the review [9] and the monograph [10]. These hierarchical low-rank tensor representations are connected with deep neural networks [6, 12]. It is the eigenvalue structure of the correlation C or equivalently the structure of the singular values of the associated linear map R in the particular splitting Eq. (16)

which determines how many terms a series representation needs to be a good reduced model with a certain accuracy.

Another frequent case is that the role of the Hilbert space $\mathcal{Q}$ is taken by a tensor product $\mathcal{W} = \mathcal{Q} \otimes \mathcal{E}$, where $\mathcal{Q}$ is as before but $\mathcal{E}$ is a *finite-dimensional* inner-product (Hilbert) space [13]. Such a situation arises when the parametric model is in $\mathcal{V} = \mathcal{U} \otimes \mathcal{E}$, where $\mathcal{U}$ is an unspecified Hilbert space as before, and one wants to see the 'small' space $\mathcal{E}$ separately. The parametric map can be defined as follows:

$$r: \mathcal{M} \to \mathcal{V} = \mathcal{U} \otimes \mathcal{E}; \quad r(\mu) = \sum_k r_k(\mu) \boldsymbol{r}_k, \tag{17}$$

where as before $r_k(\mu) \in \mathcal{U}$ and the $\boldsymbol{r}_k \in \mathcal{E}$. Typically the index k will range over the finite dimension of $\mathcal{E}$, and the $\{\boldsymbol{r}_k\}_k$ are a suitable basis. An example of such a situation is a vector field over some manifold. If at each point the vector is in the finite-dimensional space $\mathcal{E}$, e.g. the tangent space of the manifold, we model this by a tensor product $\mathcal{U} \otimes \mathcal{E}$, where $\mathcal{U}$ is some Hilbert space of scalar valued functions on the manifold.

The 'correlation' can now be given by a bilinear form. The densely defined map $C_{\mathcal{E}}$ in $\mathcal{V} = \mathcal{U} \otimes \mathcal{E}$ is defined on elementary tensors $\boldsymbol{u} = u \otimes \boldsymbol{u}$, $\boldsymbol{v} = v \otimes \boldsymbol{v} \in \mathcal{V} = \mathcal{U} \otimes \mathcal{E}$ as

$$\langle C_{\mathcal{E}} \boldsymbol{u} | \boldsymbol{v} \rangle_{\mathcal{U}} := \sum_{k,j} \langle R_k(u) | R_j(v) \rangle_{\mathcal{Q}} (\boldsymbol{u}^{\mathsf{T}} \boldsymbol{r}_k) (\boldsymbol{r}_j^{\mathsf{T}} \boldsymbol{v}) \tag{18}$$

and extended by linearity, where each $R_k : \mathcal{U} \to \mathcal{Q}$ is the map associated to $r_k(\mu)$ as before for just a single map $r(\mu)$. It may be called the *'vector correlation'*. By construction it is self-adjoint and positive. The corresponding kernel with values in $\mathcal{E} \otimes \mathcal{E}$ for the eigenvalue problem on $\mathcal{W} = \mathcal{Q} \otimes \mathcal{E}$ is

$$\varkappa_{\mathcal{E}}(\mu_1, \mu_2) = \sum_{k,j} \langle r_k(\mu_1) | r_j(\mu_2) \rangle_{\mathcal{V}} \, \boldsymbol{r}_k \otimes \boldsymbol{r}_j. \tag{19}$$

While the situation just described occurs often when $r(\mu)$ is for example the state of a system, a case which looks formally similar but allows an alternative approach happens when the vector space $\mathcal{E}$ consist of tensors of *even* degree, hence $\mathcal{E} = \mathcal{F} \otimes \mathcal{F}$ for some space of tensors $\mathcal{F}$ of half the degree. An example of such a situation is the stress or strain field of a continuum mechanics problem, where again $\mathcal{U}$ could be a space of scalar spatial functions, and $\mathcal{E}$ the space of symmetric 2nd degree tensors. Another example is the specification of a conductivity tensor field for a heat conduction problem with $\mathcal{E}$ again the space of symmetric 2nd degree tensors, or the specification of an elasticity tensor field with $\mathcal{E}$ the space of symmetric 4th degree tensors.

Such a tensor of even degree can always be thought of as a linear map from a space of tensors of half that degree into itself. Being a linear map, it can be represented as a *matrix* $\boldsymbol{A} \in \mathbb{R}^{n \times n}$, the case we shall look at here. The size of the matrix is equal to the dimension of the space $\mathcal{F}$, i.e. $n = \dim \mathcal{F}$. So the space $\mathcal{E}$ can be thought of as a space of matrices.

Often these linear maps/matrices possess some additional properties, like being symmetric positive definite—as in the examples of the conductivity or elasticity tensor—or e.g. orthogonal. One has to realise that the representation methods which have been investigated here are *linear* methods, i.e. they work best when the representation is in a *linear* manifold, essentially free from nonlinear constraints. The two examples of positive definite or orthogonal tensor fields have such nonlinear constraints. We discuss two possible approaches:

First, assume that A has to be orthogonal. It then satisfies $A^\mathsf{T} A = I = A A^\mathsf{T}$, a nonlinear constraint. But the orthogonal matrices $\mathsf{O}(n)$ and its sub-group of special orthogonal matrices $\mathsf{SO}(n)$ form compact Lie groups. One then may work in their Lie algebra $\mathfrak{o}(n) = \mathfrak{so}(n)$, the *skew* symmetric matrices, the tangent space at the group identity I. This is a *free* linear space. An element $A \in \mathsf{SO}(n)$ can be expressed with the exponential map $A = \exp(S)$ with $S \in \mathfrak{so}(n) = \mathcal{E}$. Using the exponential map from the Lie algebra to its corresponding Lie group one only has to deal with representations in the Lie algebra.

As a second example, assume that the matrix $A \in \mathsf{Sym}^+(n)$ has to be symmetric positive definite (spd). Then it can be factored as $A = G^\mathsf{T} G$ with invertible $G \in \mathsf{GL}(n)$. Both of these are nonlinear constraints. The spd matrices $\mathsf{Sym}^+(n)$ are not a linear space, but geometrically a salient open cone and a Riemannian manifold in the space of all symmetric matrices $\mathfrak{sym}(n)$. The manifold $\mathsf{Sym}^+(n)$ can be made into a Lie group in different ways. Here it is important to observe that any $A \in \mathsf{Sym}^+(n)$ can be represented again with the matrix exponential as $A = \exp(H)$ with $H \in \mathfrak{sym}(n) = \mathcal{E}$. This is a good strategy even for scalar fields, i.e. $n = 1$.

Hence, in both cases, one may investigate the representation in some *linear sub-space* $\mathfrak{g} \subseteq \mathcal{F} \otimes \mathcal{F}$, in the concrete matrix case here in a sub-space $\mathfrak{g} \subseteq \mathbb{R}^{n \times n} = \mathfrak{gl}(n)$. Such a parametric element may be represented first as $H(\mu) \in \mathcal{Q} \otimes \mathfrak{g}$ and then exponentiated:

$$H(\mu) = \sum_k \varsigma_k(\mu) H_k, \qquad H(\mu) \mapsto \exp(H(\mu)) = A(\mu). \qquad (20)$$

Hence we now concentrate on representing $H(\mu)$. The parametric map would be written analogous to Eq. (17) as

$$R(\mu) = \sum_k r_k(\mu) \otimes R_k \in \mathcal{U} \otimes \mathcal{E}, \quad \text{with} \quad R_k \in \mathfrak{g}. \qquad (21)$$

The correlation analogous to Eq. (18) may now be defined via a bilinear form on elementary tensors as a densely defined map $C_\mathcal{E}$ in $\mathcal{W} = \mathcal{U} \otimes \mathcal{F} = \mathcal{U} \otimes \mathbb{R}^n$—observe, *not* $\mathcal{U} \otimes \mathcal{E} = \mathcal{U} \otimes \mathfrak{g}$—and extended by linearity:

$$\forall (\boldsymbol{u} = u \otimes \boldsymbol{v}), (\boldsymbol{v} = v \otimes \boldsymbol{v}) \in \mathcal{W} = \mathcal{U} \otimes \mathcal{F} :$$

$$\langle C_{\mathcal{F}} \boldsymbol{u} | \boldsymbol{v} \rangle_{\mathcal{U}} := \sum_{k,j} \langle R_k(u) | R_j(v) \rangle_{\mathcal{Q}} (\boldsymbol{R}_k \boldsymbol{u})^{\mathsf{T}} (\boldsymbol{R}_j \boldsymbol{v}). \tag{22}$$

The kernel corresponding to Eq. (19) is again matrix valued,

$$\varkappa_{\mathcal{F}}(\mu_1, \mu_2) = \sum_{k,j} \langle r_k(\mu_1) | r_j(\mu_2) \rangle_{\mathcal{U}} \, \boldsymbol{R}_k^{\mathsf{T}} \boldsymbol{R}_j, \tag{23}$$

defining an eigenproblem in $\mathcal{Q} \otimes \mathcal{F}$.

For coupled systems, the approach has some similarity with the vector case $\mathcal{U} \otimes \mathcal{E}$ above. The main characteristic of a coupled system which we want to preserve is that the space state may be written as $\mathcal{U} = \mathcal{U}_1 \times \mathcal{U}_2$ with the the natural inner product $\langle u | v \rangle_{\mathcal{U}} = \langle u_1 | v_1 \rangle_{\mathcal{U}_1} + \langle u_2 | v_2 \rangle_{\mathcal{U}_2}$, where $\boldsymbol{u} = (u_1, u_2), \boldsymbol{v} = (v_1, v_2) \in \mathcal{U}$. This is for two coupled systems, labelled as '1' and '2'. The parametric map is

$$\boldsymbol{r} : \mathcal{M} \to \mathcal{U} = \mathcal{U}_1 \times \mathcal{U}_2; \quad \boldsymbol{r}(\mu) = (r_1(\mu), r_2(\mu)). \tag{24}$$

The associated linear map is

$$\boldsymbol{R} : \mathcal{U} \to \mathcal{Q}^2 = \mathcal{Q} \times \mathcal{Q}; \quad (\boldsymbol{R}(\boldsymbol{u}))(\mu) = (\langle u_1 | r_1(\mu) \rangle_{\mathcal{U}_1}, \langle u_2 | r_2(\mu) \rangle_{\mathcal{U}_2}). \tag{25}$$

As before, these $\mathbb{R}^2$ valued functions on $\mathcal{M}$ are like two problem-adapted co-ordinate systems on the joint paramcter set, one for each sub-system. From this one obtains the 'coupling correlation', again defined through a bilinear form

$$\langle C_c \boldsymbol{u} | \boldsymbol{v} \rangle_{\mathcal{U}} := \sum_{j=1}^{2} \langle R_j(u_j) | R_j(v_j) \rangle_{\mathcal{Q}}. \tag{26}$$

The kernel is then a 2×2 matrix valued function in an integral operator on $\mathcal{W} = \mathcal{Q} \times \mathcal{Q}$:

$$\varkappa_c(\mu_1, \mu_2) = \mathrm{diag}(\langle r_k(\mu_1) | r_k(\mu_2) \rangle_{\mathcal{U}_k}). \tag{27}$$

Often there is a bit more structure one wants to preserve, namely that $\mathcal{M} = \mathcal{M}_1 \times \mathcal{M}_2$, and the parameter set $\mathcal{M}_1$ is for the sub-system '1', and the set $\mathcal{M}_2$ is for sub-system '2'. We also assume that not only $\mathcal{U} = \mathcal{U}_1 \times \mathcal{U}_2$, but also $\mathcal{Q} = \mathcal{Q}_1 \times \mathcal{Q}_2$, where the scalar functions in $\mathcal{Q}_1$ depend only on $\mathcal{M}_1$, and similarly for subsystem '2'. The parametric map is hence

$$\boldsymbol{r} : \mathcal{M} = \mathcal{M}_1 \times \mathcal{M}_2 \to \mathcal{U} = \mathcal{U}_1 \times \mathcal{U}_2; \quad \boldsymbol{r}((\mu_1, \mu_2)) = (r_1(\mu_1), r_2(\mu_2)), \tag{28}$$

with the associated linear map

$$R : \mathcal{U} \to \mathcal{Q} = \mathcal{Q}_1 \times \mathcal{Q}_2; \quad (R(u))(\mu) = (\langle u_1 | r_1(\mu_1) \rangle_{\mathcal{U}_1}, \langle u_2 | r_2(\mu_2) \rangle_{\mathcal{U}_2}). \quad (29)$$

The correlation may be defined as before in Eq. (26), and also the kernel on $\mathcal{Q} = \mathcal{Q}_1 \times \mathcal{Q}_2$ is as in Eq. (27), but now the first diagonal entry is a function on $\mathcal{M}_1 \times \mathcal{M}_1$ only, and analogous for the second diagonal entry.

5 Conclusion

Parametric mappings $r : \mathcal{M} \to \mathcal{U}$ have been analysed with in a variety of settings via the associated linear map $R : \mathcal{U} \to \mathcal{Q} \subseteq \mathbb{R}^{\mathcal{M}}$, enabling the linear analysis. The RKHS setting allows a first representation, and essentially reproduces everything in $\mathcal{U}$ in the function space $\mathcal{R}$. The choice of another inner product and corresponding Hilbert space $\mathcal{Q}$ leads to measures of importance in $\mathcal{M}$, or, more precisely, in $\mathbb{R}^{\mathcal{M}}$.

It is shown that each separated representation defines an associated linear map, and that conversely under some more restrictive conditions, the normally more general notion of an associated linear map defines a representation.

Several refinements are presented to represent some additional structure in the linear map. One such structure is the information of dealing with a coupled system. This can be reflected in the structure of the associated linear map.

Acknowledgements Partly supported by the Deutsche Forschungsgemeinschaft (DFG) through SPP 1886 and SFB 880.

References

1. Benner, P., Gugercin, S., Willcox, K.: A survey of projection-based model reduction methods for parametric dynamical systems. SIAM Rev. **57**, 483–531 (2015). https://doi.org/10.1137/130932715
2. Benner, P., Ohlberger, M., Patera, A.T., Rozza, G., Urban, K. (eds.): Model Reduction of Parametrized Systems, MS&A — Modeling, Simulation and Applications, vol. 17. Springer, Berlin (2017). https://doi.org/10.1007/978-3-319-58786-8
3. Berlinet, A., Thomas-Agnan, C.: Reproducing Kernel Hilbert Spaces in Probability and Statistics. Springer, Berlin (2004). https://doi.org/10.1007/978-1-4419-9096-9
4. Chinesta, F., Keunings, R., Leygue, A.: The Proper Generalized Decomposition for Advanced Numerical Simulations. Springer, Berlin (2014). https://doi.org/10.1007/978-3-319-02865-1
5. Chinesta, F., Ladevèze, P., Cueto, E.: A short review on model order reduction based on proper generalized decomposition. Arch Comput. Methods Eng. **18**, 395–404 (2011). https://doi.org/10.1007/s11831-011-9064-7
6. Cohen, N., Sharri, O., Shashua, A.: On the expressive power of deep learning: A tensor analysis (2016). arXiv: 1509.05009 [cs.NE]
7. Courant, R., Hilbert, D.: Methods of Mathematical Physics. Wiley, Chichester (1989). https://doi.org/10.1002/9783527617234
8. Dautray, R., Lions, J.L.: Spectral Theory and Applications. Mathematical Analysis and Numerical Methods for Science and Technology, vol. 3. Springer, Berlin (1990). https://doi.org/10.1007/978-3-642-61529-0

9. Grasedyck, L., Kressner, D., Tobler, C.: A literature survey of low-rank tensor approximation techniques. GAMM-Mitteilungen **36**, 53–78 (2013). https://doi.org/10.1002/gamm.201310004

10. Hackbusch, W.: Tensor Spaces and Numerical Tensor Calculus. Springer, Berlin (2012). https://doi.org/10.1007/978-3-642-28027-6

11. Janson, S.: Gaussian Hilbert Spaces. Cambridge Tracts in Mathematics, vol. 129. Cambridge University Press, Cambridge (1997). https://doi.org/10.1017/CBO9780511526169

12. Khrulkov, V., Novikov, A., Oseledets, I.: Expressive power of recurrent neural net-works (2018). arXiv: 1711.00811 [cs.LG].

13. Krée, P., Soize, C.: Mathematics of Random Phenomena—Random Vibrations of Mechanical Structures. D. Reidel, Dordrecht (1986). https://doi.org/10.1007/978-94-009-4770-2

14. Lefrançois, E., Boufflet, J.P.: An introduction to fluid-structure interaction: Application to the piston problem. SIAM Rev. **52**, 747–767 (2010). https://doi.org/10.1137/0907583

15. Matthies, H.G.: Analysis of probabilistic and parametric reduced order models (2018). arXiv: 1807.02219 [math.NA].

16. Matthies, H.G., Litvinenko, A., Pajonk, O., Rosić, B.V., Zander, E.: Parametric and uncertainty computations with tensor product representations. In: Dienstfrey, A. Boisvert, R., (eds.) Uncertainty Quantification in Scientific Computing. IFIP Advances in Information and Communication Technology, vol. 377, pp. 139–150. Springer, Boulder (2012). https://doi.org/10.1007/978-3-642-32677-6

17. Matthies, H.G., Ohayon, R.: Analysis of parametric models linear methods and approximations (2018). arXiv: 1806.01101 [math.NA].

18. Matthies, H.G., Steindorf, J.: Strong coupling methods. In: Wendland, W., Efendiev, M., (eds.) Analysis and Simulation of Multifield Problems. Lecture Notes in Applied and Computational Mechanics, vol. 12, pp. 13–36. Springer, Berlin (2003). https://doi.org/10.1007/978-3-540-36527-3_2

Model Order Reduction of Switched Linear Systems with Constrained Switching

Ion Victor Gosea, Igor Pontes Duff, Peter Benner and Athanasios C. Antoulas

Abstract In this work, we propose a balanced truncation procedure for the reduction of large-scale switched linear systems (SLSs) which are characterized by constrained switching scenarios. To this aim, we introduce new generalized reachability and observability Gramians, related to the constrained switched sequences, which satisfy coupled Lyapunov equations. The main goal is to make use of the newly introduced Gramian matrices to develop a balancing-type procedure for model order reduction (MOR) of constrained switching SLSs. By following the classical scheme for linear time-invariant (LTI) systems, the subspaces that are related to small singular values can be truncated leading to reduced order SLSs. Finally, the efficiency of the proposed approximation is demonstrated by several numerical examples.

Keywords Switched systems · Model order reduction · Balanced truncation · Hybrid systems · Gramian matrices

I. V. Gosea (✉) · I. Pontes Duff · P. Benner · A. C. Antoulas
Max Planck Institute for Dynamics of Complex Technical Systems, Sandtorstrasse 1, 39106 Magdeburg, Germany
e-mail: gosea@mpi-magdeburg.mpg.de

I. Pontes Duff
e-mail: pontes@mpi-magdeburg.mpg.de

P. Benner
Faculty of Mathematics, Otto-von-Guericke University, Magdeburg, Germany
e-mail: benner@mpi-magdeburg.mpg.de

A. C. Antoulas
Rice University, Houston, TX, USA
e-mail: aca@rice.edu

A. C. Antoulas
Baylor College of Medicine, Houston, TX, USA

© Springer Nature Switzerland AG 2020   41
J. Fehr and B. Haasdonk (eds.), *IUTAM Symposium on Model Order Reduction of Coupled Systems, Stuttgart, Germany, May 22–25, 2018*, IUTAM Bookseries 36,
https://doi.org/10.1007/978-3-030-21013-7_3

1 Introduction

MOR aims at finding efficient computational to replace models of complex, large-scale time-varying processes by simpler and smaller models that can still capture the behavior of the original process. Such high dimensional models are often linked to the spatial discretization of partial differential equations (PDEs). The reduced order models (ROM) could be used as efficient surrogates for the original model, replacing it as a component in tasks such as simulation, analysis, or control. For details on different MOR techniques, we refer the reader to the books [1, 5] and to the surveys [4, 6].

In the context of LTI systems, balanced truncation (BT) is a common approach which was initially introduced in [16, 17]. The main idea behind BT is to transform a dynamical system to a balanced form defined in such a way that the states that are difficult to reach are also difficult to observe. For more details on BT, see [7, 14].

Switched systems, a subclass of hybrid systems, can be considered as the result of the interaction between a finite state automaton and a finite set of LTI subsystems which are also referred to as modes. Switched systems have applications in control of mechanical and aeronautical systems, power converters and also in the automotive industry. In particular, such systems can be used to model processes that are subject to known or unknown abrupt parameter variations such as synchronously switched linear systems, networks with periodically varying switchings, and sudden change of system structure. Typical examples of real world situations when switched or hybrid systems are used to model the underlying dynamical behavior include vehicle gearboxes, air conditioning systems, elevator systems or evaporation devices. For a detailed characterization of switched systems, we refer the reader to the books [13, 24, 25].

In this work, we study continuous-time switched linear systems[1] with coupling or switching matrices, i.e., matrices that scale the continuous state at the switching times. Whenever the dimension of the state space is large and, in addition, the SLS has a big number of subsystems, difficulties for certain tasks such as simulation, optimization and control might appear. To cope with this issue, the original SLS might be approximated by a reduced order SLS by applying MOR. We refer to the following contributions on model reduction of SLSs: [8, 10, 12, 15, 19–23] which present balancing-related procedures, and, [2, 3, 11] which are related to moment matching-type methods.

In general, for SLSs, the switching signal is not restricted and can follow any trajectory. In this context, the concepts of reachability and observability have been studied and described in terms of subspaces (see [26]) and Gramians (see [20]). However, in many practical applications, the switching signal follows particular sequences/patterns—examples are automatic gearbox shifting and power converters. Thus the main goal in this paper is to develop suitable MOR methods whenever the switched signal is constrained. To the best of our knowledge, the only contribution

---

[1]It is worth mentioning that, in the switched and hybrid systems literature, such classes of systems are referred to also as linear switched systems (or LSSs).

with this philosophy is [2], where the authors solve the problem of minimal realization with constrained switching.

In this paper, we propose a balanced truncation MOR procedure for reducing SLSs with constrained switching. The technique is based on defining generalized reachability and observability Gramian matrices for each discrete mode, specifically tailored to particular switching scenarios. Similar Gramians were introduced in [12], for the general switching case, i.e., with no constraints imposed on the switches. This allows us to find those states that are hard to reach and hard to observe via an appropriate transformation. Truncating such states yields reduced-order SLSs.

The paper is organized in the following way: In Sect. 2, we introduce the definition of continuous-time SLSs together with that of input–output mappings that correspond to such systems. Additionally, some notation and symbols used throughout the paper are defined in this section. In Sect. 3, we introduce the definition of energy reachability and observability Gramians of constrained switching SLSs for the case with two discrete modes. Section 4 introduces the proposed MOR balanced truncation algorithm. In Sect. 5, three numerical experiments are presented (in which we compare the performance of the new introduced method against that of others) while a summary of the findings and the conclusion are stated in Sect. 6.

2 Linear Switched Systems: Definition and Properties

A continuous time switched linear system (SLS) denoted by Σ is a control system described by the equations

$$\Sigma : \begin{cases} \dot{\mathbf{x}}(t) = \mathbf{A}_{\sigma(t)}\mathbf{x}(t) + \mathbf{B}_{\sigma(t)}\mathbf{u}(t), & \mathbf{x}(0) = \mathbf{0}, \\ \mathbf{y}(t) = \mathbf{C}_{\sigma(t)}\mathbf{x}(t), \end{cases} \tag{1}$$

where $\sigma(t) \in \Omega$ is the switching signal, $\Omega = \{1, 2, \ldots, D\}$, $D > 1$, is the set of discrete modes, $\mathbf{u}(t) \in \mathbb{R}^m$ is the input, $\mathbf{x}(t) \in \mathbb{R}^{n_q}$ is the state (depending on the current active mode q), and $\mathbf{y}(t) \in \mathbb{R}^p$ is the output. The system matrices $\mathbf{A}_q \in \mathbb{R}^{n_q \times n_q}$, $\mathbf{B}_q \in \mathbb{R}^{n_q \times m}$, $\mathbf{C}_q \in \mathbb{R}^{p \times n_q}$, where $q \in \Omega$, correspond to the linear system active in mode $q \in \Omega$.

In (1), the derivative of variable $\mathbf{x}$ with respect to time is denoted with $\dot{\mathbf{x}}$. More precisely, this denotes the derivative from the right, i.e., $\dot{\mathbf{x}}(t) = \frac{d^+}{dt}\mathbf{x}(t) = \lim\limits_{\epsilon \to 0, \epsilon > 0} \dfrac{\mathbf{x}(t + \epsilon) - \mathbf{x}(t)}{\epsilon}$. Moreover, we assume homogeneous initial condition for the first mode, i.e., $\mathbf{x}(0) = \mathbf{0}$. For more details on general properties and control of SLSs, we refer the reader to [13, 24].

For a fixed time interval $[0, T]$, the signal σ can be described using a sequence of pairs (q_i, t_i) from $\Omega \times \mathbb{R}_+$, denoted with $\mathbf{z} = (q_1, t_1)(q_2, t_2) \cdots (q_k, t_k)$ with $i, k \in \mathbb{N}$, $i \leqslant k, q_1, \ldots, q_k \in \Omega$ and $t_1, \ldots t_k \in \mathbb{R}_+$. Moreover, let $T_i = t_1 + \ldots + t_i$ and hence $T = T_k$. Then, for all $t \in [0, T]$, we have

$$\sigma(t) = \begin{cases} q_1 & \text{if } t \in [0, T_1), \\ q_i & \text{if } t \in [T_{i-1}, T_i), \ 2 \leqslant i \leqslant k. \end{cases} \qquad (2)$$

Furthermore, the transition from mode q_i to mode q_{i+1} at time T_i is made via the switching or coupling matrices $\mathbf{K}_{q_i, q_{i+1}} \in \mathbb{R}^{n_{q_{i+1}} \times n_{q_i}}$, where $q_i, q_{i+1} \in \Omega$, as follows: $\mathbf{x}_{q_{i+1}}(T_i) = \mathbf{K}_{q_i, q_{i+1}} \lim_{t \nearrow T_i} \mathbf{x}_{q_i}(t)$. This limit to the left exists because the evolution of the switched system $\mathbf{\Sigma}$ in $[T_{i-1}, T_i)$ coincides with the evolution of the mode q_i linear subsystem active in the same interval of time.

The switching matrices $\mathbf{K}_{q_i, q_{i+1}}$ allow having different dimensions for the subsystems active in different modes. If the $\mathbf{K}_{q_i, q_{i+1}}$ matrices are not explicitly given, it is considered that they are identity matrices.

For ease and clarity of presentation, we will focus our attention to the case with two switching modes. Hence, in the what follows, the presentation is tailored to the case for which $\mathbf{\Sigma}$ switches between $D = 2$ subsystems only. Let the latter be denoted by $\mathbf{\Sigma}_1$ and $\mathbf{\Sigma}_2$ and be described by the following differential equations:

$$\mathbf{\Sigma}_1 : \begin{cases} \dot{\mathbf{x}}_1(t) = \mathbf{A}_1 \mathbf{x}_1(t) + \mathbf{B}_1 u(t) \\ y(t) = \mathbf{C}_1 \mathbf{x}_1(t) \end{cases}, \quad \mathbf{\Sigma}_2 : \begin{cases} \dot{\mathbf{x}}_2(t) = \mathbf{A}_2 \mathbf{x}_2(t) + \mathbf{B}_2 u(t) \\ y(t) = \mathbf{C}_2 \mathbf{x}_2(t) \end{cases}. \qquad (3)$$

Moreover, the matrices A_q are stable for all $1 \leqslant q \leqslant D$, i.e., all eigenvalues of A_q have a strictly negative real part. This implies that all linear subsystems of the SLS are stable.

2.1 Input–Output Mapping of an SLS

The input–output behavior of an SLS can be described in time domain using the mapping $\mathbf{y} = \mathbf{f}(\mathbf{u}, \sigma)$ that can be decomposed into a *generalized kernel representation* (as presented in [18]). In the case of two modes and zero initial condition, the input–output behavior is determined by analytic functions $\mathbf{h}_{q_i, \dots, q_k} : \mathbb{R}_+^k \to \mathbb{R}^{p \times m}$ with $q_i, \dots, q_k \in \Omega$, $i, k \geqslant 1$, $k \geqslant i$. Hence, for all pairs $(\mathbf{u}, \mathbf{z})$ composed of control input $\mathbf{u}(t) \in \mathbb{R}^m$ and sequence $\mathbf{z} = (q_1, t_1)(q_2, t_2) \cdots (q_k, t_k)$, we can write:

$$\mathbf{f}(\mathbf{u}, \mathbf{z}) = \sum_{i=1}^{k} \int_0^{t_i} \mathbf{h}_{q_i, q_{i+1}, \dots, q_k}(t_i - \tau, t_{i+1}, \dots, t_k) \mathbf{u}(\tau + t_1 + \dots + t_{i-1}) \mathrm{d}\tau. \qquad (4)$$

The kernel functions $\mathbf{h}$ can be explicitly written in terms of the system's matrices for $i, k \geqslant 1$, $k \geqslant i$, as follows,

$$\mathbf{h}_{q_i, q_{i+1}, \dots, q_k}(t_i, t_{i+1}, \dots, t_k) = \mathbf{C}_{q_k} e^{\mathbf{A}_{q_k} t_k} \mathbf{K}_{q_{k-1}, q_k} e^{\mathbf{A}_{q_{k-1}} t_{k-1}} \cdots \mathbf{K}_{q_i, q_{i+1}} e^{\mathbf{A}_{q_i} t_i} \mathbf{B}_{q_i}. \qquad (5)$$

Based on the kernels in (5), introduce time-domain multivariate functionals that will be used in construction of the generalized infinite Gramian matrices for SLSs:

$$\mathbf{r}_1(t_1) = e^{\mathbf{A}_1 t_1}\mathbf{B}_1, \quad \mathbf{r}_2(t_2) = e^{\mathbf{A}_2 t_1}\mathbf{B}_2, \quad \mathbf{r}_{1,2}(t_1, t_2) = e^{\mathbf{A}_2 t_2}\mathbf{K}_{1,2}e^{\mathbf{A}_1 t_1}\mathbf{B}_1, \ldots \quad (6)$$

$$\mathbf{o}_1(t_1) = \mathbf{C}_1 e^{\mathbf{A}_1 t_1}, \quad \mathbf{o}_2(t_2) = \mathbf{C}_2 e^{\mathbf{A}_2 t_1}, \quad \mathbf{o}_{1,2}(t_1, t_2) = \mathbf{C}_2 e^{\mathbf{A}_2 t_2}\mathbf{K}_{1,2}e^{\mathbf{A}_1 t_1}, \ldots \quad (7)$$

In general, for $q_1, \ldots, q_k \in \Omega$, and based on the kernels in (5), one can write that:

$$\mathbf{h}_{q_1, \cdots, q_k}(t_1, \ldots, t_k) = \mathbf{C}_{q_k}\mathbf{r}_{q_1, \cdots, q_k}(t_1, \ldots, t_k) = \mathbf{o}_{q_1, \cdots, q_k}(t_1, \ldots, t_k)\mathbf{B}_{q_1}. \quad (8)$$

In the case where all switchings are possible, we need all the kernels in (8) in order to determine the dynamics of $\mathbf{\Sigma}$. However, if the switching pattern is constrained, then only a limited number of kernels determines the behavior of $\mathbf{\Sigma}$. Later, we use those kernels to construct the reachability and observability Gramians that are related to a constrained switching. In order to formalize the set of constrained switchings, we briefly introduce the notion of a *language*.

2.2   Constrained Switching Described by Languages

Let Ω^+ be the infinite set of non-empty sequences that can be formed with elements from $\Omega = \{1, 2\}$, i.e. $\Omega^+ = \{1, 2, 12, 21, 121, 212, \ldots\}$. Note that the kernel functions in (5) encode the input–output energy transfer for all possible switching scenarios (encoded as sequences from the set Ω^+). In what follows, we introduce the notion of a language, which encodes the restricted switching scenarios.

The elements of Ω^+ shall be referred to as words over Ω and any nonempty set $\mathscr{L} \subseteq \Omega^+$ is a language over Ω. For a word $w = q_1 q_2 \cdots q_k \in \Omega^+$, with $q_i \in \Omega$, $1 \leqslant i \leqslant k$, we consider q_i to be the ith letter of the word w. The concatenation of a word $w \in \Omega^+$ with a word $v \in \Omega^+$ will be denoted with $wv \in \Omega^+$. For example, if $w = 12$ and $v = 1$, then $wv = 121$. Also, a word $v \in \Omega^+$ is a prefix of a word $w \in \Omega^+$ if there exists $z \in \Omega^+$ so that $w = vz$. Conversely, a word $v \in \Omega^+$ is a suffix of a word $w \in \Omega^+$ if there exists $y \in \Omega^+$ so that $w = yv$.

For a language $\mathscr{L} \subseteq \Omega^+$, introduce another language $\mathscr{L}_-$ as the 1-prefix of $\mathscr{L}$ and which contains all nonempty words from $\mathscr{L}$ by omitting the last letter of each. Additionally, introduce the language $_-\mathscr{L}$ as the 1-suffix of $\mathscr{L}$, which contains all words from $\mathscr{L}$ by omitting the first letter of each.

Definition 1 A language $\mathscr{L} \subseteq \Omega^+$ is complete if the conditions $\mathscr{L}_- \subseteq \mathscr{L}$ and $_-\mathscr{L} \subseteq \mathscr{L}$ are simultaneously verified.

Thus, we conclude that the language $\mathscr{L}$ is complete if for any word $w \in \mathscr{L}$, any prefix or suffix of w also belongs to $\mathscr{L}$.

Next, denote with $|w|$ the length of a word $w \in \Omega^+$, i.e. the number of letters and also with $|\mathscr{L}|$ the length of a language $\mathscr{L} \subseteq \Omega^+$, i.e. the number of words. We say that a language $\mathscr{L}$ is finite, if $|\mathscr{L}| < \infty$.

For example, $|121| = 3$ and $|\Omega^+| = +\infty$. Consider the following language $\mathscr{L} = \{1, 2, 12, 21, 121\} \subseteq \{1, 2\}^+$. Then, it follows that $\mathscr{L}_- = \{1, 2, 12\}$, $_-\mathscr{L} = \{1, 2, 21\}$ and $|\mathscr{L}| = 5$. Note also that this language is complete.

3   Gramians for SLSs with Constrained Switching

In this section, we introduce the Gramians for SLSs with constrained switching. This is inspired by the Gramians for LTI systems. We recall that an LTI system is described by the equations $\{\dot{\mathbf{x}} = \mathbf{A}\mathbf{x} + \mathbf{B}\mathbf{u}, \mathbf{y} = \mathbf{C}\mathbf{x}$. The input–output behavior of such a system is associated with the kernel $\mathbf{h} = \mathbf{C}e^{\mathbf{A}t}\mathbf{B}$, which can be decomposed as $\mathbf{r}(t) = e^{\mathbf{A}t}\mathbf{B}$ and $\mathbf{o}(t) = \mathbf{C}e^{\mathbf{A}t}$. Using these kernels, the reachability infinite Gramian $\mathscr{P}$ and the observability infinite Gramian $\mathscr{Q}$ are defined as follows:

$$
\begin{aligned}
\mathscr{P} &= \int_0^{+\infty} e^{\mathbf{A}t}\mathbf{B}(e^{\mathbf{A}t}\mathbf{B})^T \, dt = \int_0^{+\infty} \mathbf{r}(t)(\mathbf{r}(t))^T \, dt, \\
\mathscr{Q} &= \int_0^{+\infty} (\mathbf{C}e^{\mathbf{A}t})^T \mathbf{C}e^{\mathbf{A}t} \, dt = \int_0^{+\infty} (\mathbf{o}(t))^T \mathbf{o}(t) \, dt,
\end{aligned}
\tag{9}
$$

One can show that the $\mathscr{P}$, $\mathscr{Q}$ matrices, defined in (9), satisfy Lyapunov equations:

$$
\mathbf{A}\mathscr{P} + \mathscr{P}\mathbf{A}^T + \mathbf{B}\mathbf{B}^T = \mathbf{0}, \quad \mathbf{A}^T\mathscr{Q} + \mathscr{Q}\mathbf{A} + \mathbf{C}^T\mathbf{C} = \mathbf{0}.
\tag{10}
$$

Next, we show how the results in (9) and (10) can be extended to the SLS case.

Definition 2 Let $\boldsymbol{\Sigma}$ be an SLS as in (1) for which all subsystems are stable. Let $\mathscr{L} \subseteq \Omega^+$ be a complete and finite language over $\Omega = \{1, 2\}$, which includes all allowed switching scenarios. Let $w = q_1 \cdots q_k \in \mathscr{L}$ be a word which represents a particular switching sequence. Then we can write definitions of the infinite reachability Gramian $\mathscr{P}_w$ and of the infinite observability Gramian $\mathscr{Q}_w$, associated to the word w, in terms of the functionals in (8) as

$$
\mathscr{P}_{q_1 \cdots q_k} = \int_0^{+\infty} \cdots \int_0^{+\infty} \mathbf{r}_{q_1 \cdots q_k}(t_1, \ldots, t_k)\mathbf{r}_{q_1 \cdots q_k}(t_1, \ldots, t_k)^T \, dt_1 \cdots dt_k, \tag{11}
$$

$$
\mathscr{Q}_{q_1 \cdots q_k} = \int_0^{+\infty} \cdots \int_0^{+\infty} \mathbf{o}_{q_1 \cdots q_k}(t_1, \ldots, t_k)^T \mathbf{o}_{q_1 \cdots q_k}(t_1, \ldots, t_k) \, dt_1 \cdots dt_k. \tag{12}
$$

Lemma 1 *By assuming that the conditions stated in Definition 2 hold, it follows that the Gramians introduced in (11) and (12) satisfy the following Lyapunov equations:*

$$
\mathbf{A}_{q_k}\mathscr{P}_{q_1 \cdots q_k} + \mathscr{P}_{q_1 \cdots q_k}\mathbf{A}_{q_k}^T + \mathbf{K}_{q_{k-1}, q_k}\mathscr{P}_{q_1 \cdots q_{k-1}}\mathbf{K}_{q_{k-1}, q_k}^T = \mathbf{0},
\tag{13}
$$

$$\mathbf{A}_{q_1}^T \mathscr{Q}_{q_1 \cdots q_k} + \mathscr{Q}_{q_1 \cdots q_k} \mathbf{A}_{q_1} + \mathbf{K}_{q_1,q_2}^T \mathscr{Q}_{q_2 \cdots q_k} \mathbf{K}_{q_1,q_2} = \mathbf{0}. \tag{14}$$

As a remark, note that since the language $\mathscr{L}$ is complete and $q_1 \cdots q_k \in \mathscr{L}$, it follows that both $q_1 \cdots q_{k-1}$ and $q_2 \cdots q_k$ are part of $\mathscr{L}$. Next, introduce the definition of the reachability and observability Gramians corresponding to mode $q \in \{1, 2\}$.

Definition 3 Let Σ be an SLS as in (1) and $\mathscr{L} \subseteq \Omega^+$ be a complete language over $\Omega = \{1, 2\}$. For $q \in \{1, 2\}$, let $\mathbf{P}_q$ be the reachability Gramian associated with mode q. It follows that $\mathbf{P}_q$ can be written as a summation of the reachability Gramians corresponding to all switching sequences w from $\mathscr{L}$ that end in mode q. More precisely,

$$\mathbf{P}_q = \sum_{|w|=k, w(k)=q} \mathscr{P}_w. \tag{15}$$

Additionally, let $\mathbf{Q}_q$ be the observability Gramian associated with mode q. It can also be written as a summation of the observability Gramians corresponding to all switching sequences w from $\mathscr{L}$ that start in mode q:

$$\mathbf{Q}_q = \sum_{|w|=k, w(1)=q} \mathscr{Q}_w. \tag{16}$$

Example 1 Consider the language introduced in Sect. 2, i.e. $\mathscr{L} = \{1, 2, 12, 21, 121\} \subseteq \{1, 2\}^+$. Then, we can write that:

$$\begin{cases} \mathbf{P}_1 = \mathscr{P}_1 + \mathscr{P}_{21} + \mathscr{P}_{121}, \quad \mathbf{P}_2 = \mathscr{P}_2 + \mathscr{P}_{12}, \\ \mathbf{Q}_1 = \mathscr{Q}_1 + \mathscr{Q}_{12} + \mathscr{Q}_{121}, \quad \mathbf{Q}_2 = \mathscr{Q}_2 + \mathscr{Q}_{21}. \end{cases} \tag{17}$$

In order to explicitly compute the reachability Gramians $\mathscr{P}_q$ for $q \in \mathscr{L}$, one needs to solve a series of Lyapunov equations as stated in the following:

$$\text{M1}: \begin{cases} \mathbf{A}_1 \mathscr{P}_1 + \mathscr{P}_1 \mathbf{A}_1^T + \mathbf{B}_1 \mathbf{B}_1^T = 0 \\ \mathbf{A}_1 \mathscr{P}_{2,1} + \mathscr{P}_{2,1} \mathbf{A}_1^T + \mathbf{K}_{2,1} \mathscr{P}_2 \mathbf{K}_{2,1}^T = 0 \\ \mathbf{A}_1 \mathscr{P}_{1,2,1} + \mathscr{P}_{1,2,1} \mathbf{A}_1^T + \mathbf{K}_{2,1} \mathscr{P}_{1,2} \mathbf{K}_{2,1}^T = 0 \end{cases}, \quad \text{M2}: \begin{cases} \mathbf{A}_2 \mathscr{P}_2 + \mathscr{P}_2 \mathbf{A}_2^T + \mathbf{B}_2 \mathbf{B}_2^T = 0 \\ \mathbf{A}_2 \mathscr{P}_{1,2} + \mathscr{P}_{1,2} \mathbf{A}_2^T + \mathbf{K}_{1,2} \mathscr{P}_1 \mathbf{K}_{1,2}^T = 0 \end{cases}$$

Similarly for the observability Gramians $\mathscr{Q}_q$:

$$\text{M1}: \begin{cases} \mathbf{A}_1^T \mathscr{Q}_1 + \mathscr{Q}_1 \mathbf{A}_1 + \mathbf{C}_1^T \mathbf{C}_1 = 0 \\ \mathbf{A}_1^T \mathscr{Q}_{1,2} + \mathscr{Q}_{1,2} \mathbf{A}_1 + \mathbf{K}_{1,2}^T \mathscr{Q}_2 \mathbf{K}_{1,2} = 0 \\ \mathbf{A}_1^T \mathscr{Q}_{1,2,1} + \mathscr{Q}_{1,2,1} \mathbf{A}_1 + \mathbf{K}_{1,2}^T \mathscr{Q}_{2,1} \mathbf{K}_{1,2} = 0 \end{cases}, \quad \text{M2}: \begin{cases} \mathbf{A}_2^T \mathscr{Q}_2 + \mathscr{Q}_2 \mathbf{A}_2 + \mathbf{C}_2^T \mathbf{C}_2 = 0 \\ \mathbf{A}_2^T \mathscr{Q}_{2,1} + \mathscr{Q}_{2,1} \mathbf{A}_2 + \mathbf{K}_{2,1}^T \mathscr{Q}_1 \mathbf{K}_{2,1} = 0 \end{cases}$$

4 Description of the Method

The proposed balanced truncation method is presented in Algorithm 1. Let Σ be an SLS as described by (1) with $D = 2$ modes and $\mathscr{L} \subseteq \Omega^+$ be a complete language. Denote with $\hat{\Sigma}$ the reduced order SLS obtained by applying the new proposed BT method to Σ. For $q \in \Omega$, let $(\hat{\mathbf{A}}_q, \hat{\mathbf{B}}_q, \hat{\mathbf{C}}_q)$ be the matrices corresponding to mode q of system $\hat{\Sigma}$ and let r_q be the dimension of each reduced subsystem of $\hat{\Sigma}$. Additionally $\hat{\mathbf{K}}_{q_1,q_2}$ are the reduced-order coupling matrices corresponding to $\hat{\Sigma}$.

Algorithm 1 BT using constrained Gramians

Input: $\mathbf{A}_q$ stable matrices, $\mathbf{B}_q, \mathbf{C}_q, \mathbf{K}_{q_1,q_2}$, a finite complete language $\mathscr{L}$ and order r_q.

1: Solve the Lyapunov equations as described in (13) and (14).

2: Compute the Gramians $\mathbf{P}_q$ and $\mathbf{Q}_q$ as in (15) and (16), for $q = 1, 2$.

3: Compute the Cholesky decomposition $\mathbf{P}_q = \mathbf{U}_q \mathbf{U}_q^T$ and the eigenvalue decomposition $\mathbf{U}_q^T \mathbf{Q}_q \mathbf{U}_q = \mathbf{V}_q \mathbf{\Lambda}_q^2 \mathbf{V}_q^T$, where $\mathbf{\Lambda}_q^2$ is a diagonal matrix that contains the singular values sorted in decreasing order.

4: Construct the transformation matrices $\mathbf{T}_q \in \mathbb{R}^{n_q \times n_q}$ as $\mathbf{T}_q = \mathbf{\Lambda}_q^{1/2} \mathbf{V}_q^T \mathbf{U}_q^{-1}$.

5: Compute a balanced realization of Σ as

$$\bar{\mathbf{A}}_q = \mathbf{T}_q \mathbf{A}_q \mathbf{T}_q^{-1}, \quad \bar{\mathbf{B}}_q = \mathbf{T}_q \mathbf{B}_q, \quad \bar{\mathbf{C}}_q = \mathbf{C}_q \mathbf{T}_q^{-1}, \quad \bar{\mathbf{K}}_{q_1,q_2} = \mathbf{T}_{q_2} \mathbf{K}_{q_1,q_2} \mathbf{T}_{q_1}^{-1}. \quad (18)$$

6: Truncate the matrices in (18) as:

$$\hat{\mathbf{A}}_q = \bar{\mathbf{A}}_q(1:r_q, 1:r_q), \quad \hat{\mathbf{B}}_q = \bar{\mathbf{B}}_q(1:r_q, :), \quad \hat{\mathbf{C}}_q = \bar{\mathbf{B}}_q(:, 1:r_q),$$
$$\hat{\mathbf{K}}_{q_1,q_2} = \bar{\mathbf{K}}_{q_1,q_2}(1:r_{q_2}, 1:r_{q_1}).$$

Output: reduced order matrices $\hat{\mathbf{A}}_q, \hat{\mathbf{B}}_q, \hat{\mathbf{C}}_q, \hat{\mathbf{K}}_{q_1,q_2}$.

Remark 1 For step 2 in Algorithm 1, the computation of the Gramians $\mathbf{P}_q$ and $\mathbf{Q}_q$ is performed by using the built-in standard solvers in Matlab: *lyap* and *lyapchol* (the second one directly computes the Cholesky factors needed in step 3). Both of these two direct solvers need cubic computations with respect to the dimension of the subsystems, i.e. $O(n_q^3)$. A speed-up can be achieved by using iterative solvers that only compute approximate low-rank factors of the Gramians and hence lower the computations to $O(n_q)$ (provided that the matrices are sparse).

5 Numerical Results

In this section we compare the performance of the new proposed method against that of other methods from the literature. More exactly, the initial switched linear system Σ is going to be reduced by means of five balancing methods: Balancing using averaging linear Gramians (see [15] and denoted here by BT1), balancing using generalized Gramians based on all possible switches (see [12] and denoted here by BT2), balancing using Gramians obtained by recasting an SLS as an envelope LTI (see [21] and denoted here by BT3), balancing using Gramians based on bilinear reformulation of the SLS (see [20] and denoted here by BT4), and, the proposed method in Algorithm 1 (denoted here by BT5).

5.1   First Example

For the first experiment, consider the CD player system in [9] of order 120 with two inputs and two outputs. We consider that, at any given instance of time, only one input and one output are active (the others are not functional due to mechanical failure). More exactly, consider mode j to be activated whenever the jth input and the jth output are simultaneously failing (where $j \in \{1, 2\}$). In this way, an SLS system with two operational modes is constructed (as in [12]). Both subsystems are stable SISO linear systems of order 120.

We use the signal $\mathbf{u}(t) = \cos(10t)$ as the control input and the signal $\sigma(t)$ that is initiated in mode 2 with switching times (0, 0.5, 2, 2.5, 3.5, 5.5, 6, 7.5, 8, 9, 10), as the switching signal.

Choose the truncation orders $r_1 = r_2 = 10$ for the reduced SLS using all methods. For the BT5 method, choose the language $\mathscr{L} = \{1, 2, 12, 21\}$. We compare the time domain response of the original SLS against the ones corresponding to the five reduced models. The output approximation errors are represented in Fig. 1. Note that the response of all reduced models accurately follows that of the original model. Note that the responses of the reduced models constructed with the BT2 and BT5 methods are very similar (deviation of 10^{-6} between them). Hence, in Fig. 1, the blue and magenta curves coincide. Additionally, notice that the new proposed methods provides slightly better approximation quality as compared to the others.

Fig. 1 Deviation between the original output and the ones corresponding to the reduced SLSs

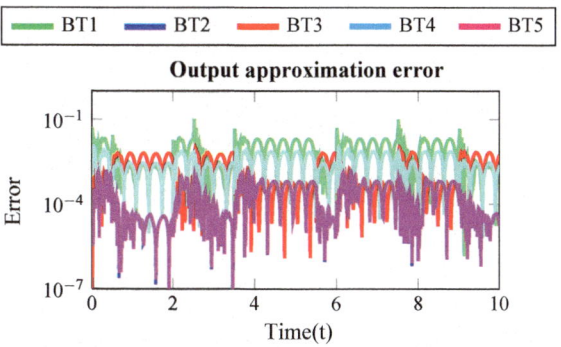

5.2   Second Example

For the second numerical experiment, let us consider an SLS with two modes of order $n_1 = n_2 = 300$ as previously introduced in [20], whose system matrices are:

$$\mathbf{A}_1 = \begin{bmatrix} -2 & 1 & & \\ 0.1 & -2 & 1 & \\ & \ddots & \ddots & \ddots \\ & & 0.1 & -2 \end{bmatrix}, \mathbf{A}_2 = \begin{bmatrix} -2 & 0.5 & & \\ 1 & -2 & 0.5 & \\ & \ddots & \ddots & \ddots \\ & & 1 & -2 \end{bmatrix}, \begin{cases} \mathbf{B}_1^T = \begin{bmatrix} 1 & 0 & \dots & 0 \end{bmatrix}, \mathbf{B}_2^T = \begin{bmatrix} 0 & \dots & 0 & 1 \end{bmatrix}, \\ \mathbf{C}_1 = \begin{bmatrix} 0 & 1 & 0 & \dots & 0 \end{bmatrix}, \mathbf{C}_2 = \begin{bmatrix} 0 & \dots & 0 & 1 & 0 \end{bmatrix} \end{cases}.$$

We use the signal $\mathbf{u}(t) = 10e^{-1/2t}\sin(20t)$ as the control input and the signal $\sigma(t)$ that is initiated in mode 1 with switching times (0, 1, 1.5, 2, 3, 4.5, 5, 6, 7.5, 8, 10), as the switching signal.

Choose the truncation order $r = 20$ for the reduced SLSs using all methods. We again compare the time domain response of the original SLS against the ones corresponding to the five reduced models. To this aim, we consider the following two languages that encode the allowed switching sequences:

1. For the first case, let $\mathscr{L}_1 = \{1, 2, 12, 21\}$.
2. For the second case, let $\mathscr{L}_2 = \{2, 1, 21, 12, 212, 121\}$.

The output approximation errors are presented in Fig. 2. Note that, for the BT5 method, the left figure corresponds to the language $\mathscr{L}_1$, while the right figure corresponds to the language $\mathscr{L}_2$. Since only BT5 is a language-dependent method, the curves corresponding to BT1, BT2, BT3 and BT4 are the same in both figures.

By inspecting Fig. 2, we observe that both BT1 and BT3 methods provide models with poor approximation quality. A possible explanation for this behavior is that BT3 is designed for low-rank switching models as stated in [21]. The model analyzed

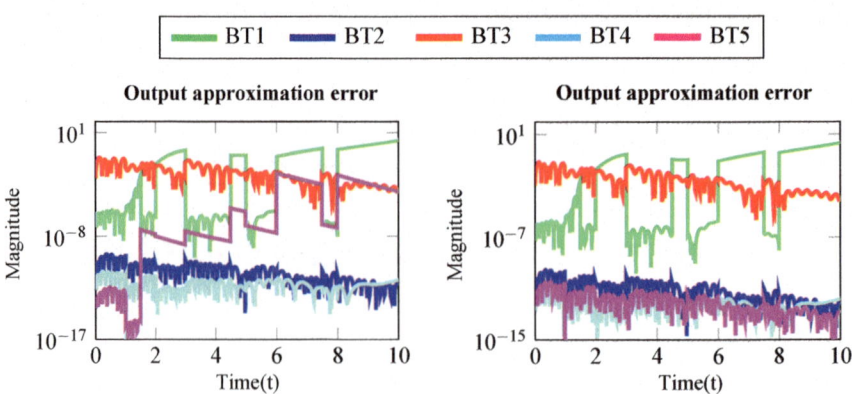

Fig. 2 Deviation between the original output and the ones corresponding to the reduced SLSs (the left figure corresponds to $\mathscr{L}_1$ while the right figure corresponds to $\mathscr{L}_2$)

in this example does not possess this property. On the other hand, note that the approximation quality of the original response using either the BT2 or BT4 methods is considerably better.

When using the language $\mathscr{L}_1$, the output approximation curve corresponding to BT5 (depicted with magenta in the left side of Fig. 2) increases after two switches. This is due to the fact that the language $\mathscr{L}_1$ does not include sufficient information.

By incorporating the sequences 121 and 212 into $\mathscr{L}_2$, we notice that the error curve (depicted with magenta in the right side of Fig. 2) stagnates at a magnitude around 10^{-12}. Hence, the new method provides comparable approximation accuracy with that of the methods BT2 and BT4 (that require solving generalized Sylvester equations and hence need longer computational time to yield a ROM).

Remark 2 The BT1 method requires the solution of a number of $2D$ standard Lyapunov equations (where D is the number of modes/subsystems of the SLS), while the BT2 method requires the solution of $2D$ generalized Lyapunov equations. Finally, the BT3 method relies on solving two standard Lyapunov equations, while the BT4 method needs the solution of two generalized Lyapunov equations.

Remark 3 The key difference between the new algorithm and the previously proposed algorithm in [12] is the definition and the computation of the Gramian matrices. With the new approach we avoid having to solve generalized Lyapunov equations, for which the solution is computational challenging and not straightforward. Instead, we relax this task by solving a number of standard Lyapunov equations which can be performed in a reliable and fast manner.

Remark 4 The computational effort of Algorithm 1 directly depends on the performance of the solver used for computing the Gramian matrices in step 2 of the algorithm. In this work, we rely on direct built-in solvers as mentioned in Remark 1. The size of the complete language directly influences the computations. In particular, the number of linear Lyapunov equations that need to be solved is equal to the number of words in the language.

6 Conclusion and Outlook

In the current work, we have proposed generalized reachability and observability Gramians for SLSs with constrained switching, which can be computed by solving a series of classical Lyapunov equations. Also, these Gramians encode the reachable and observable sets of the SLS tailored to the specific switching scenarios (described by the corresponding language). Based on the new Gramian matrices, a balancing-type procedure is proposed which enables to find local projection matrices used to construct a reduced order model. Finally, the practical applicability of the proposed method was illustrated by means of two numerical examples. Possible future research directions include establishing connections between the newly introduced Gramians and the ones introduced in [12] for switching scenarios described by infinite languages and extending this framework for SLSs with more than two modes.

References

1. Antoulas, A.C.: Approximation of Large-Scale Dynamical Systems. SIAM, Philadelphia (2005)
2. Bastug, M., Petreczky, M., Wisniewski, R., Leth, J.: Model reduction of linear switched systems by restricting discrete dynamics. In: Proceedings of the 53rd IEEE CDC (2014)
3. Bastug, M., Petreczky, M., Wisniewski, R., Leth, J.: Model reduction by nice selections for linear switched systems. IEEE Trans. Autom. Control. **61**(11), 3422–3437 (2016)
4. Baur, U., Benner, P., Feng, L.: Model order reduction for linear and nonlinear systems: a system-theoretic perspective. Arch. Comput. Meth. Eng. **21**, 331–358 (2014)
5. Benner, P., Cohen, A., Ohlberger, M., Willcox, K. (eds.): Model Reduction and Approximation. SIAM Publications, Philadelphia (2017)
6. Benner, P., Gugercin, S., Willcox, K.: A survey of projection-based model reduction methods for parametric dynamical systems. SIAM Rev. **57**(4), 483–531 (2015)
7. Benner, P., Stykel, T.: Model Order Reduction for Differential-Algebraic Equations: A Survey, Chap. 3. Surveys in Differential-Algebraic Equations IV, Part of the series Differential-Algebraic Equations Forum, pp. 107–160. Springer, Berlin (2017)
8. Birouche, A., Mourllion, B., Basset, M.: Model order-reduction for discrete-time switched linear systems. Int. J. Syst. Sci. **43**(9), 1753–1763 (2012)
9. Chahlaoui, Y., Van Dooren, P.: A collection of benchmark examples for model reduction of linear time invariant dynamical systems (2002). http://slicot.org/20-site/126-benchmark-examples-for-model-reduction
10. Gao, H., Lam, J., Wang, C.: Model simplification for switched hybrid systems. Syst. Control. Lett. **55**, 1015–1021 (2006)
11. Gosea, I.V., Petreczky, M., Antoulas, A.C.: Data-driven model order reduction of linear switched systems in the Loewner framework. SIAM J. Sci. Comput. **40**(2), B572–B610 (2018)
12. Gosea, I.V., Petreczky, M., Antoulas, A.C., Fiter, C.: Balanced truncation for linear switched systems. Adv. Comput. Math. **44**(6), 1845–1886 (2018)
13. Liberzon, D.: Switching in Systems and Control. Birkhäuser (2008)
14. Mehrmann, V., Stykel, T.: Balanced truncation model reduction for large-scale systems in descriptor form, Chap. 45. In: Benner, P., Mehrmann, V., Sorensen, D.C. (eds.), Dimension Reduction of Large-Scale Systems, pp. 83–115. Springer, Berlin (2005)
15. Monshizadeh, N., Trentelman, H.L., Camlibel, M.K.: A simultaneous balanced truncation approach to model reduction of switched linear systems. IEEE Trans. Automat. Control **57**(12), 3118–3131 (2012)
16. Moore, B.: Principal component analysis in linear systems: controllability, observability, and model reduction. IEEE Trans. Automat. Control **26**, 17–32 (1981)
17. Pernebo, L., Silverman, L.: Model reduction via balanced state space representation. IEEE Trans. Automat. Control **27**, 382387 (1982)
18. Petreczky, M., van Schuppen, J.H.: Partial-realization theory for linear switched systems - a formal power series approach. Automatica **47**, 2177–2184 (2011)
19. Petreczky, M., Wisniewski, R., Leth, J.: Balanced truncation for linear switched systems. In: Nonlinear Analysis: Hybrid Systems, Special Issue related to IFAC Conference on Analysis and Design of Hybrid Systems (ADHS 12), vol. 10, pp. 4–20 (2013)
20. Pontes Duff, I., Grundel, S., Benner, P.: New Gramians for switched linear systems: Reachability, observability, and model reduction. https://arxiv.org/abs/1806.00406, Accepted for publication in IEEE Trans. Auto. Control (2019)

21. Schulze, P., Unger, B.: Model reduction for linear systems with low-rank switching. SIAM J. Control Optim. **56**(6), 4365–4384 (2018)
22. Shaker, H.R., Wisniewski, R.: Generalized Gramian framework for model/controller order reduction of switched systems. Int. J. Syst. Sci. **42**(8), 1277–1291 (2011)
23. Shaker, H.R., Wisniewski, R.: Model reduction of switched systems based on switching generalized Gramians. Int. J. Innov. Comput., Info. Contr. **8**(7(B)), 5025–5044 (2012)
24. Sun, Z., Ge, S.S.: Switched Linear Systems: Control and Design. Springer, Berlin (2005)
25. Sun, Z., Ge, S.S.: Stability Theory of Switched Dynamical Systems. Springer, Berlin (2011)
26. Sun, Z., Ge, S.S., Lee, T.H.: Controllability and reachability criteria for switched linear systems. Automatica **38**, 775–786 (2002)

A Review on Reduced Order Modeling using DMD-Based Methods

Soledad Le Clainche and José M. Vega

Abstract This article illustrates a review on the applications of a new method that can be used either as a reduced order model or to uncover the underlying physics in spatio-temporal data. The method is based on the higher order dynamic mode decomposition (a recent extension of standard dynamic mode decomposition) of the given data, which leads to a purely data-driven, equation free approach. The high accuracy and robustness of the method makes it suitable to analyze very complex spatio-temporal data resulting from either numerical simulations or experimental measurements. The article illustrates the good performance and versatility of this new reduced order model in two specific applications: (i) speeding up numerical simulations in the wake of a circular cylinder and (ii) wind forecasting upstream wind turbines using actual experimental data databases.

Keywords Higher order dynamic mode decomposition · Data driven reduced order models · Temporal extrapolation from spatio-temporal databases · Cylinder wake · Data forecasting in wind turbines

1 Introduction

Complex, generally unsteady flows are found both in nature and in a wide range of industrial applications. For this reason, the study and understanding of the flow behaviour is a research topic of high interest. Reduced order models (ROMs) are generally used as a simple way of describing/simulating complex flows. The purpose of developing ROMs lies in several advantages for different applications. Among

S. L. Clainche (✉) · J. M. Vega
E.T.S.I. Aeronáutica y del Espacio, Universidad Politécnica de Madrid, 20040 Madrid, Spain
e-mail: soledad.leclainche@upm.es

J. M. Vega
e-mail: josemanuel.vega@upm.es

© Springer Nature Switzerland AG 2020   55
J. Fehr and B. Haasdonk (eds.), *IUTAM Symposium on Model Order Reduction of Coupled Systems, Stuttgart, Germany, May 22–25, 2018*, IUTAM Bookseries 36,
https://doi.org/10.1007/978-3-030-21013-7_4

the most important ones, it is possible to find the development of ROMs for fluid dynamics, with the aim at, for example:

a. Reducing the computational cost in numerical simulations. This task, in turn, can be performed in three different ways:

- *Preprocessed ROMs*, which are constructed by first computing a set of snapshots (following seminal ideas by Sirovich [37]) using a standard numerical solver (pre-process), then extracting a limited number of representative modes (dimension-reduction, using, e.g., POD), and finally projecting the governing equations onto the set of modes, which gives a reduced order system, whose integration is much faster than the full numerical solver. There is a variety of methods to perform these steps, including standard proper orthogonal decomposition [2], proper generalized decomposition [8], and reduced basis methods [33]. These ROMs usually exhibit a very fast online operation and in time-dependent simulations, they are useful mainly to simulate permanent dynamics, excluding transients, which can be complicated to reproduce since their dynamics are unpredictable.
- *Adaptive ROMs*. The two difficulties mentioned above in connection with preprocessed ROMs can be overcome combining along the simulation a standard numerical solver and a reduced system. The resulting method, known as POD on the Fly [36], does not require any preprocess and is able to simulate not only permanent dynamics, but also transients. The method has been used for various purposes in various fields; see, e.g., [24, 34, 40].
- *Data-driven ROMs*. Using a purely data-driven method, which does not use the underlying equations at all [21, 26] (except to generate the data when numerical data is used), although it assumes a model. One such ROM will be considered in the present paper. The benefit of this type of ROMS is that they can be applied to any type of data, even experiments (as it will be presented below).

b. Understanding the underlying physics in either numerical or experimental data, and predicting the different involved flow states [27, 29]. This is one of the tasks that will be addressed in the present paper.
c. Providing efficient tools for various applications, including flow control [13] and optimal design [30].

Describing the nonlinear dynamics underlying complex flows requires identifying both transient and permanent behaviours with a relatively good accuracy. This task may be performed by decomposing the given spatio-temporal data $\mathbf{v}(x, \mathbf{t})$ as a sum of simpler Fourier-like spatio-temporal patterns, in terms of normalized spatial modes $u_n(x)$, weighted with appropriate amplitudes a_n, and showing a temporal behavior determined by the growth rates δ_n (whose sign determines temporal growth or decay) and frequencies ω_n (which determine temporal oscillation), as

$$\mathbf{v}(x, \mathbf{t}) \simeq \sum_{n=1} \mathbf{a_n} u_n(x) \mathrm{e}^{(\delta_n + \mathrm{i}\omega_n)\mathbf{t}}. \tag{1}$$

Classical techniques can be used to obtain this approximation, based on, e.g., Fast Fourier Transform or Power Spectral Density. However, the main inconvenient of these methods is that the temporal length of the data analyzed must be very large to obtain reasonably good results. A good alternative is using some other more sophisticated techniques, such as dynamic mode decomposition (DMD) [35] or its recently introduced extension, higher order DMD (HODMD) [25], which has been successfully used in the analysis of complex flows [27], or with traditional methods for reduced order modeling [1].

Once the amplitudes, spatial modes, growth rates, and frequencies appearing in the expansion (1) have been accurately computed, transient and permanent dynamics are identified as resulting from modes with negative and zero (or very small) growth rates, respectively, namely, $\delta_n < 0$ and $\delta_n = 0$, respectively. Thus, the final attractor can be obtained from transient dynamics by just neglecting in (1) those modes such that $\delta_n < 0$. This is the basic idea behind using HODMD as a reduced order model (ROM) to predict the final permanent dynamics from transient behavior, which involves temporal extrapolation and will be considered in Sect. 3. Likewise, the spatial modes u_n appearing in (1) can be spatially interpolated or extrapolated to obtain the temporal data at spatial locations not considered in the original database, as it will be done in Sect. 4.

With the above in mind, after a brief description of the HODMD method, in Sect. 2, this article presents a review on two representative applications of the method, one dealing with its use as a data driven ROM, which is intended to diminish the computational cost in numerical simulations and will also be used for temporal extrapolation, in Sect. 3, and another concerned with the use of the method to uncovering and forecasting the dynamics behind experimental data in wind turbines, in Sect. 4. The paper ends with some concluding remarks, in Sect. 5.

2 Higher Order Dynamic Mode Decomposition

For convenience, we consider the discretized version of (1), referred to a (not necessarily structured) spatial mesh, labeled with the index j and a uniform temporal mesh, labeled with the index k, as

$$\mathbf{v}(\mathbf{x_j}, \mathbf{t_k}) \simeq \sum_n \mathbf{a_n} \mathbf{q_n}(\mathbf{x_j}) \, e^{(\delta_n + i\omega_n)(k-1)\Delta t}, \tag{2}$$

which is rewritten in the form (1) by appropriate interpolation in the x and setting $(k-1)\Delta t = t$ (which involves interpolation in the time variable). The spatial distributions for the discrete values of t are called the snapshots. Two definitions are now in order. The dimension of the span of the vector space generated by the spatial modes q_n is known as the *spatial complexity* of the expansion, while the number of terms appearing in (2) is called the *spectral complexity*, which at least equals the spatial complexity. However, the spectral complexity is larger than the spatial complexity in many problems of scientific and industrial interest; see [24].

Now, the basic idea of standard DMD and its difference with HODMD is briefly introduced in this chapter, although a better description of the algorithms can be found in [23] and [25], respectively. The DMD expansion appearing in (2) is such that

$$\mathbf{v}(\boldsymbol{x}_j, \mathbf{t}_{k+1}) \simeq \mathscr{L}\mathbf{v}(\boldsymbol{x}_j, \mathbf{t}_k), \tag{3}$$

where the finite-dimensional linear operator $\mathscr{L}$ is known as the Koopman operator. Once this operator has been computed (via, e.g., the pseudoinverse), its eigenvectors give the modes $\boldsymbol{q}_n$ appearing in (2), and their eigenvalues μ_n give the growth rates and frequencies, as $\delta_n + i\omega_n = \log(\mu_n)/\Delta t$; the amplitudes a_n may be computed via least squares fitting. However, the number of eigenvalues of $\mathscr{L}$ coincides with its size, meaning that this method can only provide the expansion (2) if the spectral complexity coincides with the spatial complexity. In the more general case, when the spectral complexity is larger than the spatial complexity, the standard DMD method does not provide good results and should be substituted by the HODMD method [25], which combines standard DMD with Takens' delay embedding theorem [39], using d index-lagged snapshots. Combination of DMD and delayed snapshots was previously suggested [41] and performed [6], in a spirit different from what is presented in this section [25]. On the other hand, in the literature it is possible to find some other variants of DMD with improved performance, although they are not related to time-delayed snapshots. Some examples are: *sparsity promoting DMD* [19] (uses a penalty to identify a smaller set of important modes using optimization techniques), *extended DMD* [43] (extends the DMD approximation with more basis functions, allowing the method to capture more complex dynamics), *optimized DMD* [7] (the expansion is computed solving an optimization problem) and some other variants suitable to reduce the level of noise in the data [9, 10, 16, 38]

The HODMD method proceeds in two steps (which are briefly summarized here, see [25] for a more detailed description and illustration of the method). A version of the code can be downloaded in [17].

1. *Dimension reduction*. To begin with, truncated singular value decomposition (SVD) [14] is applied to the snapshot matrix (whose columns are the snapshots), which gives a set of reduced snapshots. When the spatial mesh is structured, truncated SDV can be advantageously replaced by truncated higher order SVD [22, 42], which is an extension of standard SVD (which treats matrices) to deal with tensors.

2. *Generating the expansion (2)*. The standard DMD method is applied to an enlarged reduced snapshot matrix that contains, not only the reduced snapshots but also the result of applying d shifts in then index k to these snapshots. The result is an expansion of the form (2) for the reduced snapshots, which using the SVD reconstruction yields the expansion (2) for the original snapshots and completes the algorithm.

As described, the method depends on three tunable parameters, namely the index d, the tolerance ε_1 used in step 1 to truncate the SVD expansion, and a second tolerance ε_2 that is used in step 2 to neglect in (1) those modes that exhibit a too small amplitude a_n. These parameters are selected after some calibration of the method with the aim of minimizing the relative error of the reconstruction of the original data using expansion (1).

It is interesting to comment the role of errors, which are unavoidable in both numerical data (truncation and round off errors) and specially in experimental data (ambient noise and other experimental artifacts). Appropriate selection of the tolerances ε_1 and ε_2, comparable to the error size, help to filter errors. However, a more efficient error filtering method results from applying iteratively steps 1 and 2 [27].

3 Application I: Acceleration of Numerical Simulations

When computing an attractor by an unsteady numerical simulation, transient dynamics are usually found in the initial stage, where a large number of frequencies (physical and spurious) develop simultaneously. The development and evolution of the modes associated with these frequencies directly depend on the spatial and temporal discretization used to solve the numerical equations. For example, it is well known that low order schemes (i.e.: finite volumes of second order) are more dissipative than high order schemes (i.e.: high spectral methods). Thus low order schemes introduce a smaller number of frequencies, since the small amplitude frequencies are dissipated [11].

In this article, HODMD is used as a ROM to predict the final attractor from a group of data collected in the initial transient stage of a numerical simulation. Note that such aim involves temporal extrapolation. To this end, the expansion (1) will be constructed retaining only the permanent modes, in principle those with $\delta_n = 0$. However, due to numerical errors, the growth rate of permanent modes will not be exactly zero, but close to zero, defined as $|\delta_n| < \varepsilon$ (where ε is tunable). Once the permanent modes are identified, we enforce that their associated growth rates be exactly equal to zero, which converts the original expansion (1) (obtained via HODMD) into the following expansion

$$\mathbf{v}(\mathbf{x}, \mathbf{t}) \simeq \sum_{n=1}^{N} \mathbf{a_n} u_n(\mathbf{x}) e^{i\omega_n t}. \tag{4}$$

As a representative example we will show the performance of the DMD based ROM in the three-dimensional incompressible cylinder wake, which is a classical fluid dynamics problem [44, 45]. In the simplest incompressible formulation, assuming spatially uniform density and kinematic viscosity v, the nondimensional velocity and pressure fields satisfy the continuity and Navier-Stokes equations

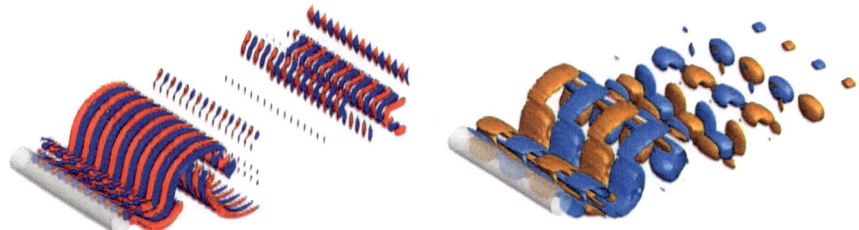

Fig. 1 Modes B (left) and A (right) of the three-dimensional cylinder wake obtained at $Re = 280$ and $L_z = 6.99$

$$\nabla \cdot \mathbf{v} = \mathbf{0}, \tag{5}$$

$$\frac{\partial \mathbf{v}}{\partial t} + (\mathbf{v} \cdot \nabla)\mathbf{v} = -\nabla \mathbf{p} + \frac{1}{\mathbf{Re}} \nabla^2 \mathbf{v}, \tag{6}$$

where the Reynolds number is defined in terms of the free stream velocity U and the cylinder diameter D as $Re = UD/\nu$. As is well known, the steady, reflection-symmetric solution for small Re exhibits a Hopf bifurcation at $Re \simeq 46$ [18, 32] that produces a two-dimensional but unsteady periodic von Karman vortex street flow. This flow remains orbitally stable up to $Re \simeq 190$, where it suffers a secondary bifurcation (Floquet multiplier $= 1$) and becomes three-dimensional [3] for some specific values of the spanwise wave length β (or the spanwise period $L_z = 2\pi/\beta$). It is possible to find two types of modes beyond this secondary instability depending on the values of Re and L_z: the synchronous periodic modes and the asynchronous quasi-periodic modes. The synchronous modes, known as modes A and B, are standing wave modes presenting different spatio-temporal symmetries. An example of modes A and B is presented in Fig. 1, as obtained via Floquet linear analysis in [31]. These modes oscillate with frequencies that are similar to that of the primary two-dimensional periodic flow [3–5]. Mode A is a long-wave mode that emerges at $Re > 189$, while mode B is a short-wave mode that appears at $Re > 259$. The asynchronous quasi-periodic modes emerge at $Re \simeq 380$ through an instability associated with a pair of complex Floquet multipliers [5].

In order to illustrate the HODMD-based ROM described above, the numerical data has been obtained upon integration of the continuity and Navier-Stokes equations (5)–(6) at $Re = 220$ and $L_z = 4$. Note that mode A is to be present for this value of the Reynolds number. The numerical solver is the open source, spectral element code Nek5000 [12], which has been applied in a sufficiently large computational domain. The boundary conditions are no-slip at the cylinder surface, periodic in the spanwise direction, and appropriately non-reflecting at the outer boundary of the computational domain; see [26] for further details.

Figure 2 shows the evolution of the spanwise velocity at a representative point of the computational domain, with coordinates $(x, y, z) = (2, 0, 0)$. The cylinder axis is the z-axis and the nondimensional cylinder diameter is 1, which means that the selected point is the near field on the cylinder flow. As can be seen, there is a

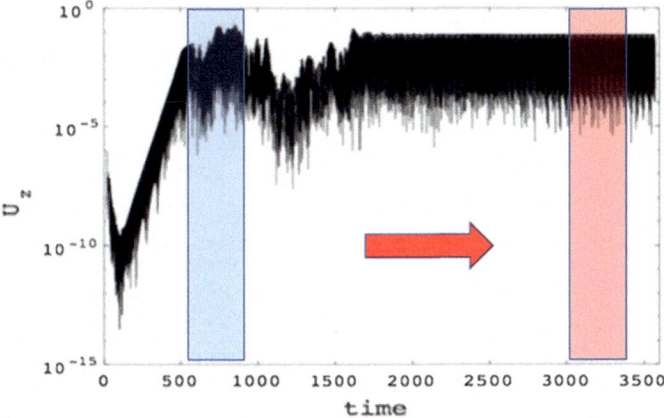

Fig. 2 Evolution of spanwise velocity in the three-dimensional numerical simulations of the cylinder wake at $Re = 280$ and $L_z = 4$ at a representative point near the cylinder. The blue rectangle represents the transient stage where the data has been extracted for the HODMD computations, and the red rectangle represents the attractor

very long transient stage until the final attractor is reached, at $t \sim 2900$. This fact makes it worthy the use of the HODMD-based ROM for guessing the attractor upon extrapolation from the transient stage, instead of continuing the integration of the Navier-Stokes equations until the attractor is reached. In fact, such extrapolation will be performed using data from the short interval $575 \leq t \leq 825$, highlighted in blue in Fig. 2, which means that the acceleration factor resulting from using the ROM will be $\sim 2900/575 = 5$.

In order to construct the HODMD-based ROM, a set of 500 snapshots have been collected in the above mentioned transient region. HODMD has been applied to these snapshots using the following values of the tunable parameters $d = 250$, $\varepsilon_1 = 10^{-4}$, and $\varepsilon_2 = 5 \cdot 10^{-3}$ (set after some calibration), for which the HODMD method retains 19 modes. The permanent modes are defined such that $|\delta_n| < 10^{-3}$. Setting $\delta_n = 0$ in these modes and neglecting the remaining modes lead to the DMD expansion (4), which is used as an approximation of the attractor. The relative root mean square (RMS) error of this approximation at $t = 2900$ is $\sim 6 \cdot 10^{-2}$. It must be noted that standard DMD (which coincides with HODMD for $d = 1$), with the same values of ε_1 and ε_2, retains only 9 modes and gives a worse approximation, with a relative RMS error ~ 0.4, which is five times larger than its counterpart obtained for $d = 250$. Retaining 19 modes in DMD only provides worse results, since the remaining modes captured by the method are related with transient or spurious artifacts (not permanents). These results are illustrated in Fig. 3, where a representative snapshot is plotted for the original flow and its reconstructions using $d = 250$ and $d = 1$. Three remarks are in order in connection with this figure:

- Plotting the level lines does not make justice to the approximation, especially at those regions where the velocity distributions are fairly flat or small.

Fig. 3 A representative snapshot in the attractor showing the streamwise (left), normal (middle), and spanwise (right) velocity components at the $z = 0$ plane for $t = 2900$, considering the original data and the HODMD extrapolations for $d = 250$ and $d = 1$

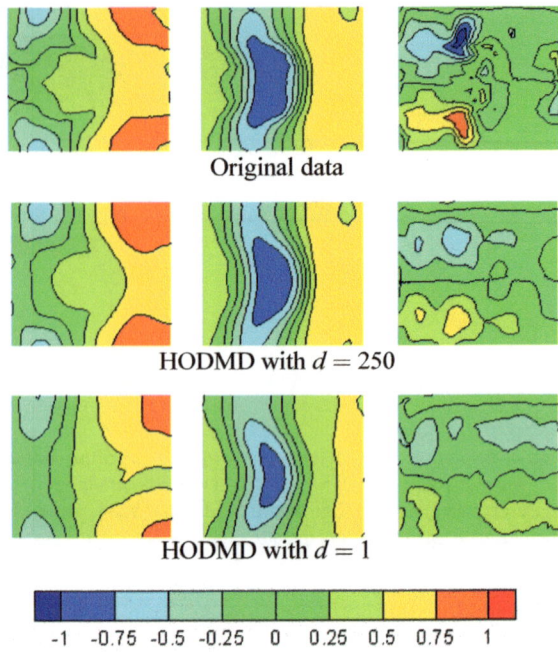

- The approximation for the spanwise velocity component is much worse than for the streamwise and normal components because the spanwise component is quite small.
- The approximation with $d = 250$ is much better than with $d = 1$, especially for the streamwise velocity component. In particular, the up-down symmetry in the original data is preserved with $d = 250$, but not with $d = 1$.

Summarizing the above, the HODMD-based ROM gives fairly good reconstruction of the final attractor, with a computational cost that is five times smaller than its counterpart using the full numerical solver.

4   Application II: Data Forecasting in Wind Turbines

As another interesting application of HODMD-based ROMs, we consider data forecasting based on experimental data. A case of particular interest is linked to the field of renewable energies. Light detection and ranging (LiDAR) [15, 20] measurements is a experimental a method usually employed in the wind energy for the remote measurement of the line-of-sight component of the wind speed. LiDAR measurements are based on detection of the Doppler shift for light backscattered from natural aerosols transported by the wind in the atmosphere. This technique is usually employed in the wind energy industry with different goals. LiDAR measurements offer time depen-

Fig. 4 Plane of measurements upstream the wind turbine in the LiDAR experimental campaign. Data are available in the six planes that are closer to the wind turbine, while the data on the seventh plane are to be predicted

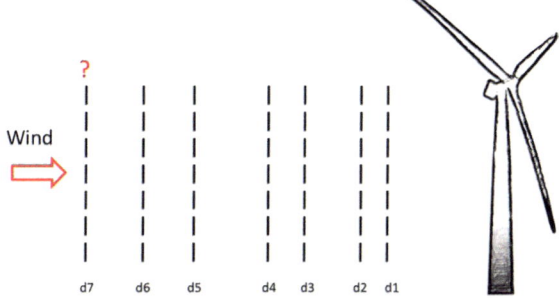

dent signals that present high noise levels (15–20 %). The LiDAR method usually gives various signals upstream of the wind turbine, at the left and right sides of the wind turbine, and at various distances that must be located between 20 and 200 m from the measurement device, which is a restriction in certain applications. Here, we consider six distances between 33 and 201 m. The aim of the HODMD-based ROM in the present case is to obtain the time-dependent signals at a seventh distance, namely 228 m, upstream the measurement device, which is not accessible to LiDAR measurements; see Fig. 4. To this end, we collect the experimental data such that each snapshot collects the two (left and right) measurements at the six available distances. Thus, the spatial complexity is 12, which is much smaller than the very large number of involved frequencies, namely the spectral complexity, meaning that HODMD, with $d > 1$, is needed to obtain good approximations of the dominant frequencies.

In order to obtain a HODMD-based ROM, we first use a set of data collected during 24 h in the six different locations presented. The HODMD method is applied to these data with tolerances $\varepsilon_1 = \varepsilon_2 = 10^{-4}$ and various values of d in the interval $30 \leq d \leq 40$ (set after some calibration). The obtained results are fairly insensitive to the selected value of d in this interval, which illustrates the robustness of the method. With the results obtained, a reliable and accurate ROM is constructed as in Eq. (1). Such expansion has been used to predict the wind velocity at the unknown seventh distance (during the same 24 h period), by simply extrapolating in space the spatial modes appearing in the expansion (1). To this end, both linear and quadratic spatial interpolation is used. Figure 5 compares the original data with the predictions using both linear and quadratic spatial extrapolation.The data at the unknown seventh distance are predicted with relative RMS errors ∼2% [29]. The computational cost for predicting these measurements is negligible, making possible to easily update the model in real-time.

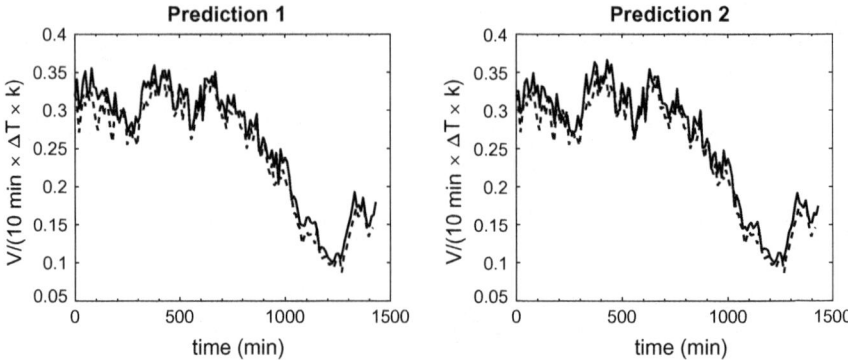

Fig. 5 Original (solid lines) and HODMD-predictions (dashed lines) of the wind velocity upstream the wind turbine at $x = 228$ m using a linear (left) and quadratic (right) approximation model

5 Conclusions

The HODMD method has been briefly described and applied to construct purely data-driven ROMs, whose performance has been illustrated in two applications. Namely,

- The temporal extrapolation to obtain the final attractor in the cylinder wake from transient data, which is useful to decrease the computational cost of obtaining the attractor. In this case, the transient stage data has been obtained from numerical simulation.
- The spatial extrapolation, to obtain the temporal evolution of data not provided beforehand. In this case, highly noisy experimental data has been used.

As a main conclusion, the data-driven ROM considered in this paper seems to be a fairly useful tool in several tasks of scientific and industrial interest. Several extensions of the presented methods include the substitution of HODMD by the more sophisticated spatio-temporal modal decomposition [28]. This method decomposes the spatio-temporal data, not only in temporal Fourier-like modes, but in spatio-temporal traveling waves.

Acknowledgements This work was supported by the Spanish Ministry of Economy and Competitiveness, under gran TRA2016-75075-R.

References

1. Alla, A., Kutz, N.: Nonlinear model order reduction via dynamic mode decomposition. SIAM J. Sci. Comp. **39**(5), 778–796 (2018)
2. Alonso, D., Vega, J.M., Velazquez, A., de Pablo, V.: Reduced-order modeling of three-dimensional external aerodynamic flows. J. Aerosp. Eng. **25**(4), 588–599 (2012)

3. Barkley, D., Henderson, R.: Three-dimensional Floquet stability analysis of the wake of a circular cylinder. J. Fluid Mech. **322**, 215–241 (1996)
4. Barkley, D., Tuckerman, L.S., Golubitsky, M.: Bifurcation theory for three-dimensional flow in the wake of a circular cylinder. Phys. Rev. E **61**, 5247–5252 (2000)
5. Blackburn, H.M., Marques, F., Lopez, J.M.: Symmetry breaking of two-dimensional time-periodic wakes. J. Fluid Mech. **522**, 395–411 (2005)
6. Brunton, S.L., Brunton, B.W., Proctor, J.L., Kaiser, E., Kutz, J.N.: Chaos as an intermittently forced linear system. Nat. Commun. **8**(19), 1–9 (2017)
7. Chen, K.K., Tu, J.H., Rowley, C.W.: Variants of dynamic mode decomposition: boundary condition, Koopman and Fourier analyses. J. Nonlinear Sci. **22**, 8871–8875 (2012)
8. Chinesta, F., Keunings, R., Leygue, A.: The proper generalized decomposition for advanced numerical simulations. In: SpringerBriefs in Applied Sciences and Technology. Springer-Verlag, Berlin (2014)
9. Dawson, S.T.M., Hemati, M.S., Williams, M.O., Rowley, C.W.: Characterizing and correcting for the effect of sensor noise in the dynamic mode decomposition. Exp. Fluids 57–42 (2016)
10. Dicle, C., Mansour, H., Tian, D., Benosman, M., Vetro, A.: Robust low rank dynamic mode decomposition for compressed domain crowd and traffic flow analysis. IEEE Int. Conf. Multimedia Expo (ICME) 1–6 (2016)
11. Ferrer, E., Willden, R.H.: Blade-wake interactions in cross-flow turbines. Int. J. Marine Energy **11**, 71–83 (2015)
12. Fischer, P.F., Lottes, J.W., Kerkemeier, S.G.: Nek5000 Web page (2008). http://nek5000.mcs.anl.gov
13. Gao, C., Zhang, W., Kou, J., Liu, Y., Ye, Z.: Active control of transonic buffet flow. J. Fluid Mech. **824**, 312–351 (2017)
14. Golub, G.H., van Loan, G.T.: Matrix Computations. John Hopkins Univ Press (1996)
15. Harris, M., Hand, M., Wright, A.: A Lidar for turbine control. Technical Report NREL/TP-500-39154, National Renewable Energy Laboratory (NREL). Golden, CO, USA (2006)
16. Hemati, M.S., Rowley, C.W., Deem, E.A., Cattafesta, L.N.: De-biasing the dynamic mode decomposition for applied Koopman spectral analysis of noisy datasets. Theoret. Comput. Fluid Dyn. 349–368 (2017)
17. https://github.com/LeClaincheVega/HODMD
18. Jackson, C.P.: A finite-element study of the onset of vortex shedding in ow past variously shaped bodies. J. Fluid Mech. **182**, 23–45 (1987)
19. Jovanović, M.R., Schmid, P.J., Nichols, J.W.: Sparsity-promoting dynamic mode decomposition. Phys. Fluids **26**, 024103 (2014)
20. Karlsson, C.J., Olsson, F.A.A., Letalick, D., Harris, M.: All-fiber multifunction continuous-wave coherent laser radar at 1.55 mm for range, speed, vibration, and wind measurements. Appl. Opt. **39**, 3716–3726 (2000)
21. Katznelson, Y.: An Introduction to Harmonic Analysis. Dover Publications, Inc., New York (1976)
22. Kolda, T.G., Bader, B.W.: Tensor decompositions and applications. SIAM Rev. **51**, 455–500 (2009)
23. Kutz, J.N., Brunton, S.L., Brunton, B.W., Proctor, J.L.: Data-driven modeling of complex systems. Soc. Indus. Appl. Math. (2016)
24. Le Clainche, S., Varas, F., Vega, J.M.: Accelerating reservoir simulations using POD on the fly. Int. J. Num. Meth. Eng. **28**, 79–100 (2017)
25. Le Clainche, S., Vega, J.M.: Higher orderd ynamic mode decomposition. SIAM J. Appl. Dyn. Syst. **16**(2), 882–925 (2017)
26. Le Clainche, S., Vega, J.M.: Higher order dynamic mode decomposition to identify and extrapolate flow patterns. Phys. Fluids **29**, 084102 (2017)
27. Le Clainche, S., Vega, J.M., Soria, J.: Higher order dynamic mode decomposition of noisy experimental data: the flow structure of a zero-net-mass-flux jet. Exp. Therm. Fluid Sci. **88**, 336–353 (2017)

28. Le Clainche, S., Vega, J.M.: Spatio-temporal Koopman decomposition. J. Nonlin. Sci. **28**(3), 1–50 (2018)
29. Le Clainche, S., Lorente, L., Vega, J.M.: Wind predictions upstream wind turbines from a LiDAR database. Energies **11**(3), 543–558 (2018)
30. Park, K.H., Jun, S.O., Baek, S.M., Cho, M.H., Yee, K.J., Lee, D.H.: Reduced-order model with an artificial neural network for aerostructural design optimization. J. Aircr. **50**(4), 1106–1116 (2013)
31. Perez, J.M., Le Clainche, S., Vega, J.M.: Alternative three-dimensional instability analysis of the wake of a circular cylinder. In: Proceedings of 8th AIAA Theoretical Fluid Mechanics Conference, AIAA paper 2017-4021 (2017)
32. Provanasal, M., Mathis, C., Boyer, L.: Bénard-von Kármán instability: transient and forcing regimes. J. Fluid Mech. **182**, 1–22 (1987)
33. Quarteroni, A., Manzoni, A., Negri, F.: Reduced Basis Methods for Partial Differential Equations. Springer-Verlag, Berlin (2016)
34. Rapun, M.L., Terragni, F., Vega, J.M.: Adaptive POD-based low dimensional modeling supported by residual estimates. Int. J. Num. Meth. Eng. **104**, 844–868 (2015)
35. Schmid, P.J.: Dynamic mode decomposition of numerical and experimental data. J. Fluid Mech. **656**, 5–28 (2010)
36. Sirisup, S., Karniadakis, G.E., Xiu, D., Kevrekidis, I.G.: Equations-free/Galerkin-free POD assisted computation of incompressible flows. J. Comput. Phys. **207**, 568–587 (2005)
37. Sirovich, L.: Turbulence and the dynamic of coherent structures. Q. Appl. Math. **45**(3), 561–571 (1987)
38. Takeishi, N., Kawahara, Y., Tabei, Y., Yairi, T.: Bayesian dynamic mode decomposition. In: Proceedings of the 26th Intl Joint Conference on Artificial Intelligence (IJCAI), pp. 2814–2821 (2017)
39. Takens, F.: Detecting strange attractors in turbulence. In: Rand, D.A., Young, L.-S. (eds.) Lecture Notes in Mathematics. Springer-Verlag, pp. 366–381 (1981)
40. Terragni, F., Vega, J.M.: Construction of bifurcation diagrams using POD on the fly. SIAM J. Appl. Dyn. Syst. **13**, 339–365 (2014)
41. Tu, J.H., Rowley, C., Luchtemburg, D.M., Brunton, S.L., Kutz, J.N.: On dynamic mode decomposition: theory and applications. J. Comp. Dyn. **1**(2), 391–421 (2014)
42. Tucker, L.R.: Some mathematical notes on three-mode factor analysis. Psikometrica **31**, 279–311 (1996)
43. Williams, M.O., Kevrekidis, I.G., Rowley, C.W.: A data driven approximation of the Koopman operator: extending dynamic mode decomposition. J. Nonlin. Sci. **25**, 1307–1346 (2015)
44. Williamson, C.H.K.: Oblique and parallel modes of vortex shedding in the wake of a circular cylinder at low Reynolds number. J. Fluid Mech. **206**, 579–627 (1989)
45. Williamson, C.H.K.: Vortex dynamics in the cylinder wake. Annu. Rev. Fluid. Mech. **28**, 477–539 (1996)

An Adaptive Way of Choosing Significant Snapshots for Proper Orthogonal Decomposition

Steffen Kastian and Stefanie Reese

Abstract In structural engineering problems, the resulting partial differential equations (PDEs) are often solved using the finite element method (FEM). The number of degrees of freedom (DOF) and hence the computational time increases depending upon the complexity of the problem (linear/nonlinear) and discretization of space and time. The proper orthogonal decomposition (POD) yields a valuable set of vector bases which can be used in model order reduction (MOR) techniques to reduce the computational time. However, for nonlinear problems the trade-off with respect to accuracy to gain speedup is still high. In this context, we present an adaptive method to choose snapshots, which leads to a POD technique with either increased accuracy or increased speed-up for a fixed accuracy compared to the classical POD.

Keywords Model order reduction · Proper orthogonal decomposition · Adaptivity

1 Introduction

Normally, calculation of engineering problems requires many simulations. In order to get accurate results, the simulations include many degrees of freedom (DOF) which make the calculation expensive. Model order reduction (MOR) can help to reduce the cost. Due to the fact that optimization or uncertainty quantification require the solution of many similar problems which only differ by slight changes of parameters, the use of MOR might lead to a significant improvement in computational cost. Among others, proper orthogonal decomposition (POD) has been shown to be a suitable method for this purpose [12, 19]. The POD method is already used in many different fields e.g. in stochastics [3], transport phenomena [18], turbulent flows [15, 21],

S. Kastian (✉) · S. Reese
Institute of Applied Mechanics, RWTH Aachen, Mies-van-der-Rohe-Str. 1,
52074 Aachen, Germany
e-mail: steffen.kastian@rwth-aachen.de

S. Reese
e-mail: stefanie.reese@rwth-aachen.de

© Springer Nature Switzerland AG 2020
J. Fehr and B. Haasdonk (eds.), *IUTAM Symposium on Model Order Reduction
of Coupled Systems, Stuttgart, Germany, May 22–25, 2018*, IUTAM Bookseries 36,
https://doi.org/10.1007/978-3-030-21013-7_5

acoustics [22], image or signal analysis [6]. POD has been extensively studied in the last decades in the fields of dynamics and solid mechanics [4, 11, 14]. Since nonlinear simulations are still expensive with POD the CPU time can be additionally reduced by using the discrete empirical interpolation method (DEIM) [17]. The system of equations of a FE problem in general has a dimension as high as the number of DOF. The POD projects this high dimension system of equations into an equation system of smaller dimension. Due to this reduction, an error is generated. In literature, different kinds of approaches for the estimation of this error are proposed [8]. Volkwein et al. show approaches for a posteriori error estimators [20, 24] and Chaturantabut et al. introduce an error estimate for POD-DEIM [5]. If the expected error is too high, most approaches add more basis functions. Radermacher and Reese introduce the selective POD where POD is only applied to regions with approximately elastic behaviour [16, 17]. A hp certified reduced basis (RB) was introduced by Eftang et al. [7] in which a POD/Greedy [9] sampling procedure is used for the initial partition of the parameter domain (h-refinement). Another interesting approach was shown by Kunisch et al. [13] which focuses on the allocation of possible additional snapshots. Haasdonk et al. [10] proposed a method which starts with an empty projection matrix and extends this basis with the worst resolved parameter combination. This procedure enriches the projection matrix until a certain error tolerance is fulfilled. The focus of their paper is to find a good training set for parameterized problems. Local basis vectors for problems with different physical regimes and parameter variations where shown by Amsallem et al. [1].

In contrast to the existing approaches, the present one incorporates a projection matrix which adapts itself to the state of deformation The adaptive algorithm enables a selection of a suitable snapshot set, while meaningless snapshots are sorted out. During the course of a simulation different snapshots might become important to obtain the best possible basis functions, while other snapshots are not important and can be neglected. This approach is especially interesting for nonlinear problems, since it allows to capture the change in the behaviour of a system adaptively.

The paper is organized as follows: Section 2 describes the main equations and ideas of the POD method. The new adaptive approach for obtaining adjusted projection matrices for each time step is developed in Sect. 3. It also includes an iterative approach to select the number of modes. In Sect. 4, we present the investigated nonlinear 3D example. In the context of this example, various aspects, for instance the number of considered snapshots as well as the change of the POD modes, are discussed. The paper finishes with a conclusion in Sect. 5.

2 Proper Orthogonal Decomposition

POD is based on precalculations, where so called snapshots are saved. In our case, the snapshots are solution vectors for different time steps. Depending on the problem, it might be necessary to carry out several precalculations. Here, the solution vectors are the displacement vectors. The snapshots are stored in a snapshot matrix $D \in \mathbb{R}^{n \times l}$

with $D = [u_1, ..., u_l]$ where n describes the number of degrees of freedom and l the number of collected snapshots. In the case of a single precalculation, the number of collected snapshots l is equal to the number of time steps j of the precalculation. The solution vector of one time step in one precalculation is the displacement vector u with the dimension $n \times 1$. The goal of POD is to use the snapshot matrix D to find an orthonormal basis such that the distance with respect to the projection of D onto the subspace defined by Φ is minimized. The singular value decomposition of D can be used to solve this problem:

$$D = S \Sigma V^T = \sum_{k=1}^{l} \sigma_k s_k v_k^T. \tag{1}$$

In the latter equation the upper block of Σ is a diagonal matrix, which contains singular values in decreasing order, and the lower matrix contains a null matrix:

$$\Sigma = \begin{pmatrix} \sigma_1 & & \\ & \ddots & \\ & & \sigma_n \\ & \mathbf{0} & \end{pmatrix} \text{ with } \sigma_1 \geq ... \geq \sigma_n \tag{2}$$

The matrices $S = (s_1, ..., s_l)$ and $v = (v_1, ..., v_l)$ are orthonormal. The vectors s_i corresponding to the m highest singular values are used to span the projection matrix

$$\Phi = [s_1, ..., s_m] \quad \in \mathbb{R}^{n \times m}. \tag{3}$$

with the dimension $n \times m$. The number of modes is represented by m which should be much smaller than the number of degrees of freedom ($m \ll n$). This projection matrix Φ can project the system of equations into a system of equations of smaller dimension and hence reduce the cost.

3 Adaptive Proper Orthogonal Decomposition

In classical POD, the modes are selected once and stay constant throughout the simulation. This means that the behaviour is not expected to change significantly over time. Nonlinear problems usually require more basis vectors than linear problems to obtain accurate results. In this paper we introduce an approach to overcome the issues which appear in a nonlinear regime. The main difference between the classical POD and the new adaptive proper orthogonal decomposition (APOD) is that the projection matrix Φ is not constant for all time steps. The projection matrix $\bar{\Phi}(\bar{u}_{max})$ depends on the maximum deformation

$$\bar{u}_{max} = ||\bar{u}||_\infty \tag{4}$$

which is defined as the infinity norm of the displacement vector $\bar{u}$ of the previous time step. This will be called current state in the following. Other versions based on an angle computation or a projection are currently under investigation. We believe that this method is also applicable to problems of other fields and is not limited to quasi-static problems. The infinity norm was chosen for two reasons. First of all, its computation can be done by low effort. Secondly, we believe that the norm represents the current state in a good manner. The displacement vectors in the snapshot matrix have to be ordered by their infinity norm,

$$||u_1||_\infty \le ||u_2||_\infty \le \dots \le ||u_l||_\infty, \tag{5}$$

and with these ordered snapshots the total snapshot matrix is given by

$$D = [u_1, \dots, u_l]. \tag{6}$$

It includes all snapshots from one or more precomputations. For more than one precomputation, the snapshots in the snapshot matrix can be mixed up and only need to be ordered by their infinity norm. The current maximum deformation is compared to the infinity norm of the snapshot set to find snapshots which can describe the current deformation in the best way. We find the position of the snapshot matrix D with the minimum difference between the infinity norm of each snapshot and the maximum current deformation $\bar{u}_{max}$ with

$$b(\bar{u}_{max}) = \underset{i \in \{1,\dots,l\}}{\operatorname{argmin}} \left(|(||u_i||_\infty - \bar{u}_{max})| \right) \tag{7}$$

where b is the subscripted number of the best fitting snapshot. Additionaly $a \in \mathbb{R}$ snapshots are chosen in the neighborhood of the best fitting snapshot. Using these snapshots we obtain a new snapshot matrix $\bar{D}$ for the current time step

$$\bar{D} = [u_{\lceil b-\frac{a}{2}\rceil} \ \dots \ u_b \ \dots \ u_{\lceil b+\frac{a}{2}\rceil}]. \tag{8}$$

With this current snapshot matrix $\bar{D}$ one can carry out singular value decomposition analogously to Eq. 1 and compute the current projection matrix $\bar{\Phi}$ analogously to Eq. 3. This is the projection matrix for the next time step. It is possible to precompute the basis for each possible best fitting snapshots b in an offline step. This yields to up to l different bases.

3.1 Iterative Number of Modes

It can be a challenging task to choose the number of modes a priori. In previous papers [2, 23, 24], it was shown that the estimated error correlates with the decay of the singular values. This knowledge is used here to define an iterative scheme to determine the required number of modes. The minimum number of modes is defined as m_{min}, which is the start value for the first iteration step with $m = m_{min}$. This is followed by the procedure of the APOD method described in Sect. 3. However, after calculating the singular values, the possible quality of the reduction is verified. For this verification the decay of the singular values is observed. If the decay of the singular values fulfills

$$\frac{\sigma_m}{\sum_{i=1}^{m} \sigma_i} \leq C_{tol} \tag{9}$$

then, the number of modes m is sufficient. Here, σ_i represent the singular values with a decreasing sequence. Otherwise the procedure has to be repeated with an increased m

$$m \leftarrow m + 1 \tag{10}$$

until the criterion in Eq. 9 is fullfilled. Equation (9) can be dissatisfactory if C_{tol} is too small. Then all snapshots are taken into account. Alternatively, the tolerance value C_{tol} could be increased. This iterative process could also take place in an offline step, so that only the corresponding basis has to be selected online. In the following a flow chart of the algorithm is shown:

Offline:

(1) Collect snapshots in precomputations
(2) Order the existing snapshots by their infinity norm
(3) Create a basis for each possibly selected snapshot and store them with the corresponding number of the best fitting snapshot (b)

Online:
loop time steps

(1) Determine current maximum displacement u_{max}
(2) Select corresponding basis $\boldsymbol{\Phi}$
(3) Galerkin projection
(4) Solve reduced system next time step

4 Numerical Example

In Sects. 2 and 3, POD and the new APOD were introduced. A comparison of these two methods regarding accuracy and efficiency will be presented in Sect. 4.3.

We are solving the weak form

$$\int_{\Omega} \rho \, \ddot{\boldsymbol{u}} \cdot \boldsymbol{w} \, dV + \int_{\Omega} \boldsymbol{\sigma} : (\boldsymbol{\Delta w}) dV = \int_{\Omega} \rho \, \boldsymbol{b} \cdot \boldsymbol{w} dV + \int_{\partial \Omega_t} \boldsymbol{t} \cdot \boldsymbol{w} dA$$

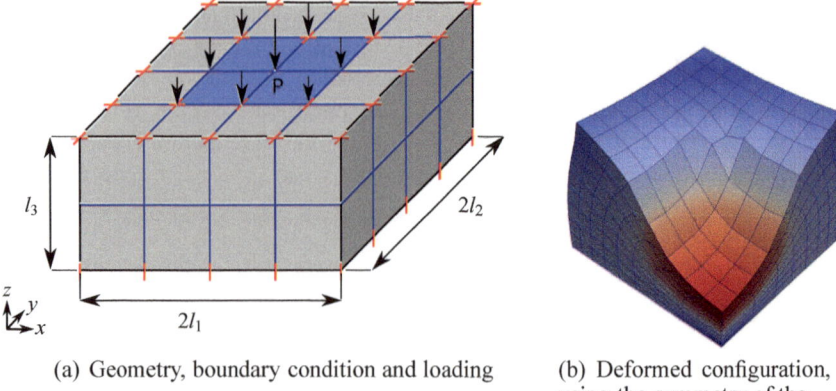

(a) Geometry, boundary condition and loading

(b) Deformed configuration, using the symmetry of the system

Fig. 1 Cube under compression

of the balance of linear momentum where w represents a test function, $\ddot{u}$ the acceleration vector, σ the stress and ρb the body force density. We incorporate finite strains and describe the material behaviour by means of a Neo-Hooke model. The equation is discretized in space by means of the finite element method. Here we use isoparametric trilinear shape functions.

A cube under compression, as shown in Fig. 1a, is chosen as an example. Due to symmetry it is sufficient to consider only a quarter of the cube for the simulation. The nodes at the bottom, at coordinate $z = 0$, are fixed during the whole simulation, such that they can only move in x- and y-direction. Additionally the nodes at the top, with initial z-coordinate $z = l_3$ are fixed in x- and y-direction, i.e. they can only move in z-direction. The symmetric cube is discretized by 8 elements in each direction which yields a total of 8^3 elements. This is a 3D example with a Neo-Hookean material law, which means that large deformations are taken into account. One deformed system is plotted in Fig. 1b.

The Neo-Hookean material law includes two parameters. The Poisson's ratio ν is set to 0.4999, which represents almost incompressible material behaviour. The Young's modulus varies between $E \in [100, 2000] \, \mathrm{N/mm^2}$ and the geometry of the cube undergoes small changes. A nondimensional length is introduced as $l_i = L_i/L_0$. Therefore, the side lengths of the cube vary between $l_1, l_2, l_3 \in [0.95, 1.05]$. In many engineering problems one has one or a small number of quantities of interest. The displacement of the degree of freedom which is maximal is called u_{max}. For the following, this maximum deformation will be considered to be the quantity of interest.

4.1 Considered Snapshots

In order to apply POD, either the number of modes or a threshhold C_{tol} has to be chosen. For the APOD, the number of considered snapshots a needs to be chosen

Fig. 2 Error plotted over the considered snapshots with $m = 5$ modes, one precomputation with 100 available snapshots

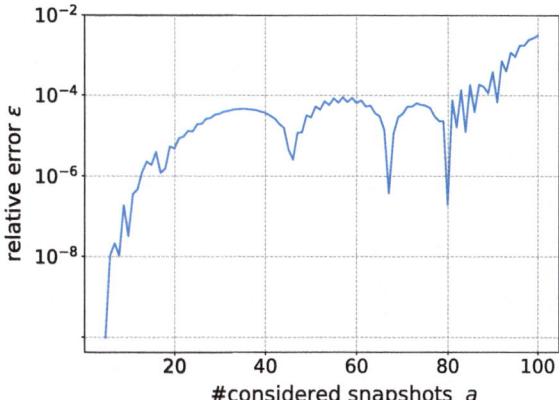

additionally. The number of considered snapshots should depend on the number of precomputations and the number of used modes. To ensure that enough possible basis vectors are available, the number of considered snapshots has to be bigger than the number of modes:

$$a \geq m. \tag{11}$$

The first example in Fig. 2 is calculated with a fixed length, $l_1, l_2, l_3 = 1$, and a fixed Young's modulus $E = 100$ N/mm² for the precalculation as well as the reduced calculation. In the following, the error between the unreduced model and (A)POD is compared. The relative error is defined as

$$\varepsilon = \frac{|u_{max, \, full} - u_{max,(A)POD}|}{|u_{max, \, full}|},$$

where $u_{max, \, full}$ is the maximum displacement of the unreduced reference system and $u_{max,(A)POD}$ the maximum displacement of the reduced simulation. In Fig. 2 the error over the considered snapshots is shown for a constant number of modes ($m = 5$) regarding the example shown in Sect. 4. In this example we have a fixed pool of 100 snapshots. For each time step we select a fixed number $a \leq 100$ of snapshots. This is the number of considered snapshots shown on the x-axis. Each time step only considers the amount of snapshots which are selected as shown in Sect. 3. All other snapshots for one time step are neglected. It can be seen that in the case of one precomputation, a smaller number of considered snapshots yields better results. The smallest error is achieved if the number of considered snapshots is approximately equal to the number of modes $a = m$. This leads to the additional advantage that the CPU time is reduced. This is due to the fact that for each time step an SVD of smaller dimension is required, which results in faster calculations. Considering all available snapshots leads to the classic POD and yields the same results.

Allowing a variation in the size of the cube $l_1, l_2, l_3 \in [0.95, 1.05]$ results in a more complex problem and requires a higher amount of precomputations. The number of

Table 1 Precomputations to capture the influence of the cube for different geometries

l_1	l_2	l_3	E [N/mm^2]
0.95	0.95	0.95	200
1.05	0.95	0.95	200
0.95	1.05	0.95	200
0.95	0.95	1.05	200

Fig. 3 Error plotted over the considered snapshots with $m = 10$ modes, four precomputations and 400 available snapshots

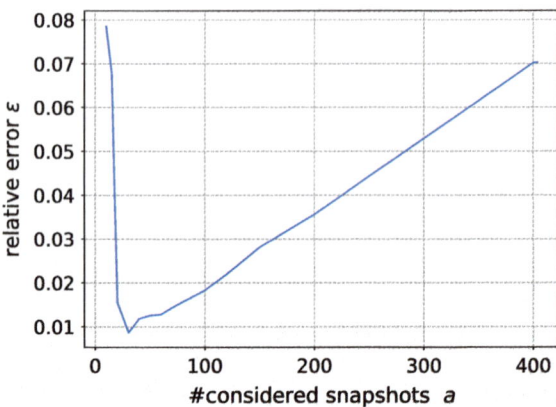

precomputations is called p. For this example $p = 4$ precomputations are chosen. Table 1 shows the variations in the size of the cube upon which the precomputation are based. The ranges for the different lengths are chosen in such a way that the nonlinearity coming from the variation of geometry can be captured.

Figure 3 shows the error over the number of considered snapshots. Ten modes are used in this example. The size of the cube was set to $l_1 = 1.05$, $l_2 = 1.00$, $l_3 = 0.95$ to ensure that the simulation differs from the available sets of snapshots. For a small number of considered snapshots ($s < 20$) the reduced model does not approximate the original system very well. On the other hand, using a higher number of considered snapshots, the reduced system contains too much information which does not describe the problem appropriately. Thus, the error increases with the number of considered snapshots.

According to Figs. 2 and 3 and other calculated examples (not shown in this paper) the multiplication of the number of modes m times the number of precomputations p

$$a = m\, p \qquad (12)$$

is a good estimate for the number of snapshots (a) to be considered.

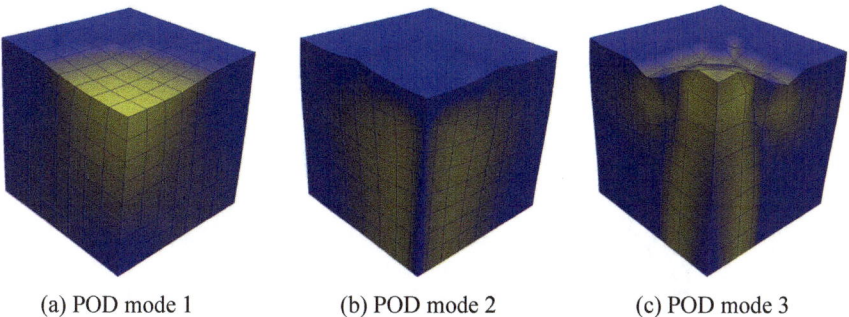

(a) POD mode 1   (b) POD mode 2   (c) POD mode 3

Fig. 4 POD modes for the cube under compression with one precalculation and the classic POD

4.2 Comparison of Modes

In the classical POD the modes are calculated once and stay constant for the whole simulation. Fig. 4a–c depict the first three modes for the cube under compression. These are the most significant modes which are physically meaningful and have the highest singular values.

Mode 1 in Fig. 4a captures the main structure of the problem. For a linear problem, mode 1 would be sufficient. Modes 2 and 3 are important to capture partly nonlinear behaviour of the cube.

In comparison to the physically meaningful modes there are modes that correspond to singular values which are 10^{15} times smaller than the meaningful modes. These modes have a negligible influence on the system, which is the reason why only a small number of modes needs to be considered in order to obtain accurate results. The non-meaningful modes are mainly interpreted as some kind of noise or numerical inaccuracy. The 100th mode is pictured in Fig. 5 as an example of a non-physical mode.

For the APOD new modes are calculated at each timestep. This enables the method to capture a good response for nonlinear simulations. In Figs. 6 and 7 the mode 1 for different timesteps and the corresponding displacement are shown. Mode 1 corresponds to the eigenvector related to the highest singular value. For each timestep other snapshots are considered, which results in a changing mode 1 over time. While for the first timestep mode 1 for APOD is qualitatively similar to mode 1 of POD, one can see that the behaviour of the cube changes in the 90th timestep. For large deformations (see e.g. Fig. 7c), the response of the cube in the corner where the loading is applied becomes stiffer. This can be seen in mode 1 of the corresponding timestep in Fig. 6c.

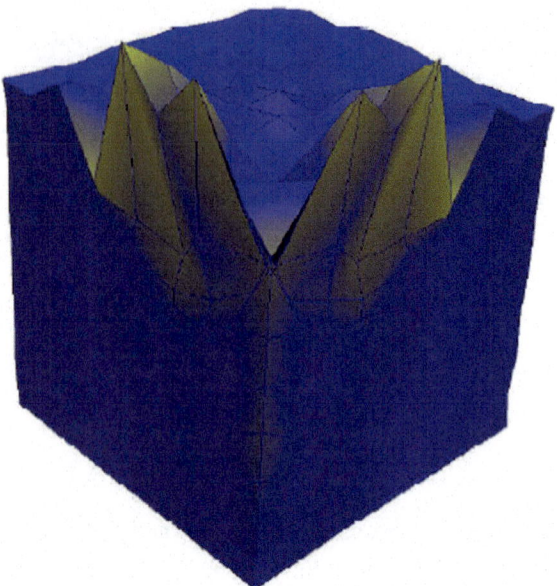

Fig. 5 POD mode 100

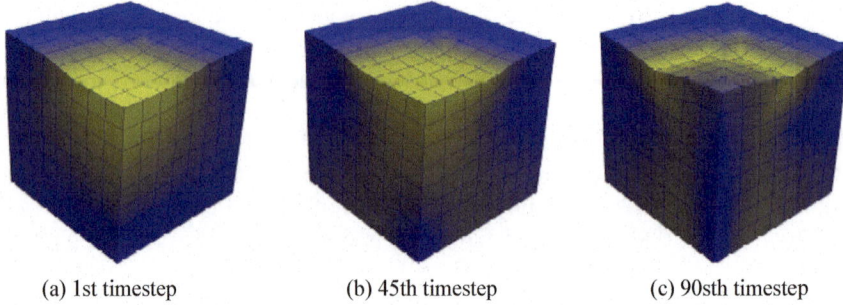

(a) 1st timestep   (b) 45th timestep   (c) 90sth timestep

Fig. 6 APOD mode 1 for the cube under compression for different timesteps

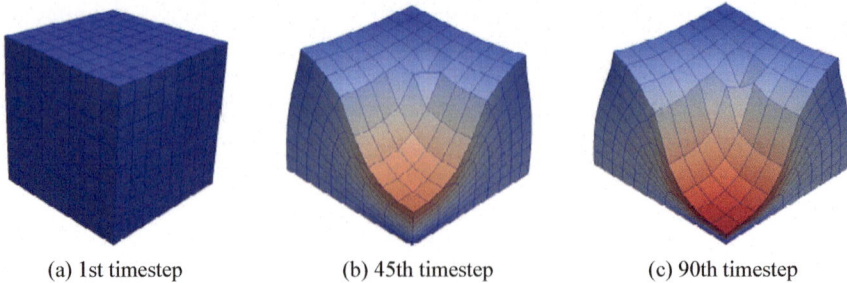

(a) 1st timestep   (b) 45th timestep   (c) 90th timestep

Fig. 7 Displacements for the cube under compression for different timesteps

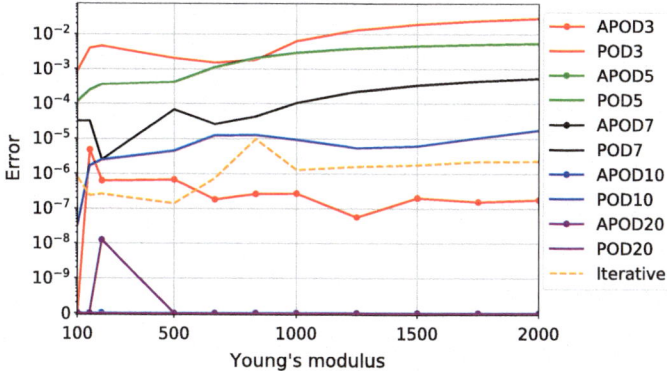

Fig. 8 Comparison of the error between POD and APOD for a range of Young's moduli

4.3 Results

In the previous chapter, the error over the number of considered snapshots and some modes where shown. Finally, we can compare the APOD with the POD method regarding accuracy and efficiency. The cube under compression shown in Fig. 1a is investigated for this purpose. A Galerkin projection is used to reduce the model. Sections 2 and 3 describe how to set up the projection matrix. To compare the different methods, the size of the cube is fixed to $l_1, l_2, l_3 = 1.0$. All reduced calculations were done based on only one precalculation with a Young's modulus of $E = 100\,\text{N/mm}^2$. In order to carry out simulations with properties different from the precomputation we allow a variation of the Young's modulus $E \in [100, 2000]\,\text{N/mm}^2$. The error is defined in Eq. 1. The system is described in detail in Sect. 4. The number of considered snapshots is equal to the number of modes ($a = m$). For the iterative approach the tolerance was set to $C_{tol} = 10^{-15}$.

In Fig. 8 the number after the method represents the number of used modes, e.g. APOD3 depicts the APOD method with $m = 3$ modes. It can be seen that even for a small number of modes APOD has accurate results while the classic POD needs a higher number of modes to achieve equally accurate results. Comparing the computational time in Fig. 9, one observes that for the same number of modes the APOD is slightly slower than POD but has significantly better accuracy. For the similar level of accuracy, it is clearly seen that APOD is faster than POD. APOD with an iterative number of modes has the advantage that there is no need to choose the number of modes m a priori. In addition, the method is error controlled. However, it is slower than when choosing the number of modes a priori which is based on the fact that for each time step several SVDs are required until the tolerance is fulfilled.

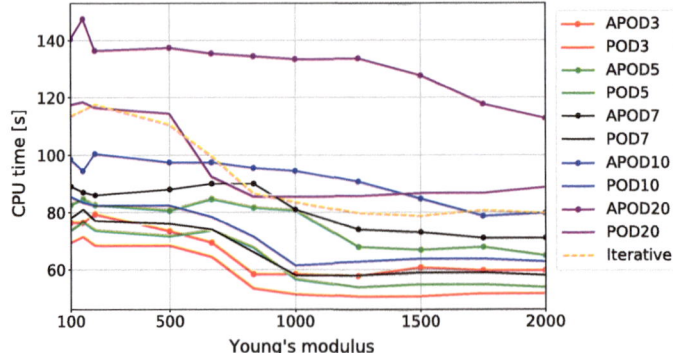

Fig. 9 Comparison of the CPU time of POD and APOD for a range of Young's moduli

5 Conclusion

We developed a new method which is able to select snapshots from an existing pool of snapshots to create an adaptive basis for each timestep. This method is able to capture nonlinearities better than the classic POD method. The adaptive proper orthogonal decomposition can be coupled with other features of model order reduction methods like error estimators or methods which enrich the pool of basis vectors.

Furthermore, it was shown that the APOD achieves more accurate results with similar costs. It is seen that APOD has better stability than the classical POD for a small number of modes.

Yet, the method might be limited to problems with a clear structure or one significant value which can be used to compare the snapshots with the current state in an appropriate way. Nevertheless, the possibility of a more general formulation as well as the investigation of more complex problems should be in the centre of further research.

Acknowledgements We would like to thank the German Research Foundation (Deutsche Forschungsgemeinschaft(DFG)) for the financial support of the SPP 1886 with the title "Polymorphic uncertainty modelling for the numerical design of structures".

References

1. Amsallem, D., Zahr, M.J., Farhat, C.: Nonlinear model order reduction based on local reduced order bases. Int. J. Numer. Meth. Eng. **92**(10), 891–916 (2012)
2. Antoulas, A.C., Danny, C.S., Yunkai, Z.: On the decay rate of Hankel singular values and related issues. Syst. Control Lett. **46.5**, 323–342 (2002)
3. Bastine, D., Vollmer, L., Wächter, M., Peinke, J.: Stochastic wake modelling based on POD analysis energies **11**(3), 612 (2018)

4. Bolzon, G., Buljak, V.: An effective computational tool for parametric studies and identification problems in materials mechanics. Comput. Mech. **48**, 675–687 (2011). https://doi.org/10.1007/s00466-011-0611-8
5. Chaturantabut, S., Sorensen, D.S.: A state space error estimate for POD-DEIM nonlinear model reduction. SIAM J. Numer. Anal. **50**(1), 46–63 (2012)
6. Chatterjee, A.: An introduction to the proper orthogonal decomposition. Curr. Scir. **78**(7), 808–817 (2000)
7. Eftang, J.L., Knezevic, D.J., Patera, A.T.: An hp certified reduced basis method for parametrized parabolic partial differential equations. Math. Comput. Modell. Dyn. Syst. **17**(4), 395–422 (2011)
8. Haasdonk, B., Ohlberger, M.: Efficient reduced models and a posteriori error estimation for parametrized dynamical systems by offline/online decomposition. Math. Comput. Modell. Dyn. Syst. **17**(2), 145–161 (2011)
9. Haasdonk, B.: Convergence rates of the pod-greedy method ESAIM. Math. Modell. Numer. Anal. **47**(3), 859–873 (2013)
10. Haasdonk, B., Dihlmann, M., Ohlberger, M.: A training set and multiple bases generation approach for parameterized model reduction based on adaptive grids in parameter space. Math. Comput. Modell. Dyn. Syst. **17**(4), 423–442 (2011)
11. Kerschen, G., Golinval, J.C., Vakakis, A., Bergman, L.: The method of proper orthogonal decomposition for dynamical characterization and order reduction of mechanical systems: an overview. Nonlinear Dyn. **41**(1–3), 147–169, 2005. https://doi.org/10.1007/s11071-005-2803-2
12. Kunisch, K., Volkwein, S.: Proper orthogonal decomposition for optimality systems ESAIM. Math. Modell. Numer. Anal. **42**(1), 1–23 (2008)
13. Kunisch, K., Volkwein, S.: Optimal snapshot location for computing POD basis functions ESAIM. Math. Modell. Numer. Anal. **44**(3), 509–529 (2010)
14. Lenaerts, V., Kerschen, G., Golinval, J., Chevreuils, C.D.: Proper orthogonal decomposition for model updating of nonlinear mechanical systems. In: Golinval 2001 Mechanical Systems and Signal Processing, pp. 31–41 (2001)
15. Lumley, J.L., Holmes, P., Berkooz, G.: Turbulence. Coherent structures. In: Dynamical Systems and Symmetry. Cambridge University Press, Cambridge (1996)
16. Radermacher, A., Reese, S.: Model reduction in elastoplasticity: proper orthogonal decomposition combined with adaptive sub-structuring. Comput. Mech. **54**(3), 677–687 (2014)
17. Radermacher, A., Reese, S.: POD—based model reduction with empirical interpolation applied to nonlinear elasticity. Int. J. Numer. Meth. Eng. **107**(6), 477–495 (2016)
18. Reiss, J., Schulze, P., Sesterhenn, J., Mehrmann, V.: The shifted proper orthogonal decomposition: a mode decomposition for multiple transport phenomena. SIAM J. Sci. Comput. **40**(3), A1322–A1344 (2018)
19. Sachs, E.W., Volkwein, S.: POD-Galerkin approximations in PDE-constrained optimization. GAMM-Mitteilungen **33**(2), 194–208 (2010)
20. Tonn, T., Urban, K., Volkwein, S.: Comparison of the reduced-basis and POD a posteriori error estimators for an elliptic linear-quadratic optimal control problem. Math. Comput. Modell. Dyn. Syst. **17**(4), 355–369 (2011)
21. Meyer, R.D., Tan, G.: Use of proper orthogonal decomposition and linear stochastic estimation technique to investigate real-time detailed airflows for building ventilation. Indoor Built Environ. **25**(2), 378–389 (2016)
22. Nagarajan, K.K., Singha, S., Cordier, L., Airiau, C.: Open-loop control of cavity noise using proper orthogonal decomposition reduced-order model. Comput. Fluids **160**, 1–13 (2018)
23. Penzl, T.: Eigenvalue decay bounds for solutions of Lyapunov equations: the symmetric case. Syst. Control Lett. **40**.2, 139–144 (2000)
24. Volkwein, S., Tröltzsch, F.: POD a-posteriori error estimates for linear-quadratic optimal control problems. Comput. Optim. Appl. **44**, 83115 (2009)

Fully Online ROMs and Collocation Based on LUPOD

Maria-Luisa Rapún, Filippo Terragni and José M. Vega

Abstract This paper deals with the acceleration of time-dependent solvers for non-linear dissipative systems. The governing equations are Galerkin-projected onto a set of modes, which are obtained by applying proper orthogonal decomposition (POD) to a set of snapshots calculated by a standard numerical solver. The advantage of this approach is that the online operation of the resulting Galerkin system should be much faster than the standard numerical solver. The basic version of such reduced order model uses snapshots computed in a preprocess that is usually very computationally expensive. This difficulty can be overcome by an adaptive combination of the standard numerical solver and the Galerkin system along the simulation, using a method called POD on the Fly, which will be illustrated in a representative application. In addition, Galerkin projection can be performed using only a suitable set of collocation points, which decreases even further the computational cost. In this context, an efficient collocation method called LUPOD will be described and tested in various applications, including its combination with Galerkin projection.

Keywords Reduced order models · Collocation methods · Proper orthogonal decomposition · LU decomposition · Galerkin projection

M.-L. Rapún · J. M. Vega (✉)
E.T.S.I. Aeronáutica y del Espacio, Universidad Politécnica de Madrid,
Plaza Cardenal Cisneros 3, 28040 Madrid, Spain
e-mail: josemanuel.vega@upm.es

M.-L. Rapún
e-mail: marialuisa.rapun@upm.es

F. Terragni
Department of Materials Science and Engineering, G. Millán Institute
for Fluid Dynamics, Nanoscience and Industrial Mathematics,
Universidad Carlos III de Madrid, 28911 Leganés, Spain
e-mail: fterragn@ing.uc3m.es

© Springer Nature Switzerland AG 2020   81
J. Fehr and B. Haasdonk (eds.), *IUTAM Symposium on Model Order Reduction*
of Coupled Systems, Stuttgart, Germany, May 22–25, 2018, IUTAM Bookseries 36,
https://doi.org/10.1007/978-3-030-21013-7_6

1 Introduction

Reducing development time and computational costs by means of reduced order models (ROMs) is a challenging research field that has received an increasing attention in the last two decades. The interest is twofold: ROMs help in the analysis and understanding of the underlying physical problems, and facilitate their industrial application.

In this paper, we first introduce in Sect. 2 some of our recent works [14, 16] on the design of *adaptive* online ROMs for the acceleration of time-dependent numerical solvers for dissipative problems, using a combination of an inexpensive proper orthogonal decomposition (POD)-based Galerkin system (GS) with short runs of a standard numerical solver (NS). The latter provides the snapshots needed for the construction of the POD modes, which form the linear subspace the governing equations are Galerkin-projected onto. Alternation between running the NS and integrating the GS is an 'on the fly' process, in which the underlying POD modes are dynamically adapted over time to the local dynamics. Thus, this method is called *POD on the Fly*.

In Sect. 3, we discuss a *sampling* technique that, given a set of snapshots, selects convenient collocation points retaining the 'essence' of the data, expressed, e.g., by concentrated spatio-temporal complexity. In essence, this collocation strategy is based on two steps:

(i) Truncated (with a given accuracy) Gauss elimination (or LU decomposition) with double pivoting is performed on the full snapshot matrix, which selects the N most linearly independent snapshots and a set of N spatial points that better account for such linear independency. The number N is comparable to the number of POD modes that would be selected by applying POD to the full snapshot matrix.

(ii) POD is applied to the selected snapshots using only the selected points, and Galerkin projection is performed to obtain the ROM using only the values of the obtained POD modes at the selected points.

Since it is a combination of LU and POD decompositions, the method is referred to as *LUPOD method* [15]. Let us note here that LUPOD is different to related methods in the literature, such as the so-called Q-DEIM [9], which in turn is an improvement of plain DEIM [7]. Specifically, in Q-DEIM the first step consists in applying truncated POD to the full snapshot matrix, which selects a number of POD modes, to which a QR decomposition with pivoting is then applied to select a number of spatial points to be used in Galerkin projection.

The LUPOD method can be synergically combined with Galerkin projection to improve the performance of the resulting ROM. In fact, LUPOD both enhances the quality of the POD modes and decreases the required computational resources. Section 4 illustrates the combination of LUPOD with *preprocessed* ROMs, in which snapshots and modes are computed offline just once, to describe attractors. It is worth mentioning that the final goal of these ideas would be constructing adaptive ROMs in which the LUPOD-selected collocation points (together with the modes) may change over time to adapt to the transient local dynamics that are simulated. This is left as future research; nonetheless simulations presented in Sect. 4, in which the efficiency and robustness of LUPOD-based preprocessed ROMs are tested in some challenging test cases, are an important step towards the mentioned extension of the method.

2   ROMs Based on POD on the Fly

Let us consider an evolution dissipative problem of the form

$$\partial_t \mathbf{q} = \mathscr{L}\mathbf{q} + \mathbf{f}(\mathbf{q}, t) \tag{1}$$

for the state vector $\mathbf{q}(\mathbf{x}, t)$, where $\mathscr{L}$ is a linear operator ($\mathscr{L}$ typically involves spatial derivatives and is elliptic) and $\mathbf{f}$ is nonlinear. A discussion on convenient assumptions for such operators can be found in [14] and references therein. After spatial discretization of the operators $\mathscr{L}$ and $\mathbf{f}$ (for which we will keep the same notation), time discretization is performed by using, for instance, the Crank-Nicolson plus Adams-Bashforth scheme with time step Δt (see [5]), yielding

$$\frac{2}{\Delta t}(\mathbf{q}^{k+1} - \mathbf{q}^k) = \mathscr{L}(\mathbf{q}^{k+1} + \mathbf{q}^k) + 3\mathbf{f}(\mathbf{q}^k, t_k) - \mathbf{f}(\mathbf{q}^{k-1}, t_{k-1}), \tag{2}$$

where $t_k = k\Delta t$ and $\mathbf{q}^k \in \mathbb{C}^J$ stands for the spatial distribution of the state variable $\mathbf{q}$ at the J grid points of the used spatial mesh. To generate a low dimensional approximation of (2), we study the combination of POD plus Galerkin projection. The idea is as follows. Consider the $J \times K$ snapshot matrix $\mathbf{S} = [\mathbf{s}_1, \ldots, \mathbf{s}_K]$ whose columns are spatial distributions of the state variable $\mathbf{q}$ in the spatial mesh with J grid points at K different time instants $\tau_1 < \ldots < \tau_K$, computed via the NS defined by (2). Standard POD [6, 20] applied to $\mathbf{S}$ gives an orthonormal set of $M < K$ POD modes $\mathbf{u}_1, \ldots, \mathbf{u}_M \subset \mathbb{C}^J$ such that $\mathbf{s}_k \approx \sum_{m=1}^{M} \langle \mathbf{s}_k, \mathbf{u}_m \rangle \mathbf{u}_m$, where $\langle \cdot, \cdot \rangle$ is the inner product associated to the spatial mesh, namely

$$\langle \mathbf{u}, \mathbf{v} \rangle = \sum_{j=1}^{J} u_j \overline{v_j}, \quad \text{for } \mathbf{u}, \mathbf{v} \in \mathbb{C}^J, \tag{3}$$

where overbars stand hereafter for the complex conjugate. POD modes are optimal in the sense that they provide the best approximation with M modes of the whole snapshot set with respect to the root mean square (RMS) error. We consider then the linear expansion $\mathbf{q}_{GS} = \sum_{m=1}^{M} a_m(t)\mathbf{u}_m$ and substitute $\mathbf{q}_{GS}$ into the spatially discretized counterpart of (1) to perform a Galerkin projection onto the POD modes. Thus, the mode amplitudes $a_1, \ldots, a_M$ obey the GS

$$\frac{d}{dt}a_i = \sum_{j=1}^{M} \mathcal{L}_{ij}^{GS} a_j + f_i^{GS}(a_1, \ldots, a_M, t), \quad i = 1, \ldots, M, \tag{4}$$

where

$$\mathcal{L}_{ij}^{GS} = \langle \mathbf{u}_i, \mathcal{L}\mathbf{u}_j \rangle, \quad f_i^{GS} = \left\langle \mathbf{u}_i, \mathbf{f}\left(\sum_{m=1}^{M} a_m \mathbf{u}_m, t\right)\right\rangle.$$

Then, the system (4) of ODEs is integrated using the same time discretization as for the original NS. When choosing the Crank-Nicolson plus Adams-Bashforth scheme, the mode amplitude vector $\mathbf{a} = [a_1, \ldots, a_M]^\top$ is obtained from

$$\frac{2}{\Delta t}(\mathbf{a}^{k+1} - \mathbf{a}^k) = \mathcal{L}^{GS}(\mathbf{a}^{k+1} + \mathbf{a}^k) + 3\mathbf{f}^{GS}(\mathbf{a}^k, t_k) - \mathbf{f}^{GS}(\mathbf{a}^{k-1}, t_{k-1}).$$

This kind of combination of POD with Galerkin projection has been widely used for evolution problems, for instance in [1, 11]. In principle, time integration of the GS involves a much smaller computational effort than the integration of the original system. However, the GS can exhibit spurious dynamics, which seems to be due to the non-invariance of the POD subspace under the true dynamics [17]. To avoid this problem, when the GS is designed to approach attractors, [8, 19] proposed to correct either the GS or the POD subspace. In previous papers coauthored by some of us [14, 16, 21], we suggested a different approach to provide good approximations not only of attractors, but also of transients. The method, called *local POD plus Galerkin projection* or *POD on the Fly*, can be summarized as follows.

- **Step 1**. Snapshots are calculated by the original NS in an initial time interval, I_{NS}.
- **Step 2**. POD is applied to identify the most relevant POD modes. The number M of retained modes to guarantee an approximation within accuracy ε is selected in terms of the relative RMS error, estimated as

$$RMSE_M^R = \sqrt{\sum_{j=M+1}^{R} (\sigma_j)^2 \Big/ \sum_{j=1}^{R} (\sigma_j)^2},$$

where the positive scalars σ_j are the POD singular values and R is the rank of the snapshot matrix. In principle, M could be selected by imposing $RMSE_M^R < \varepsilon$. However, to anticipate drifts in the system, we keep M larger than necessary; furthermore, an even larger number of modes, $M_1 > M$, is retained to monitor the mode truncation error (see step 3). These numbers are selected such that [16, 21]

$$RMSE_M^R < \varepsilon/100, \quad RMSE_{M_1}^R < \varepsilon/10000.$$

- **Step 3**. The governing equations are Galerkin-projected onto the POD modes and the resulting GS is integrated in a new interval, I_{GS}, until the Galerkin approximation is no longer accurate. To detect when the integration fails, we proposed in [16, 21] to monitor (i) the mode truncation error by means of the estimate

$$E_M^{M_1} = \sqrt{ \sum_{j=M+1}^{M_1} |a_j|^2 \Big/ \sum_{j=1}^{M_1} |a_j|^2 }, \qquad (5)$$

where a_j are the amplitudes appearing in (4), and (ii) possible mode instabilities by an estimate based on the simultaneous integration of two GSs of different dimension. The latter method was improved in [14] by considering residual estimates.

- **Step 4**. A few new snapshots are computed in a new I_{NS} interval using the NS to update the POD subspace. This is done by mixing some of the old (weighted) POD modes with the most relevant (weighted) ones calculated from the new snapshots (see [14, 16] for further details). Then, the process is repeated from step 3.

For illustration, we consider the complex Ginzburg-Landau equation in 1D,

$$\partial_t q = (1 + i\alpha)\partial_{xx}^2 q + \mu q - (1 + i\beta)|q|^2 q, \quad \text{with } q = 0 \text{ at } x = 0, 1, \qquad (6)$$

with parameters $\alpha = 2$, $\beta = -3.5$ and $\mu = 80$, discretized using centered second-order finite differences in a uniform grid of $J = 1000$ points and using for the time discretization the Crank-Nicolson plus Adams-Bashforth scheme [5] with time step $\Delta t = 5 \cdot 10^{-5}$. The initial condition $q(x, 0) = i \sin(2\pi x) + (1 + i) \sin(3\pi x)$ is selected to avoid the dynamics to be restricted to an invariant subspace. In Fig. 1a we show the temporal evolution of $|q|$ at the spatial points $x = 0.25$, $x = 0.5$ and $x = 0.75$ for $0 \le t \le 3$. We observe that the dynamics completely changes around $t = 0.8$. The spatio-temporal color map of $|q|$ is shown in Fig. 1b. When applying the POD on the Fly method, the initial NS interval, where the original NS is run, is selected as $I_{NS} = [0, 0.12]$ and the first GS interval is $I_{GS} = [0.12, 0.817]$. The integration of the GS stops precisely at $t = 0.817$, where the method detects that the approximation is going to fail. To visualize this, we have represented in Fig. 1c both the E_2 error, defined as $E_2 = \|\mathbf{q} - \mathbf{q}_{GS}^M\| / \|\mathbf{q}\|$ (where $\mathbf{q}$ is the 'exact' solution computed by the NS, $\mathbf{q}_{GS}^M$ is the GS solution when considering M modes and $\| \cdot \|$ is the

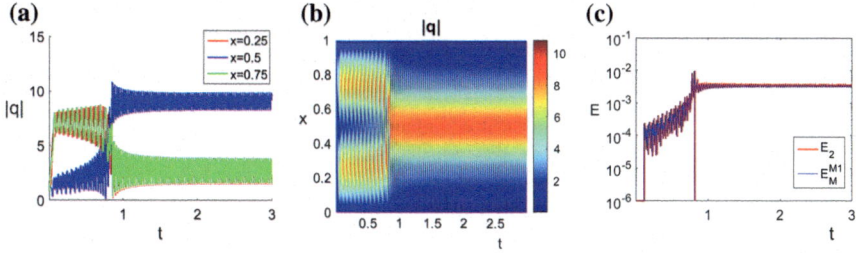

Fig. 1 **a** Temporal evolution of $|q|$ at points $x = 0.25$, 0.5 and 0.75. **b** Spatio-temporal color map of $|q|$. **c** E_2 and $E_M^{M_1}$ errors for $\varepsilon = 0.01$

norm associated with the inner product (3)), and the estimated truncation error (5). By observing these errors, we identify the first I_{NS} interval because both errors are set to zero and the first I_{GS} interval where both errors are almost plot-indistinguishable. We also observe that at $t = 0.817$, the estimate $E_M^{M_1}$ equals the desired accuracy, which is set to $\varepsilon = 0.01$. The numbers of modes are $M = 10$ and $M_1 = 16$. Then, a very short NS interval is required to update the POD basis, $I_{NS} = [0.817, 0.819]$. Note that the new POD subspace, with $M = 10$ and $M_1 = 14$, is sufficiently accurate to describe the dynamics until the final time $T = 3$. To measure the efficiency of the method, we compute the acceleration factor, defined as

$$C = \frac{\text{CPU time (NS)}}{\text{CPU time (ROM)}}. \tag{7}$$

In this example, it turns out that $C = 3.20$.

In order to improve the efficiency of the method, we can either (i) reduce the length of the initial I_{NS} interval, which was the topic of our previous papers [13, 14], where we proposed an efficient way to use previous (offline) information, or (ii) reduce the cost of the POD computation and the GS integration by using information of the snapshots at a reduced number $N < J$ of collocation points, $x_{j_1}, \ldots, x_{j_N}$, namely by considering the inner product

$$\langle \mathbf{u}, \mathbf{v} \rangle = \sum_{n=1}^{N} u_{j_n} \overline{v_{j_n}}, \quad \text{for } \mathbf{u}, \mathbf{v} \in \mathbb{C}^J, \tag{8}$$

instead of (3), which was associated to the full spatial mesh. Notice that, by doing so, the new POD modes will be orthogonal with respect to the inner product (8), and Galerkin projection will be also carried out by using (8). Obviously, the simplest choice is to select $j_1, \ldots, j_N$ equispaced. This was the strategy explored in [14, 16]. When repeating the numerical experiment of Fig. 1 by considering $N = 100$ or $N = 30$ equispaced points, the acceleration factor increases to $C = 4.97$ and $C = 6.25$, respectively. The counterpart plots are indistinguishable from those in Fig. 1 and the numbers of POD modes for the two GSs are unchanged. We have to emphasize that a further reduction of the number of points promotes a deterioration of the approximation. Furthermore, in [16] we calibrated, by trial and error, that the number of equispaced points should satisfy $N > 2M_1$. Another possibility is to select the collocation points concentrated in the spatial regions that are known to be the most significant, as done in [21] for the unsteady lid-driven cavity problem, where points were located near the upper and lateral boundaries, where the solution exhibits a richer spatial structure. In that case, N was calibrated to satisfy $N > 3M_1$. More efficient selections can be performed by using more sophisticated sampling strategies, such as missing point estimation [3], empirical [4, 10] and discrete empirical [7, 9, 12] interpolations, and hyper-reduction [2, 18]. However, the computational cost to obtain the collocation points by using those methods could offset the advantage of using the inner product (8). In the next section, we describe a very promising alternative that selects a number of points comparable to the number of retained modes and whose computational cost is comparable to standard POD.

3 Collocation via LU and the LUPOD Method

In our recent paper [15], we proposed a sampling technique based on an incomplete LU decomposition (i.e., Gauss elimination with double pivoting) to select both an appropriate set of collocation points and a convenient set of snapshots from a snapshot matrix. The algorithm, called *LUPOD method*, is as follows.

- **Step 1**. Given a snapshot matrix $\mathbf{S} = [\mathbf{s}_1, \ldots, \mathbf{s}_K]$, where $\mathbf{s}_k \in \mathbb{C}^J$, we first identify the element $S_{j_1 k_1}$ of $\mathbf{S}$ with the largest absolute value. The indices j_1 and k_1 define the first collocation point and the first selected snapshot, respectively.
- **Step 2**. Using $S_{j_1 k_1}$ as pivot, we perform Gauss elimination by columns to set to zero the j_1-th row of $\mathbf{S}$. The k_1-th column of $\mathbf{S}$ is then removed to define the first modified snapshot matrix $\mathbf{S}_1$. The first two steps are sketched in Fig. 2.
- **Step 3**. Steps 1 and 2 are iteratively performed for the modified matrices $\mathbf{S}_1, \mathbf{S}_2, \ldots$ to identify new collocation points (indices $j_2, j_3, \ldots$) and modified snapshots (indices $k_2, k_3, \ldots$) until the pivot is smaller than a given small threshold $\varepsilon_{LU} > 0$, namely until $\|\mathbf{S}_{N+1}\|_{\max}/\|\mathbf{S}\|_{\max} < \varepsilon_{LU}$.

Once the indices $\{j_1, \ldots, j_N\}$ and $\{k_1, \ldots, k_N\}$ are identified, we proposed in [15] to perform POD (via truncated SVD) on the matrix $\widetilde{\mathbf{S}} = [\mathbf{s}_{k_1}, \ldots, \mathbf{s}_{k_N}]$ using the inner product (8) based on the selected collocation points. Notice that POD is applied to the original snapshots $\mathbf{s}_{k_1}, \ldots, \mathbf{s}_{k_N}$, not to the modified snapshots that appear during the Gauss elimination process. Observe also that all the original snapshots can be approximated as linear combinations of the selected snapshots, using information at the collocation points only, within a relative maximum error of size ε_{LU}. We select a set of POD modes $\mathbf{u}_1, \ldots, \mathbf{u}_M$, where the number M is found as in the standard POD.

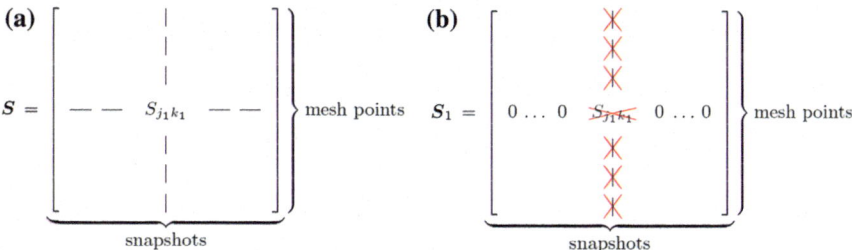

Fig. 2 First two steps in the LUPOD collocation method

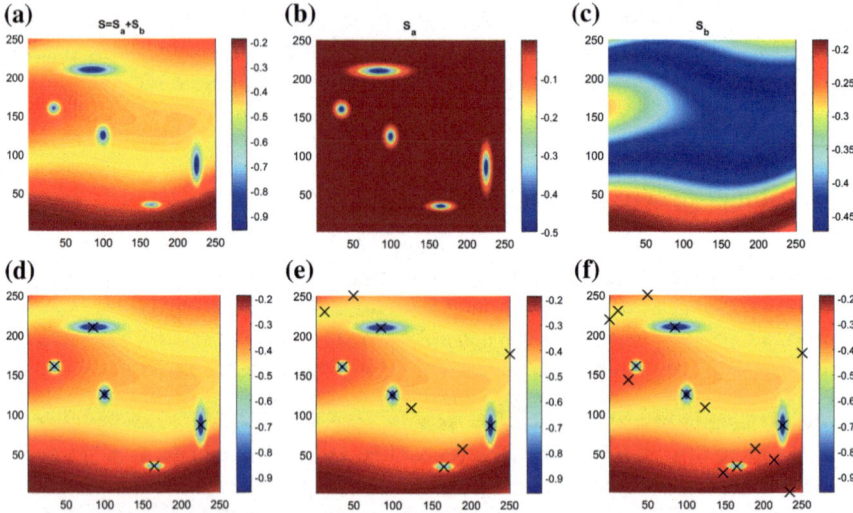

Fig. 3 First row: **a** Snapshot matrix $\mathbf{S} = \mathbf{S}_a + \mathbf{S}_b$. **b** Matrix $\mathbf{S}_a$. **c** Matrix $\mathbf{S}_b$. Second row: Location of the first N snapshots and collocation points, with **d** $N = 5$, **e** $N = 10$, **f** $N = 15$

To illustrate the performance of the LUPOD method, we consider an academic toy model inspired in one of those proposed in [15], where the snapshot matrix has concentrated complexity associated with localized peaks in the snapshots, which will promote spatial concentration of the LUPOD points in those regions. It is worth mentioning that, as explained in [15], concentrated complexity is not only related to steep changes in the behavior of the snapshots, but also to changes of their first or higher order derivatives. The considered matrix $\mathbf{S}$ of size 250×250 (represented in Fig. 3a) has been obtained by adding two snapshot matrices, $\mathbf{S} = \mathbf{S}_a + \mathbf{S}_b$. The matrix $\mathbf{S}_a$ is equal to zero everywhere, except for five small elliptical regions (see Fig. 3b), while $\mathbf{S}_b$ corresponds to a smooth function obtained by combination and composition of transcendental functions (see Fig. 3c). Columns and rows in these matrices correspond to snapshots and points, respectively.

In Fig. 3d–f, we represent by ×−marks the first five, ten and fifteen snapshots and collocation points selected by the LUPOD method. We observe that the first five are located at the centers of the ellipses, while the subsequent snapshots and collocation points are related to the complexity associated with the matrix $\mathbf{S}_b$.

In Fig. 4a we illustrate the accuracy of the LUPOD reconstruction of $\mathbf{S}$ when varying the number N of collocation points/snapshots and retaining $M = N$ POD modes, in terms of both the RMS and the maximum errors. We observe that errors decay exponentially. For a further illustration, we compare in Fig. 4b the RMS and maximum errors when fixing $N = 25$ and varying the number M of retained modes, while Fig. 4c compares both errors when fixing $M = 10$ and varying the number N of selected collocation points/snapshots. We observe that, for an optimal performance of the method, N and M should be comparable. Finally, Fig. 4d compares

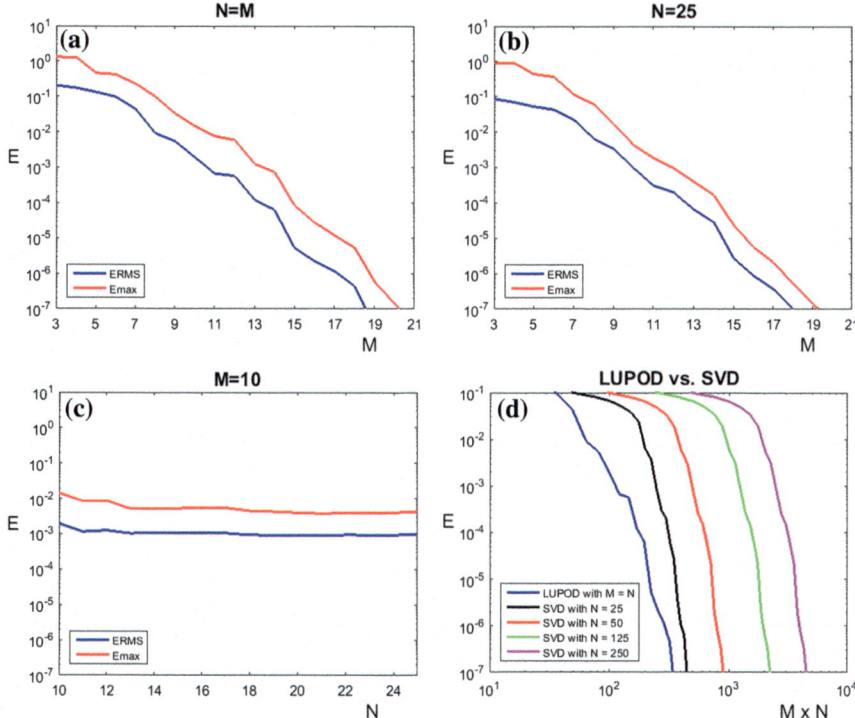

Fig. 4 **a** Reconstruction errors versus the number N of collocation points/snapshots when selecting $M = N$ modes. **b** Reconstruction errors versus the number M of modes when selecting $N = 25$ collocation points/snapshots. **c** Reconstruction errors versus the number N of collocation points/snapshots when selecting $M = 10$ modes. **d** RMS error when using LUPOD with $M = N$ (blue line) and SVD with N collocation points (remaining lines) versus $M N$

the RMS error versus $M N$ ($M N$ is the computational complexity associated with a Galerkin projection in a POD-based ROM when using N collocation points and M POD modes), when using LUPOD and standard SVD with uniform distributions of N points and retaining $M \leq N$ POD modes. This plot evidences that LUPOD outperforms in comparison with SVD for uniform mesh grids.

4 LUPOD for Preprocessed ROMs

In this section, we show how to use the LUPOD method for POD-based ROMs. The final challenge would be to apply the strategy in the POD on the Fly method using LUPOD instead of POD. However, the full adaptation of the strategy is still work in progress, therefore we will devote this section to the application of the LUPOD plus Galerkin projection to approximate attractors, namely the POD subspace will not be adaptively changed. Nonetheless, it is worth remarking that testing the ability of a LUPOD-based ROM to capture concentrated complexity in a nonlinear evolution problem, as in the simulations below, is an essential step before the mentioned adaptation may be developed. More precisely, we will consider the integration of the problem (1) in a time interval $[T_0, T]$ and proceed as follows. First, we use the NS in an interval $[T_0, T_1]$ to compute the snapshots, then we apply the LUPOD method to identify the collocation points and compute the POD modes by using the inner product (8) based on such collocation points, and finally we Galerkin-project the equations by means of the same inner product to approximate the dynamics in the interval $[T_1, T]$.

For illustration, we consider the integration of the complex Ginzburg-Landau equation (6) with the same parameter values and the same discretization scheme as in Sect. 2. The original NS is used to integrate problem (6) in the interval $[0, 1.12]$. We then consider 240 equispaced snapshots in the interval $[1, 1.12]$ to apply LUPOD. The modulus of these snapshots at 1000 equispaced spatial points is shown in Fig. 5a. Notice that this plot corresponds to the restriction to the time interval $[1, 1.12]$ of the spatio-temporal color map represented in Fig. 1b. When applying LUPOD with a tolerance $\varepsilon_{LU} = \varepsilon/100 = 10^{-4}$, only seven snapshots/collocation points are selected, which are represented in Fig. 5a by $\times$-marks. The location of the collocation points on the spatial interval $[0, 1]$ is shown in Fig. 5b to further emphasize their concentration in $[0.25, 0.75]$, where q is more complex. The LUPOD method selects $M = 5$ and $M_1 = 7$ POD modes. The reduced model is finally integrated in the GS interval $[1.12, 4]$, providing an approximation within a relative error smaller than the prescribed tolerance $\varepsilon = 0.01$, as can be seen in Fig. 6b. The acceleration factor corresponding to the LUPOD plus Galerkin method in the interval $[1,4]$ is $C = 9.09$. For comparison with our previous POD plus Galerkin method, we applied POD to the same initial snapshots, considering $N = 15$ uniformly distributed spatial points for the reduced inner product (8), which yields $M = 5$ and $M_1 = 10$ modes. The acceleration factor is $C = 7.52$. The corresponding E_2 and $E_M^{M_1}$ errors are shown in Fig. 6c. Therefore, in this case, the acceleration factor obtained via LUPOD is larger than with standard POD. Furthermore, we tested in [15] that the acceleration factors in 2D problems when using LUPOD can be impressive.

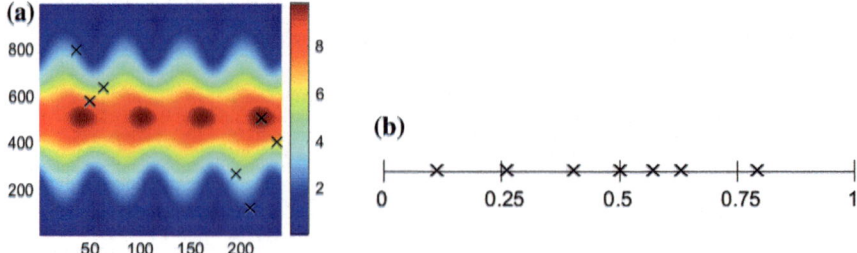

Fig. 5 **a** Snapshot matrix corresponding to the integration of equation (6) in the time interval $[1, 1.12]$ when considering a spatial mesh with 1000 points. The columns are the modulus of the snapshots at the time instants $t_k = 1 + 0.0005k$, for $k = 1, \ldots, 240$. **b** Location of the selected collocation points

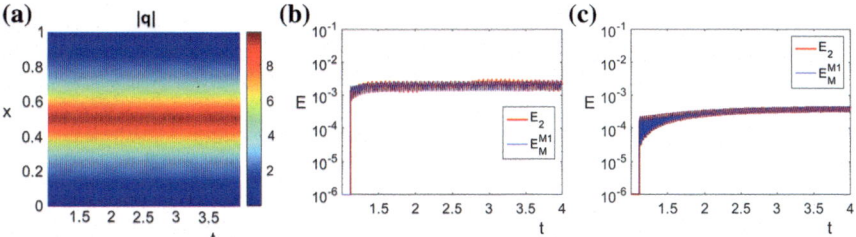

Fig. 6 a Spatio-temporal color map of $|q|$ in the interval $[1,4]$. **b** E_2 and $E_M^{M_1}$ errors for the LUPOD plus Galerkin method. **c** E_2 and $E_M^{M_1}$ errors for the POD plus Galerkin method

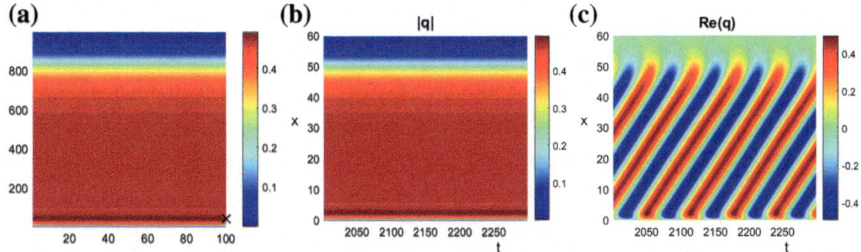

Fig. 7 a Modulus of 100 equispaced snapshots in the time interval $[2000, 2001]$; the snapshot/collocation point selected by the LUPOD method is indicated by an ×-mark. **b** Spatio-temporal color map of $|q|$ in the interval $[2000, 2300]$. **c** Spatio-temporal color map of $Re(q)$ in the interval $[2000, 2300]$

To end this section, let us show even further the incredible ability of the LUPOD method to identify the relevant features of the underlying dynamics. With this aim, we consider an example extracted from [22], where the complex Ginzburg-Landau equation with drift

$$\partial_t q = (1 + i\alpha)\partial_{xx}^2 q + c\partial_x q + \mu q - (1 + i\beta)|q|^2 q, \quad \text{with } q = 0 \text{ at } x = 0, 60,$$

with parameters $\alpha = 0.45$, $\beta = -2$, $\mu = 0.3$ and $c = 1$, is taken into account. The original NS is the same as for the complex Ginzburg-Landau equation without drift (adding centered finite differences for the first space derivative), with a spatial mesh of $J = 1000$ points and a time step $\Delta t = 5 \cdot 10^{-4}$. After eliminating the transient, 100 snapshots are selected in the time interval $[2000, 2001]$, which are processed by the LUPOD method with a tolerance $\varepsilon_{LU} = \varepsilon/100 = 10^{-4}$. The strategy selects just one collocation point and one snapshot, which are indicated in Fig. 7a by an ×-mark. Only one POD mode is then kept. Integration of the GS on the interval $[2001, 2300]$, considering just this point and this mode, provides an approximation within an RMS error $E_2 < 10^{-6}$. The modulus and the real part of the solution on such interval are plotted in Fig. 7b, c. It is rather impressive that the LUPOD method is able to capture the dynamics by means of just one collocation point.

5 Conclusions and Future Work

We have considered two ingredients that highly improve the performance of standard ROMs. On the one hand, combining along the simulation a standard numerical solver with a much faster Galerkin system leads to a POD on the Fly method, which is a fully online method that improves standard (preprocessed) ROMs in which the POD modes need to be computed from the outset. On the other hand, we have presented a very efficient sampling method to select a good set of collocation points to perform Galerkin projection. This method is called LUPOD because it is based on a synergic combination of an LU decomposition of the snapshot matrix and the application of POD. The technique has been combined with Galerkin projection to improve the performance of preprocessed ROMs for the approximation of attractors exhibiting concentrated spatio-temporal complexity. The combination of POD on the Fly and the LUPOD method is very promising but it is beyond the scope of the paper and will be the object of future research.

Acknowledgements This work was supported by the Spanish Ministry of Economy and Competitiveness, under grants TRA2016-75075-R and MTM2017-84446-C2-2-R.

References

1. Alonso, D., Velázquez, A., Vega, J.M.: A method to generate computationally efficient reduced order models. Comput. Meth. Appl. Mech. Eng. **198**, 2683–2691 (2009)
2. Amsallem, D., Zahar, M.J., Washabaugh, K.: Fast local reduced basis updates for the efficient reduction of nonlinear systems with hyper-reduction. Adv. Comput. Math. **41**, 1187–1230 (2015)
3. Astrid, P., Weiland, S., Willcox, K., Backx, T.: Missing point estimation methods in models described by proper orthogonal decomposition. IEEE Trans. Autom. Control **53**, 2237–2250 (2008)
4. Barrault, M., Maday, Y., Nguyen, N.C., Patera, A.T.: An 'empirical interpolation' method: application to efficient reduced-basis discretization of partial differential equations. C. R. Math. Acad. Sci. Paris **339**, 667–672 (2004)
5. Cebeci, T.: Convective Heat Transfer. Springer, Berlin (2002)
6. Chatterjee, A.: An introduction to the proper orthogonal decomposition. Current Sci. **78**, 808–817 (2000)
7. Chaturantbut, S., Sorensen, D.C.: Nonlinear model reduction via discrete empirical interpolation. SIAM J. Sci. Comput. **32**, 2737–2764 (2010)
8. Couplet, M., Basdevant, C., Sagaut, P.: Calibrated reduced-order POD-Galerkin system for fluid flow modelling. J. Comput. Phys. **207**, 192–220 (2005)
9. Drmac, Z., Gugercin, S.: A new selection operator for the discrete empirical interpolation method—improved a priori error bound and extensions. SIAM J. Sci. Comput. **38**, A631–A648 (2016)
10. Drohmann, M., Haasdonk, B., Ohlberger, M.: Reduced basis approximation for nonlinear parametrized evolution equations based on empirical operator interpolation. SIAM J. Sci. Comput. **34**, A937–A969 (2012)
11. LeGresley, P., Alonso, J.: Investigation of non-linear projection for POD based reduced order models for Aerodynamics. AAIA paper 2001-0926 (2001)

12. Peherstorfer, B., Butnark, D., Willcox, K., Bungartz, H.J.: Localized discrete empirical interpolation method. SIAM J. Sci. Comput. **36**, A168–A192 (2014)
13. Rapún, M.-L., Terragni, F., Vega, J.M.: Mixing snapshots and fast time integration of PDEs. In: Papadrakakis, M., Oñate, E., Schrefler B. (eds.) IV International Conference on Computational Methods For Coupled Problems in Science and Engineering, pp. 861–872 (2011)
14. Rapún, M.-L., Terragni, F., Vega, J.M.: Adaptive POD-based low-dimensional modeling supported by residual estimates. Int. J. Numer. Meth. Eng. **104**, 844–868 (2015)
15. Rapún, M.-L., Terragni, F., Vega, J.M.: LUPOD: collocation in POD via LU decomposition. J. Comput. Phys. **335**, 1–20 (2017)
16. Rapún, M.-L., Vega, J.M.: Reduced order models based on local POD plus Galerkin projection. J. Comput. Phys. **229**, 3046–3063 (2010)
17. Rempfer, D.: On low-dimensional Galerkin models for fluid flow. Theor. Comput. Fluid Dyn. **14**, 75–88 (2000)
18. Ryckelynck, D.: Hyper-reduction of mechanical models involving internal variables. Int. J. Numer. Meth. Eng. **77**, 75–89 (2009)
19. Sirisup, S., Karniadakis, G.E., Xiu, D., Kevrekidis, I.G.: Equations-free/Galerkin-free POD assisted computation of incompressible flows. J. Comput. Phys. **207**, 568–587 (2005)
20. Sirovich, L.: Turbulence and the dynamics of coherent structures. Q. Appl. Math. **45**, 561–590 (1987)
21. Terragni, F., Valero, E., Vega, J.M.: Local POD plus Galerkin projection in the unsteady lid-driven cavity problem. SIAM J. Sci. Comput. **33**, 3538–3561 (2011)
22. Tobias, S.M., Proctor, M.R.E., Knobloch, E.: Convective and absolute instabilities of fluid flows in finite geometry. Physica D **113**, 43–72 (1998)

A Posteriori Error Estimation in Model Order Reduction of Elastic Multibody Systems with Large Rigid Motion

Ashish Bhatt, Jörg Fehr, Dennis Grunert and Bernard Haasdonk

Abstract We consider the equation of motion of an elastic multibody system in absolute coordinate formulation (ACF). The resulting nonlinear second order DAE of index two has a unique solution and is reduced using the strong POD-greedy method. The reduced model is certified by deriving a posteriori error estimators, which are independent of the model order reduction (MOR) method used to obtain the projection basis. The first error estimation technique, which we establish in this paper, is a first order linear integro-differential equation. It relies on the gradient of a function and can be integrated along with the reduced simulation (in-situ). The second error estimation technique is hierarchical and requires a more enriched basis in order to estimate the error in the solution due to a coarser basis. To verify and illustrate the efficacy of the estimators, reproductive and predictive numerical experiments are performed on a coupled elastic multibody system consisting of a double elastic pendulum.

Keywords EMBS · Nonlinear PDAE · A posteriori error estimation · In-situ · Hierarchical

A. Bhatt (✉)
IIT ISM Dhanbad, 2nd Floor, Science Block, Jharkhand 826004, India
e-mail: ashish.bhatt@mathematik.uni-stuttgart.de

J. Fehr · D. Grunert
University of Stuttgart, ITM, Pfaffenwaldring 5, 70569 Stuttgart, Germany
e-mail: joerg.fehr@itm.uni-stuttgart.de

D. Grunert
e-mail: dennis.grunert@itm.uni-stuttgart.de

B. Haasdonk
University of Stuttgart, IANS, Pfaffenwaldring 57, 70569 Stuttgart, Germany
e-mail: haasdonk@mathematik.uni-stuttgart.de

© Springer Nature Switzerland AG 2020
J. Fehr and B. Haasdonk (eds.), *IUTAM Symposium on Model Order Reduction of Coupled Systems, Stuttgart, Germany, May 22–25, 2018*, IUTAM Bookseries 36,
https://doi.org/10.1007/978-3-030-21013-7_7

1   Introduction

Material properties of a body with linear elasticity can be described by its density, Young's modulus, and Poisson's ratio. The material properties and geometric properties, such as length and breadth, along with external forces can comprise the parameter space of the mechanical problem. The motion of an elastic multibody system with equality constraints is governed by the principle of virtual work, which results in a nonlinear *parameterized* partial differential-algebraic equation (PDAE) with a second order differential part describing the evolution of the unknown displacement and an algebraic part describing the constraints. This PDAE is accompanied by initial conditions for the displacement and the velocity.

The motion of an elastic body can be decomposed into small elastic and relatively large rigid body motion. When deriving the equations of motion, one can choose between the floating frame of reference formulation (FFRF) and the absolute coordinate formulation (ACF). The former expresses the elastic displacement in the local frame assigned to the body and couples it with the global frame for the entire system. This FFRF procedure results in a state-dependent nonlinear mass matrix and has been widely used [10, 11, 28]. The latter ACF procedure measures the wholesome displacement of the body in a global (inertial) frame and therefore results in a linear mass matrix with a co-rotated stiffness matrix [5]. We will consider the latter approach in this work, see [13, 14] for a comparison of the two approaches. The FFRF and ACF solutions can be transformed into each other [28]. We derive the equations of motion from first principles for the ACF in Sect. 2.

Spatial discretization of the parameterized PDAE results in a high dimensional parameterized differential-algebraic equation (DAE). It is often imperative to simulate the resulting high dimensional system repeatedly for varying parameters which calls for a model order reduction (MOR) method. There exist a number of MOR methods, e.g. balanced truncation, rational interpolation, and reduced basis methods [1, 2, 15, 16, 20, 26]. In this work, we will use the strong POD-greedy method. MOR methods are frequently also accompanied by hyperreduction techniques [6, 7, 9] to speed up the computation of high dimensional nonlinear terms, which we do not include in the current study. We set up the reduced model in Sect. 3.

MOR methods often introduce an approximation error. For reduced basis type methods of time-dependent problems, several error estimators have been proposed, most of which are residual based. A space-time residual is computed, or a spatial residual is integrated in order to compute such estimators [18, 20, 21, 26]. We take a different route in this paper and deduce an in-situ error estimator in Sect. 4 which is the solution of a linear integro-differential equation. This novel approach is compared against and blended with hierarchical error estimators derived in the same section. Finally, an illustrative experiment is presented in Sect. 5, and we close the discussion with conclusions in Sect. 6.

2   Equations of Motion of an Elastic Multibody System

When an elastic body, $\Omega \in \mathbb{R}^d$, at rest is perturbed on a time interval $[0, T]$ with an external force, it reacts by transferring energy into elastic and kinetic components. As a result of attempting to attain the equilibrium among all the forces, the system is displaced, which results in elastic and rigid displacements. Proceeding as in [5], the principle of virtual work [23] results in a PDAE that reads in the weak form, with Lagrange multiplier $\lambda(x, t) : \Omega \times (0, T) \rightarrow \mathbb{R}^d$,

$$\int_{\Omega} \rho \ddot{u} \cdot \delta u \, d\Omega + \int_{\Omega} \widetilde{E} : \mathscr{E} : \delta \widetilde{E} \, d\Omega + (\nabla C(u))^{\mathsf{T}} \lambda \cdot \delta u$$

$$= \int_{\Omega} g_{s} \cdot \delta u \, d\Omega + \int_{\partial\Omega} g_{\Gamma} \cdot \delta u \, d\Omega, \quad (1)$$

$$C(u) = 0.$$

Here ρ is the constant mass density of the elastic body, $g_s(x, t)$, $g_{\Gamma}(x, t) : \Omega \times (0, T) \rightarrow \mathbb{R}^d$ are the source and traction terms, respectively, $C : \mathbb{R}^d \rightarrow \mathbb{R}^m$ denotes the accompanying linear constraints (e.g. a pivot or joints in a multibody system), over dot denotes a time derivative as usual, δu denotes variation in the unknown displacement $u(x, t) : \Omega \times (0, T) \rightarrow \mathbb{R}^d$, $\partial\Omega$ is the boundary of the region $\Omega \subset \mathbb{R}^d$ occupied by the system, and we have omitted space and time dependence in the interest of brevity of exposition. The *reduced* strain tensor $\widetilde{E}$ in absolute coordinate formulation (ACF) reads

$$\widetilde{E}(x, t) = \mathrm{Sym}(R^{\mathsf{T}} \nabla u_{\mathrm{flex}}).$$

Here R is a rotation matrix, ∇ denotes the gradient w.r.t. the spatial coordinate x, displacements $u_{\mathrm{flex}}(x, t)$, $u_0(x, t) : \Omega \times (0, T) \rightarrow \mathbb{R}^d$ are the flexible and rigid parts of the total displacement u such that $u = u_{\mathrm{flex}} + u_0$, and Sym denotes the symmetric part of a matrix. The fourth order elasticity tensor $\mathscr{E}$ is obtained from the constitutive equation

$$\mathscr{E} = \lambda_L I \otimes I + \mu_L \mathscr{I} \quad (2)$$

with Lamé constants $\lambda_L, \mu_L \in [a, b] \subset \mathbb{R}$, identity matrix I, tensor product $\otimes$, and the fourth order symmetric identity tensor

$$(\mathscr{I})_{ijkl} = \delta_{ik}\delta_{jl} + \delta_{il}\delta_{jk}.$$

Here δ_{ij} are the Kronecker delta functions and the Lamé constants depend on the Young's modulus and Poisson's ratio of the elastic material (Fig. 1).

We now spatially discretize Eq. (1) to derive the semi-discretized system of equations assuming $\Omega \subset \mathbb{R}^d$ has smooth boundary $\partial\Omega$. To this end, substituting the finite element ansatz

Fig. 1 Elastic and rigid
body displacements u_{flex} and
u_0, respectively, of an elastic
body. The total displacement
is measured w.r.t. the global
frame X in ACF

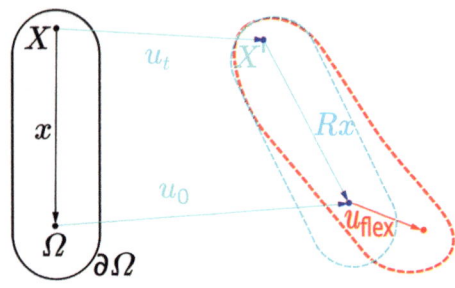

$$u(x, t) \approx u_h(x, t) = N(x)q(t)$$

into the equation, where $N(x) \in \mathbb{R}^{d \times n}$ is the finite element shape function matrix
and $q(t) \in \mathbb{R}^n$ is the unknown solution vector, and supplying initial conditions, we
obtain the discretized system

$$M\ddot{q} + R_h(q)K R_h^\mathsf{T}(q)q_{\text{flex}} + B^\mathsf{T}\lambda_h = f(q),$$
$$Bq = 0, \tag{3}$$
$$q(0) = q_0, \ \dot{q}(0) = \dot{q}_0,$$

with

$$M = \int_\Omega \rho N^\mathsf{T} N \, d\Omega, \ \ K = \int_\Omega \text{Sym}(\nabla N) : \mathscr{E} : \text{Sym}(\nabla N) \, d\Omega, \ \text{and } B = \nabla C \, N.$$

Here M and K are constant mass and stiffness matrices dependent on the material
parameters ρ, λ_L, and μ_L, respectively, $R_h(q)$ is the rotation matrix, $f(q)$ is a
nonlinear function comprising of source and traction forces and the rotation matrix
gradient, $q_{\text{flex}} \in \mathbb{R}^n$ is the unknown flexible displacement such that $q = q_{\text{flex}} + q_{\text{rig}}$
for the rigid displacement $q_{\text{rig}} \in \mathbb{R}^n$, $B \in \mathbb{R}^{m \times n}$, $m \ll n$, is the constraint matrix
resulting from the linear constraint function C of Eq. (1), and $\lambda_h \in \mathbb{R}^m$ is the Lagrange
multiplier. The constraint matrix B can model a variety of constraints including fixed
and conjoined nodes in the FEM mesh of a multibody system. The constants q_0 and
$\dot{q}_0$ are initial conditions. Consistency of the initial conditions $Bq_0 = 0, \ B\dot{q}_0 = 0$ is
tacitly assumed here. For details of the derivation of Eq. (3) from Eq. (1), see [5, 14]
for instance.

Existence of the unique solution of Eq. (3) follows from [25]. Indeed, all the terms
of the equation are smooth, $m < n$, the matrix B has full row rank, and the mass
matrix M is positive definite. This ensures that the matrix

$$\begin{bmatrix} M & B^\mathsf{T} \\ B & 0 \end{bmatrix}$$

is an isomorphism. Furthermore, DAE (3) with $\lambda_h = 0$ has index two via consistency of the initial condition $B\dot{q}_0 = 0$.

Remark 1 If additionally, the dynamics is consistent with the constraints

$$BM^{-1}(-R_h(q)KR_h^\top(q)q_{\text{flex}} + f(q)) = 0,$$

then it follows that $B\ddot{q} = 0$ by Eq. (3) and thus via integration $B\dot{q} = 0$ and $Bq = 0$ for all times. This implies that $\lambda_h = 0$ and q is equivalently a solution of the (unconstrained) IVP

$$M\ddot{q} + R_h(q)KR_h^\top(q)q_{\text{flex}} = f(q),$$
$$q(0) = q_0, \ \dot{q}(0) = \dot{q}_0. \tag{4}$$

Solving the constrained problem Eq. (3) seems preferable as the constraint satisfaction is expected to be numerically more accurate as compared to Eq. (4).

Numerical solution of (3) can be obtained by discretizing the equation in time and then solving the resulting nonlinear system of equations iteratively. To this end, let us define

$$Corr\,K(q) = R_h(q)KR_h^\top(q)q, \ \overline{f}(q) = f(q) + R_h(q)KR_h^\top(q)q_{rig},$$

where we have used the splitting $q = q_{flex} + q_{rig}$. Using this notation, (3) can be rewritten as

$$M\ddot{q} + Corr\,K(q) + B^\top\lambda_h = \overline{f}(q),$$
$$Bq = 0, \tag{5}$$
$$q(0) = q_0, \ \dot{q}(0) = \dot{q}_0,$$

Let us introduce the trapezoidal approximation, using j as the time-stepping index and $\tau = t_{j+1} - t_j$ as the uniform time step,

$$\lambda_h \approx \lambda_{j+1}, \ \ddot{q}(t_{j+1}) \approx a_{j+1}, \ q(t_{j+1}) \approx q_{j+1} = \underbrace{q_j + \tau v_j + \frac{\tau^2}{4}a_j}_{\tilde{q}_{j+1}} + \frac{\tau^2}{4}a_{j+1},$$

$$\overline{f}(q(t_{j+1})) \approx \overline{f}(q_{j+1}), \ R_h(q(t_{j+1})) \approx R_h(q_{j+1}).$$

Here v is the velocity and a is the acceleration. Substituting this discretization in (5) yields

$$\tilde{M}(q_{j+1})a_{j+1} + B^\top\lambda_{j+1} = \overline{f}(q_{j+1}) + d(q_{j+1}),$$
$$B\,a_{j+1} = e(q_{j+1}). \tag{6}$$

with

$$\tilde{M}(q_{j+1}) = R_h(q_{j+1}) \left(M + \frac{\tau^2}{4} K \right) R_h^{\mathsf{T}}(q_{j+1}),$$

$$d(q_{j+1}) = -Corr\, K(q_{j+1})\tilde{q}_{j+1},$$

$$e(q_{j+1}) = -\frac{4}{\tau^2} B(\tilde{q}_{j+1}).$$

Here we have used invariance of the mass matrix under R_h [13].

Equation (6) is a system of nonlinear equations which needs to be solved iteratively in every time step. For one fixed time t_j the iteration suggested by [14] reads

1. set $k = j$
2. initialize $q^{k+1} = q_k$
3. compute $R_h(q^{k+1})$, $\overline{f}(q^{k+1})$, $e(q^{k+1})$, $d(q^{k+1})$
4. compute λ^{k+1} using (8)
5. compute a^{k+1} using (7)
6. compute $q^{k+1} = \tilde{q}^{k+1} + \frac{\tau^2}{4} a^{k+1}$, $v^{k+1} = v_k + \frac{\tau}{2}(a_k + a^{k+1})$
7. go to step 2 until convergence.
8. set $\lambda_{j+1} = \lambda^{k+1}$, $q_{j+1} = q^{k+1}$, $v_{j+1} = v^{k+1}$, $a_{j+1} = a^{k+1}$.

The acceleration is computed with

$$a_{j+1} = \tilde{M}(q_{j+1})^{-1} \left(\overline{f}(q_{j+1}) + d(q_{j+1}) - B^{\mathsf{T}} \lambda_{j+1} \right), \tag{7}$$

$$\tilde{M}(q)^{-1} = R_h(q(t)) \left(M + \frac{\tau^2}{4} K \right)^{-1} R_h^{\mathsf{T}}(q(t)),$$

and the Lagrange multiplier λ_{j+1} is calculated with the Schur-complement

$$B\tilde{M}(q_{j+1})^{-1} B^{\mathsf{T}} \lambda_{j+1} = B\tilde{M}(q_{j+1})^{-1} \left(\overline{f}(q_{j+1}) + d(q_{j+1}) \right) - e(q_{j+1}). \tag{8}$$

Let us emphasize that this process does not involve inverting any time-dependent matrix.

3 Reduced Order Model of DAE (3)

MOR works by replacing the high dimensional space in which solutions reside with an approximating space of much lower dimension via appropriate projection. Proper orthogonal decomposition (POD) aims to minimize this projection error in an appropriate norm. POD of differential and differential-algebraic equations has been studied extensively [4, 29]. Here, we obtain the projection basis using the strong POD-greedy algorithm [16, 17]. This algorithm starts with an initial basis and the parameter value which maximizes the given selection criterion. The solution trajectories corresponding to this parameter are then computed. Orthogonal trajectories are evaluated with respect to the POD projection operator and the basis is enlarged with a POD basis of their image space. This process continues until the selection criterion is less than the given tolerance. We choose the Euclidean norm of the POD projection error to be the selection criterion. Projection bases of various sizes can be obtained by varying the tolerance. We orthonormalize the resulting projection bases for better numerical stability.

For given projection matrices V, $W \in \mathbb{R}^{n \times k}$, $k \ll n$, obtained via strong POD-greedy such that $q \approx q_r = Vz$, $q_{\text{flex}} \approx q_{r,\text{flex}} = Vz_{\text{flex}}$ with reduced states z, $z_{\text{flex}} \in \mathbb{R}^k$ and $W^\mathsf{T} V = I \in \mathbb{R}^{k \times k}$, Eq. (3) can be reduced to the DAE

$$W^\mathsf{T} M V \ddot{z} + W^\mathsf{T} R_h(Vz) K R_h^\mathsf{T}(Vz) V z_{\text{flex}} + W^\mathsf{T} B^\mathsf{T} \lambda_r = W^\mathsf{T} f(Vz),$$
$$B V z = 0,$$
$$z(0) = W^\mathsf{T} q_0, \quad \dot{z}(0) = W^\mathsf{T} \dot{q}_0, \tag{9}$$

with Lagrange multiplier $\lambda_r \in \mathbb{R}^m$. This reduced system can also be solved with the same numerical solver used to solve the full system as mentioned in the previous section. If we assume that the bases V, W are consistent with the constraints i.e. $BV = 0$, $BW = 0$, then Eq. (9) simplifies to the unconstrained ODE system, which is simply the projection of (4),

$$W^\mathsf{T} M V \ddot{z} + W^\mathsf{T} R_h(Vz) K R_h^\mathsf{T}(Vz) V z_{\text{flex}} = W^\mathsf{T} f(Vz),$$
$$z(0) = W^\mathsf{T} q_0, \quad \dot{z}(0) = W^\mathsf{T} \dot{q}_0. \tag{10}$$

Consistency of the bases is ensured by POD as the columns of V are a linear combination of snapshots of q, all of which satisfy the constraint. Due to orthogonality, we have $W = V$, which ensures consistency of W also.

After reduction, linear problems can be efficiently simulated to get the reduced solution without further downsizing. However for nonlinear systems it is often necessary to further reduce the nonlinearity (hyperreduction) in order to gain economy. Offline-online decomposition [16] of parameter-separable forms is a common technique to improve the economy. For non-affine nonlinearities, one can use empirical interpolation (EIM) [3], discrete empirical interpolation (DEIM) [30], Gauss-Newton with approximated tensors (GNAT) [6], energy-conserving sampling and weighting (ECSW) [9], etc. The reduced Eq. (9) can be discretized in time using, e.g., trapezoidal approximations.

4 A Posteriori Error Estimation

The reduced model (10), although efficient, induces an error compared to the full solution and therefore error estimation is needed for certification. One may be interested in the error in a low dimensional output quantity of interest $y = h(q)$ or the error in the solution itself. Many of the error estimators depend on evaluation of intermediate solution dependent quantities whose values and economy influences the efficacy of the estimator. Moreover, it is often desirable for the error estimator to have the same structure as the reduced ODE (9) in order to be able to solve the error ODE *in-situ* along integration of the system (10) by expanding it. We here assume that the dynamics of the full problem lies on the constrained manifold, hence Remark 1 applies, $\lambda_r = 0$, and Eqs. (9) and (10) are equivalent.

4.1 In-Situ Error Estimator

We begin by taking the difference of the exact and reconstructed solutions and then differentiate w.r.t. time to obtain the second order error ODE. This ODE is then integrated in time to get an integro-differential equation for the error. This equation is used to find the equation for the norm of the error. For succinctness, let us define

$$g(q) = M^{-1}(f(q) - R_h(q)K R_h^{\mathsf{T}}(q)q_{\text{flex}})$$

then the full (4) and reduced system (10) read

$$\ddot{q} = g(q), \quad \ddot{z} = W^{\mathsf{T}}g(Vz), \tag{11}$$

respectively. Defining the error $e(t) = q(t) - Vz(t)$, we deduce

$$\ddot{e} = \ddot{q} - V\ddot{z} = g(q) - VW^{\mathsf{T}}g(Vz). \tag{12}$$

For linear systems, Eq. (12) can be solved for the error e using variation of parameters formula. This approach was introduced in [18] for general first order systems where an online/offline decomposition of parameterized quantities is used to accelerate the online phase. It was later applied to linear mechanical systems in [27]. For second order nonlinear systems, we proceed to integrate in time

$$
\begin{aligned}
\dot{e}(t) &= \dot{e}(0) + \int_0^t g(q(s)) - VW^{\mathsf{T}}g(Vz(s))ds \\
&= \dot{e}(0) + \int_0^t (I - VW^{\mathsf{T}})g(Vz(s)) + \nabla g(Vz(s))e(s) + \mathscr{O}(\|e(s)\|^2)ds
\end{aligned}
$$

where we have used first order multidimensional Taylor series expansion [8] for the function g around Vz. Ignoring the second order error term, using triangular inequality, and $\frac{d}{dt}\|e(t)\| \leq \|\dot{e}(t)\|$, we get the scalar integro-differential equation

$$\frac{d}{dt}\|e(t)\| \leq \|\dot{e}(0)\| + \int_0^t \|(I - VW^\mathsf{T})g(Vz(s))\| + \|\nabla g(Vz(s))\|\|e(s)\|ds$$

The solution $\|e(t)\|$ of this inequality satisfies $\|e(t)\| \leq \|r(t)\|$ [22, 24] for the solution $\|r(t)\|$ of the corresponding equation

$$\frac{d}{dt}\|r(t)\| = \|\dot{e}(0)\| + \int_0^t \|(I - VW^\mathsf{T})g(Vz(s))\| + \|\nabla g(Vz(s))\|\|r(s)\|ds \tag{13}$$

when $\|e(0)\| = \|r(0)\|$.

Equation (13) can be solved using a numerical method [12] and the solution thereof is denoted by the error estimator

$$\Delta_1^k(t) := \|r(t)\|. \tag{14}$$

Here the superscript k on the estimator Δ_1 denotes its dependence on size k of the reduced system. Clearly, the efficacy of this estimator depends on the quantities $\|(I - VW^\mathsf{T})g(Vz(s))\|$ and $\|\nabla g(Vz(s))\|$. In order to integrate the estimator in-situ, one needs to supply it with the initial condition $\|r(0)\| = \|q(0) - Vz(0)\|$, which is zero if $q(0) \in \mathrm{colspan}(V)$. The quantity $\|\dot{e}(0)\|$ is given by $\dot{e}(0) = \dot{q}(0) - V\dot{z}(0)$ whose approximation results in different values of $\|e(0)\|$ and $\|r(0)\|$.

Remark 2 Another approach to error estimation is to turn Eq. (11) into a first order ODE and then use Grönwall's lemma [30]. To this end, let us rewrite the equation as

$$\dot{\eta} = h(\eta).$$

Here $\eta = [q^\mathsf{T}\dot{q}^\mathsf{T}]^\mathsf{T}$ and $h = [\dot{q}^\mathsf{T}g(q)^\mathsf{T}]^\mathsf{T}$. Then, as before,

$$\langle e, \dot{e} \rangle \leq L(h)\|e\|^2$$

where $L(h)$ is the upper logarithmic Lipschitz constant, which can be approximated by the largest eigenvalue of the symmetric part of the matrix $\nabla h(Vz)$ up to an error $\mathcal{O}(\|e\|)$. For a linear spring-mass system with stiffness $k \in \mathbb{R}$ and mass $m \in \mathbb{R}$ such that $0 < k/m < 1$, it holds

$$h(\eta) = \begin{bmatrix} 0 & 1 \\ -\frac{k}{m} & 0 \end{bmatrix}, \quad \text{and} \quad L(h) = \frac{1}{2}\sqrt{(1 - \frac{k}{m})} > 0.$$

This results in an exponentially growing error bound and renders it impractical for our second order mechanicals systems.

4.2 Hierarchical Error Bound

Hierarchical error estimators have been successfully used in the context of a non-stationary problem using space-time FEM formulation in [21] and later in [19] on a stationary problem by ensuring a saturated projection basis. Here, we adopt the finite element error estimates proposed in [21] for parametric elastodynamics problems. Given two reduced bases V_{k_1} and V_{k_2} such that $k_1 < k_2$ and the corresponding reduced solutions $z_1 \in \mathbb{R}^{k_1}$, $z_2 \in \mathbb{R}^{k_2}$, it holds

$$
\begin{aligned}
\|e^{k_1}(t)\| &:= \|q(t) - V_{k_1} z_1(t)\| \\
&= \|q(t) - V_{k_2} z_2(t) - V_{k_1} z_1(t) + V_{k_2} z_2(t)\| \\
&\leq \|q(t) - V_{k_2} z_2(t)\| + \|V_{k_1} z_1(t) - V_{k_2} z_2(t)\| =: \Delta_0^{k_1}(t).
\end{aligned}
\tag{15}
$$

Here $\|\cdot\|$ is the Euclidean norm and the last equation defines the hierarchical estimator $\Delta_0^{k_1}(t)$. The motivation is that the term $\|q(t) - V_{k_2} z_2(t)\|$ will have a very small value, assuming k_2 is large and the strong POD-greedy method converges, and the second term can be evaluated exactly, i.e.,

$$
\Delta_0^{k_1}(t) \approx \|V_{k_1} z_1(t) - V_{k_2} z_2(t)\|.
$$

Hence the bound promises to be very precise although it is impractical for practical use because it requires the exact solution $q(t)$. In order to get a practical error estimator, we seek a bound for the term $\|q(t) - V_{k_2} z_2(t)\|$ instead of using a saturation assumption on the reduced bases. Such a bound can be provided by the estimator $\Delta_1^k(t)$ of Eq. (14) i.e.

$$
\begin{aligned}
\Delta_0^{k_1}(t) &:= \|q(t) - V_{k_2} z_2(t)\| + \|V_{k_1} z_1(t) - V_{k_2} z_2(t)\| \\
&\leq \Delta_1^{k_2}(t) + \|V_{k_1} z_1(t) - V_{k_2} z_2(t)\| =: \Delta_2^{k_1}(t).
\end{aligned}
\tag{16}
$$

Remark 3 The choice of the inner product depends on the underlying physics of the problem. In certain applications, it may be more appropriate to use a Grammian-weighted norm than the Euclidean norm to estimate the errors.

Remark 4 Even though we have used strong POD-greedy method to obtain the reduced system (9), we emphasize that none of the error estimators introduced in this section are dependent on the model order reduction method used to obtain the projection basis. As such, we are free to chose other projection based reduction methods as well.

Fig. 2 Double pendulum
example: initial (blue stars)
and final (red circles)
position

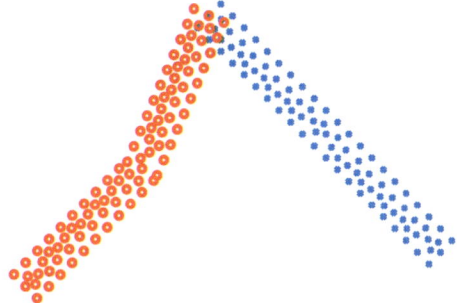

5 Double Elastic Pendulum

We now present some initial results using error estimators of the previous section.
To this end, consider an elastic pendulum with another elastic pendulum attached
to its end. Under the effect of gravity and Neumann forces, such a system can be
represented by a nonlinear PDAE such as (1) and then reduced as in (9). See Fig. 2
for a double elastic pendulum displaced from its initial position. To compare the
performance of the error estimators, we use the formula

$$\eta_i^k(t) := \frac{\Delta_i^k(t)}{\|e^k(t)\|}; \ i \in \{0, 1, 2\}; \ k \in \mathbb{N}$$

to measure *effectivities* of the estimators for reduced basis size k. Here $\Delta_i^k(t)$ and
$\|e^k(t)\|$ are the estimated and the exact errors, respectively, in the state $q(t)$. Sub-
scripts $i = 0, 1, 2$ correspond to the estimated errors and the effectivities, for various
bases sizes k, due to the estimators $\Delta_0^k(t)$, $\Delta_1^k(t)$, $\Delta_2^k(t)$ of (15), (14), (16), respectively. Although Δ_0^k depends on the exact solution, limiting its usefulness, it is a good
indicator of the optimal performance of the estimators based on Eq. (15) nonetheless.
In the following, the first experiment estimates the reproduction error and while the
second one estimates the prediction error.

In this reproduction experiment, we simulate the double elastic pendulum with
system parameters and other user-defined input values enlisted in Table 1. These
simulation results are used to obtain strong POD-greedy bases of varying sizes k
enlisted in the table. We employ the error estimators developed in the last section
and plot the estimates and effectivities in Figs. 3 and 4. Figure 3 shows the exact
error, the estimate, and the effectivities corresponding to three different values of k
for the estimator Δ_1. Figure 4 shows only two lines for each of the estimators $\Delta_0^k(t)$
and $\Delta_2^k(t)$ as they require a more enriched basis ($k = 24$). We observe decreasing
estimates as the basis size increases but the exact error begins to saturate after $k = 18$
for this reproduction case. The estimator Δ_0^k eclipses the exact errors and has the
best effectivities, all close to 1 and indistinguishable from one-another on the log
scale (Fig. 4). This is as we expected from Eq. (15) and the discussion that follows

Table 1 System parameters and other user-defined input values for reproducing simulation

Parameter	Range/value in MKS units
Young's modulus	2.00003E+11
Poisson's ratio	0.4
Density ρ	1520
Neumann forces[a,b]	−10
Gravitational acceleration	9.81
Geometry Ω	Rectangle
Dimensions $\partial\Omega$	5 × 1
Time step τ	0.01
Time span $[0, T]$	[0, 1]
d, m, n	2, 4, 192
Strong POD-greedy tolerance	{1E−7, 1E−9, 1E−11}
Basis size k	{11, 18, 24}

[a] At the end of beam 1 in the x direction
[b] At the end of beam 2 in the x direction

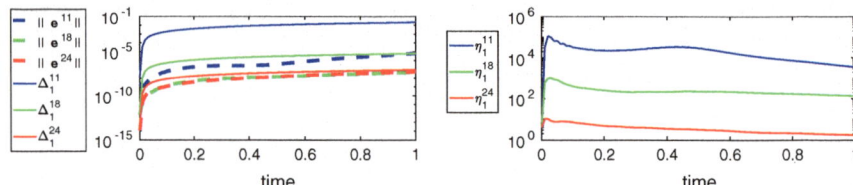

Fig. 3 In-situ error estimator results for different basis sizes: $\|e^k\|$ and Δ_1^k (left) and corresponding effectivities η_1^k (right) versus time

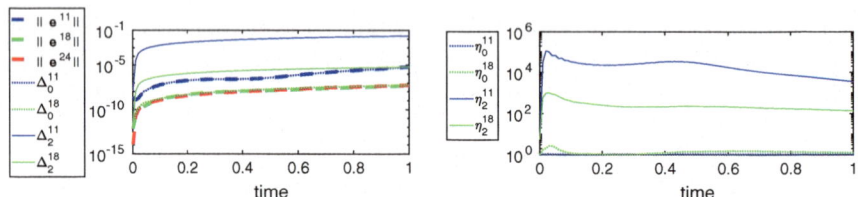

Fig. 4 Hierarchical error estimator results for different basis sizes: $\|e^k\|$, Δ_0^k, and Δ_2^k (left) and corresponding effectivities η_0^k, η_2^k (right) versus time. Δ_0^k almost entirely eclipse $\|e^k\|$

immediately after. The effectivities of the estimator Δ_1^k, which does not rely on the exact solutions, are the next best (Fig. 3).

In another experiment, we start with the same set-up but estimate the error in simulating the model for a parameter point not in the training parameter set, i.e., we now perform a prediction experiment instead of reproduction. Table 2 shows the parameter space and other user-defined input used in this experiment. The dimension of the parameter space is 5 and the number of training parameter points are 32. The

Table 2 Five dimensional parameter space, delineated for clarity, other user-defined input values, and derivative quantities for predicting simulation of a randomly selected parameter point from the space

Parameter	Range/value in MKS units
Young's modulus	(2.0001E+11, 2.0003E+11)
Poisson's ratio	(0.3, 0.4)
Density ρ	(1520, 1530)
Neumann force[a,b]	$(-100, -10)$
Gravitation acceleration	9.81
Geometry Ω	Rectangle
Dimensions $\partial\Omega$	5×1
Time step τ	0.01
Time span $[0, T]$	$[0, 0.51]$
d, m, n	2, 4, 192
Training parameter points	32
Strong POD-greedy tolerance	{1E−6, 1E−8, 1E−10}
Basis size k	{10, 17, 29}

[a] At the end of beam 1 in the x direction
[b] At the end of beam 2 in the x direction

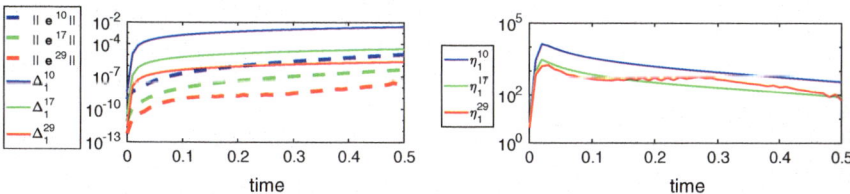

Fig. 5 In-situ error estimator results for different basis sizes: $\|e^k\|$ and Δ_1^k (left) and corresponding effectivities η_1^k (right) versus time

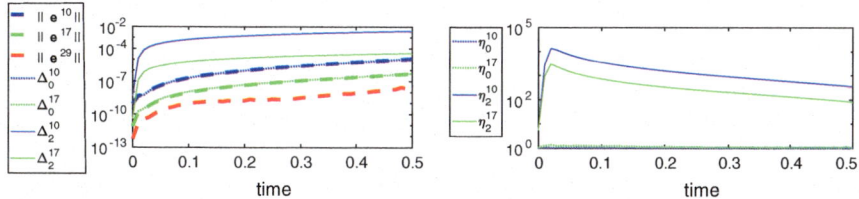

Fig. 6 Hierarchical error estimator results for different basis sizes: $\|e^k\|$, Δ_0^k, and Δ_2^k (left) and corresponding effectivities η_0^k, η_2^k (right) versus time. Δ_0^k almost entirely eclipse $\|e^k\|$

basis size k is dependent on the POD-greedy tolerance. We plot errors, estimates, and effectivities in Figs. 5 and 6 due to the in-situ and the hierarchical estimators. As before, one can see the steadily decreasing error with increasing basis size and near perfect effectivities of the estimators Δ_0^k as was expected from Eq. (15) and the discussion that follows immediately afterwards. Since hierarchical estimators require a more enriched basis ($k = 29$ in this case), only two lines are drawn for Δ_2 and Δ_0 both.

6 Conclusions

A second order nonlinear PDAE system is discretized with FEM in space and trapezoidal rule in time and reduced using strong POD-greedy algorithm. The resulting full and reduced constrained nonlinear systems of equations are solved with the method of Lagrange multipliers. A very stiff material (e.g. steel) only has very few dominant singular values because of almost no elastic motion whereas a less stiff material has more gradually decreasing singular values of the POD basis. An in-situ error estimator for the state variable is established, compared against, and blended with existing hierarchical estimators. The in-situ estimator derived in this work shows promising results both for the reproduction and prediction cases. Unlike many estimators suggested in earlier works, it does not require rewriting the mechanical system as a first order system and estimates only the state and not the enlarged state-velocity space. It also does not require the exact solution or a more enriched basis unlike hierarchical estimators. Future work comprises of a POD-greedy basis generation using the estimators established here, hyper-reduction, and application to FFRF multibody simulators.

Acknowledgements The authors gratefully acknowledge the support of DFG grants FE1583/2-1 and HA5821/5-1. We also thank Andreas Schmidt for helpful discussions during the preparation of this manuscript.

References

1. Antoulas, A.: Approximation of Large-Scale Dynamical Systems. SIAM Publications, Philadelphia, PA (2005)
2. Banagaaya, N., Alì, G., Schilders, W.: Index-aware model order reduction. In: Index-Aware Model Order Reduction Methods, pp. 61–70. Springer (2016)
3. Barrault, M., Maday, Y., Nguyen, N., Patera, A.: An "empirical interpolation" method: application to efficient reduced-basis discretization of partial differential equations. Comptes Rendus de l'Académie des Sciences, Series I(339), 667–672 (2004)
4. Benner, P., Stykel, T.: Model order reduction for differential-algebraic equations: a survey. In: Surveys in Differential-Algebraic Equations IV, pp. 107–160. Springer (2017)
5. Bhatt, A., Fehr, J., Haasdonk, B.: Model order reduction of an elastic body under large rigid motion. In: Proceedings of ENUMATH 2017, Voss, Norway (2018)

6. Carlberg, K., Farhat, C., Cortial, J., Amsallem, D.: The GNAT method for nonlinear model reduction: effective implementation and application to computational fluid dynamics and turbulent flows. J. Comput. Phys. **242**, 623–647 (2013)
7. Chaturantabut, S., Sorensen, D.C.: Discrete empirical interpolation for nonlinear model reduction. SIAM J. Sci. Comput. **32**(5), 2737–2764 (2010)
8. Dattorro, J.: Convex Optimization & Euclidean Distance Geometry. Meboo Publishing USA (2010)
9. Farhat, C., Avery, P., Chapman, T., Cortial, J.: Dimensional reduction of nonlinear finite element dynamic models with finite rotations and energy-based mesh sampling and weighting for computational efficiency. Int. J. Numer. Meth. Eng. **98**(9), 625–662 (2014)
10. Fehr, J.: Automated and error-controlled model reduction in elastic multibody systems. Ph.D. thesis, University of Stuttgart (2011)
11. Fehr, J., Grunert, D., Bhatt, A., Haasdonk, B.: A sensitivity study of error estimation in reduced elastic multibody systems. In: Proceedings of MATHMOD 2018, Vienna, Austria (2018)
12. Gelmi, C.A., Jorquera, H.: IDSOLVER: a general purpose solver for nth-order integro-differential equations. Comput. Phys. Commun. **185**(1), 392–397 (2014)
13. Gerstmayr, J., Ambrósio, J.: Component mode synthesis with constant mass and stiffness matrices applied to flexible multibody systems. Int. J. Numer. Meth. Eng. **73**(11), 1518–1546 (2008)
14. Gerstmayr, J., Schöberl, J.: A 3D finite element method for flexible multibody systems. Multibody Syst. Dyn. **15**(4), 305–320 (2006)
15. Grundel, S., Jansen, L., Hornung, N., Clees, T., Tischendorf, C., Benner, P.: Model order reduction of differential algebraic equations arising from the simulation of gas transport networks. In: Progress in Differential-Algebraic Equations, pp. 183–205. Springer (2014)
16. Haasdonk, B.: Reduced basis methods for parametrized PDEs—a tutorial introduction for stationary and instationary problems. In: Benner, P., Cohen, A., Ohlberger, M., Willcox, K. (eds.) Model Reduction and Approximation: Theory and Algorithms, pp. 65–136. SIAM, Philadelphia (2017)
17. Haasdonk, B., Ohlberger, M.: Reduced basis method for finite volume approximations of parametrized linear evolution equations. ESAIM: M2AN. **42**(2), 277–302 (2008)
18. Haasdonk, B., Ohlberger, M.: Efficient reduced models and a posteriori error estimation for parametrized dynamical systems by offline/online decomposition. Math. Comput. Model. Dyn. Syst. **17**(2), 145–161 (2011)
19. Hain, S., Ohlberger, M., Radic, M., Urban, K.: A hierarchical a-posteriori error estimator for the reduced basis method. arXiv preprint arXiv:1802.03298 (2018)
20. Hesthaven, J., Rozza, G., Stamm, B.: Certified reduced basis methods for parametrized partial differential equations. In: SpringerBriefs in Mathematics (2016)
21. Hughes, T.J., Hulbert, G.M.: Space-time finite element methods for elastodynamics: formulations and error estimates. Comput. Meth. Appl. Mech. Eng. **66**(3), 339–363 (1988)
22. Lakshmikantham, V., Leela, S.: Differential and integral inequalities. Acad. Press, New York (1969)
23. Lanczos, C.: The Variational Principles of Mechanics. Courier Corporation (2012)
24. Petrovitsch, M.: Sur une manière d'étendre le théorème de la moyence aux équations différentielles du premier ordre. Math. Ann. **54**(3), 417–436 (1901)
25. Rheinboldt, W.C.: On the existence and uniqueness of solutions of nonlinear semi-implicit differential-algebraic equations. Nonlinear Anal.: Theory Meth. Appl. **16**(7–8), 647–661 (1991)
26. Rozza, G., Huynh, D.B.P., Patera, A.T.: Reduced basis approximation and a posteriori error estimation for affinely parametrized elliptic coercive partial differential equations. Arch. Comput. Meth. Eng. **15**(3), 229–275 (2008)
27. Ruiner, T., Fehr, J., Haasdonk, B., Eberhard, P.: A-posteriori error estimation for second order mechanical systems. Acta Mechanica Sinica **28**(3), 854–862 (2012)
28. Shabana, A.A.: Dynamics of Multibody Systems. Cambridge University Press (2013)

29. Volkwein, S.: Model reduction using proper orthogonal decomposition. In: Lecture Notes. University of Konstanz (2011)
30. Wirtz, D., Sorensen, D., Haasdonk, B.: A posteriori error estimation for DEIM reduced nonlinear dynamical systems. SIAM J. Sci. Comput. **36**(2), A311–A338 (2014)

A Reduced Order Approach for the Embedded Shifted Boundary FEM and a Heat Exchange System on Parametrized Geometries

Efthymios N. Karatzas, Giovanni Stabile, Nabil Atallah, Guglielmo Scovazzi and Gianluigi Rozza

Abstract A model order reduction technique is combined with an embedded boundary finite element method with a POD-Galerkin strategy. The proposed methodology is applied to parametrized heat transfer problems and we rely on a sufficiently refined shape-regular background mesh to account for parametrized geometries. In particular, the employed embedded boundary element method is the Shifted Boundary Method (SBM), recently proposed in Main and Scovazzi, J Comput Phys [17]. This approach is based on the idea of shifting the location of true boundary conditions to a surrogate boundary, with the goal of avoiding cut cells near the boundary of the computational domain. This combination of methodologies has multiple advantages. In the first place, since the Shifted Boundary Method always relies on the same background mesh, there is no need to update the discretized parametric domain. Secondly, we avoid the treatment of cut cell elements, which usually need particular attention. Thirdly, since the whole background mesh is considered in the reduced basis construction, the SBM allows for a smooth transition of the reduced modes across the immersed domain boundary. The performances of the method are verified in two dimensional heat transfer numerical examples.

E. N. Karatzas · G. Stabile · G. Rozza (✉)
SISSA, International School for Advanced Studies, Mathematics Area, mathLab,
34136 Trieste, Italy
e-mail: grozza@sissa.it

E. N. Karatzas
e-mail: efthymios.karatzas@sissa.it

G. Stabile
e-mail: gstabile@sissa.it

N. Atallah · G. Scovazzi
Civil and Environmental Engineering, Duke University, Durham, NC 27708, USA
e-mail: nabil.atallah@duke.edu

G. Scovazzi
e-mail: guglielmo.scovazzi@duke.edu

© Springer Nature Switzerland AG 2020
J. Fehr and B. Haasdonk (eds.), *IUTAM Symposium on Model Order Reduction of Coupled Systems, Stuttgart, Germany, May 22–25, 2018*, IUTAM Bookseries 36,
https://doi.org/10.1007/978-3-030-21013-7_8

Keywords Unfitted mesh · Reduced basis methods · Shifted boundary method ·
Immersed/embedded finite element method · Reduced order modeling

1 Introduction

In this work we present a reduced order modeling strategy for parametrized geometries, starting from an embedded boundary method solver. The main idea in the
current manuscript is to exploit the advantages of embedded methods and in particular of the Shifted Boundary Method (SBM), [17, 18, 25], in a reduced order
modeling setting. Embedded methods, as full order conformal finite element methods, discretize the original set of equations into a usually high dimensional system of
algebraic equations. When a large number of different system configurations need to
be tested, or a large reduction in computational cost is the goal, the resolution of such
high dimensional system of equations becomes unfeasible. Reduced Order Methods
(ROM) have demonstrated to be a viable way to limit the computational burden [4, 6,
12, 19]. In this particular case, the attention is focused on parametrized geometries.
The methodology is tested on a simple heat transfer problem which will serve as a
base for future more complex scenarios such as flow problems [15]. The manuscript
is organized as follows: in Sect. 2 we introduce the mathematical problem and its full
order discretization; in Sect. 3 we present the reduced order model formulation and
its main features and differences with respect to a standard setting; finally in Sect. 4
numerical results are reported, and in Sect. 5 conclusions and perspectives for future
improvements are given.

2 Full Order Model Approximation

We start recalling, by a sketch description, the continuous strong formulation of
the problem and the weak formulation used for the full-order discretizaton of the
problems under consideration. The discrete SBM formulation will be used for the Full
Order Method (FOM) simulation during the offline stage. The ROM is constructed
using a Proper Orthogonal Decomposition (POD) Galerkin approach following what
is reported in Sect. 3.

2.1 The Thermal-Heat Exchange Model

Given a k—dimensional parameter space $\mathscr{P}$ and the parameter vector $\mu \in \mathscr{P} \subset \mathbb{R}^k$,
let $\mathscr{D}(\mu) \subset \mathbb{R}^d$, $d = 2, 3$ be a bounded parametrized domain depending on μ, with
boundary $\Gamma(\mu)$. We consider the following model problem in $\mathscr{D}(\mu)$:

Find the temperature $T(\mu) : \bar{\mathscr{D}}(\mu) \times \mathscr{P} \to \mathbb{R}^d$ such that in $\mathscr{P}$ we have

$$
\begin{aligned}
-\Delta T(\mu) &= f(\mu) && \text{in } \mathscr{D}(\mu), \\
T(\mu) &= g_D(\mu) && \text{on } \Gamma_D(\mu),
\end{aligned}
\tag{1}
$$

where $\Gamma_D(\mu)$ is the boundary onto which a Dirichlet boundary condition is applied, and the imposed forces $f(\mu), g_D(\mu)$ are given functions in $\mathscr{D}(\mu)$ and on the boundary $\Gamma_D(\mu)$, respectively.

2.2 Weak SBM Formulation

In this subsection we briefly recall the SBM formulation which was originally presented in [17, 18, 25]. In what follows, we denote by $\tilde{\Gamma}$ the surrogate boundary composed of the edges/faces of the mesh that are the closest to the true boundary Γ. The closest faces/edges of $\tilde{\Gamma}$ to Γ are detected by means of a closest-point projection algorithm.

The surrogate boundary $\tilde{\Gamma}$ encloses the surrogate domain $\tilde{\mathscr{D}}$. Furthermore, $\tilde{\mathbf{n}}$ indicates the unit outward-pointing normal to the surrogate boundary $\tilde{\Gamma}$, and it differs from the outward-pointing normal $\mathbf{n}$ of Γ (see Fig. 1).

Notice also that the closest-point projection, in spite of the segmented/faceted nature of the surrogate boundary $\tilde{\Gamma}$ is actually a smooth mapping $\mathbf{M}$ from points $\tilde{x}$ on $\tilde{\Gamma}$ to points x on Γ, namely,

$$
\mathbf{M} : \tilde{\mathbf{x}}|_{\tilde{\Gamma}} \to \mathbf{x}|_{\Gamma},
$$

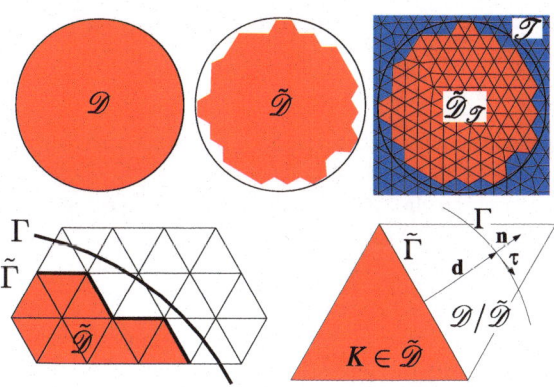

Fig. 1 Example of the SBM mesh on a disc. In the first row, from left to the right: the real geometry; the SBM surrogate geometry and the background mesh together with the surrogate SBM discretized geometry. In the second row, from left to right: a zoom of the surrogate SBM mesh/ surrogate boundary and the normal and distance vector considering one element

which also defines a distance vector function:

$$\mathbf{d} \equiv \mathbf{d_M}(\tilde{\mathbf{x}}) = \mathbf{x} - \tilde{\mathbf{x}} = [\mathbf{M} - \mathbf{I}](\tilde{\mathbf{x}}).$$

The distance vector, as seen in Fig. 1, is oriented along the normal to the true boundary, that is $\mathbf{d} = ||\mathbf{d}||\mathbf{n}$, as a consequence of the use of the closest point projection. Between the normal $\mathbf{n}$ to the true boundary and the normal $\tilde{\mathbf{n}}$ to the surrogate boundary, the minimal grid resolution assumption $\mathbf{n} \cdot \tilde{\mathbf{n}} \geq 0$ is made. The unit normal vector $\mathbf{n}$ and the unit tangential vectors $\boldsymbol{\tau}_i$ ($1 < i < d - 1$) to the boundary Γ, can be easily extended to the boundary $\tilde{\Gamma}$ since $\bar{\mathbf{n}}(\tilde{\mathbf{x}}) \equiv \mathbf{n}(\mathbf{M}(\tilde{\mathbf{x}}))$, $\bar{\boldsymbol{\tau}}_i(\tilde{\mathbf{x}}) \equiv \boldsymbol{\tau}_i(\mathbf{M}(\tilde{\mathbf{x}}))$. Here we denote by $\bar{\mathbf{n}}, \bar{\boldsymbol{\tau}}_i$ the extensions to $\tilde{\Gamma}$ of $\mathbf{n}, \boldsymbol{\tau}_i$, which are defined on Γ. In the following, whenever we write $\mathbf{n}(\tilde{\mathbf{x}})$ we actually mean $\bar{\mathbf{n}}(\tilde{\mathbf{x}})$ at a point $\tilde{\mathbf{x}} \in \tilde{\Gamma}$, and similarly for $\boldsymbol{\tau}_i(\tilde{\mathbf{x}})$ and $\bar{\boldsymbol{\tau}}_i(\tilde{\mathbf{x}})$. Moreover, the above constructions are the key ingredients when building an extension $\bar{g}_D$ of the Dirichlet boundary condition g_D to the boundary $\tilde{\Gamma}$ of the surrogate domain.

Now we can introduce the Shifted Boundary (SB) variational formulation. The SBM weak discrete formulation for the heat exchange system, with non-homogeneous Dirichlet boundary conditions, reads:

Find $T \in V_h = \left\{ \upsilon \in C^0(\tilde{\mathscr{D}}(\mu)) : \upsilon|_K \in P^1(K), \forall K \in \tilde{\mathscr{D}}_{\mathscr{T}}(\mu) \right\}$, with number of degrees of freedom equal to dim $V_h = N_h < \infty$ for all $h > 0$ such that

$$a(T, w) = \ell(w), \ \forall w \in V_h, \tag{2}$$

with

$$a(T, w) = (\nabla w, \nabla T)_{\tilde{\mathscr{D}}} - \langle w + \nabla w \cdot \mathbf{d}, \nabla T \cdot \tilde{\mathbf{n}} \rangle_{\tilde{\Gamma}_D} - \langle \nabla w \cdot \tilde{\mathbf{n}}, T + \nabla T \cdot \mathbf{d} \rangle_{\tilde{\Gamma}_D}$$
$$+ \langle \nabla w \cdot \mathbf{d}, (\mathbf{n} \cdot \tilde{\mathbf{n}})/||\mathbf{d}||\nabla T \cdot \mathbf{d} \rangle_{\tilde{\Gamma}_D} + \langle \alpha/h^{\perp}(w + \nabla w \cdot \mathbf{d}), T + \nabla T \cdot \mathbf{d} \rangle_{\tilde{\Gamma}_D},$$
$$\ell(w) = (w, f)_{\tilde{\mathscr{D}}} - \langle \nabla w \cdot \tilde{\mathbf{n}}, \bar{g}_D \rangle_{\tilde{\Gamma}_D} - \langle \nabla w \cdot \mathbf{d}, (\nabla \bar{g}_D \cdot \boldsymbol{\tau}_i)\boldsymbol{\tau}_i \cdot \tilde{\mathbf{n}} \rangle_{\tilde{\Gamma}_D}$$
$$+ \langle \alpha/h^{\perp}(w + \nabla w \cdot \mathbf{d}), \bar{g}_D \rangle_{\tilde{\Gamma}_D},$$

where α is the Nitsche penalty parameter, $h^{\perp}$ is a characteristic length of the elements in the direction orthogonal to the boundary and $\mathbf{d}$, $\mathscr{T}$, $\tilde{\mathscr{D}}_{\mathscr{T}}$ are the distance vector, the background mesh and the discretized surrogate geometry respectively, see e.g Fig. 1. Finally, the standard notation $(\cdot, \cdot)_{\tilde{\mathscr{D}}}$, $\langle \cdot, \cdot \rangle_{\tilde{\Gamma}_D}$ have been used for the $L^2(\tilde{\mathscr{D}})$ and $L^2(\tilde{\Gamma}_D)$ inner products onto the surrogate geometry $\tilde{\mathscr{D}}$ and $\tilde{\Gamma}_D$, respectively.

The idea of the Shifted Boundary method is to enforce the Dirichlet boundary conditions weakly on the surrogate domain and to modify the value of the boundary conditions to be imposed by means of a second-order accurate Taylor expansion, that is $T + \nabla T \cdot \mathbf{d} \approx \bar{g}_D$, with the purpose of maintaining overall second-order accuracy with a piecewise linear discretization.

The SBM weak formulation can be transformed in a system of linear equations and rewritten in matrix form:

$$A(\mu)T(\mu) = F_g(\mu), \tag{3}$$

where $A(\mu) \in \mathbb{R}^{N_h \times N_h}$ corresponds to the bilinear form $a(\cdot, \cdot)$, $T(\mu) \in \mathbb{R}^{N_h \times 1}$ is the vector of the unknowns and $F_g(\mu) \in \mathbb{R}^{N_h \times 1}$ corresponds to the linear form $\ell(\cdot)$.

3 Reduced Order Method by a POD-Galerkin Technique

In this section we briefly recall the POD-Galerkin technique used to generate the reduced order model and we highlight its peculiarities with respect to standard approaches. In general a ROM is a simplification of a FOM that preserves essential behavior and dominant effects, for the purpose of reducing solution time or storage capacity. In particular here we employ a projection-based reduced order model which consists of the projection of the governing equations onto the reduced basis space.

In the recent past, RB methods were applied to linear elliptic equations in [21], to linear parabolic equations in [10] and to non-linear problems in [9, 26]. Although the number of works on reduced order models is now considerable (see e.g. [12] and references therein), to the best of the authors' knowledge, only very few research works [1] can be found dealing with embedded boundary methods and ROM.

From a reduced order modeling point of view, our aim is to investigate how ROMs are applied to the SBM and, more generally, to embedded boundary methods. Our main interest is to generate ROMs on parametrized geometries. The SBM unfitted/surrogate mesh finite element method is used to apply parametrization and reduced order techniques considering Dirichlet boundary conditions.

An important objective is also to test the efficiency of a geometrically parametrized reduced order method without the usage of the transformation to reference domains, which can be an important advantage of embedded methods relying on fixed background meshes.

Before going into the details, we just remind the basics of the reduced basis method. The first step is the generation of a set of full order solutions of the parametrized problem under the variation of the parameter values. The final goal of RB methods is to approximate any member of this solution set with a low number of basis functions and is based on a two stage procedure, the offline and the online stage, [11, 19, 23].

Offline Stage

In this stage one performs a certain number of full order solves in order to use the solutions for the construction of a low dimensional reduced basis that approximates any member of the solution set to a prescribed accuracy. It is then possible to perform a Galerkin projection of the full order differential operators, describing the governing equations, onto the reduced basis space in order to create a reduced system of equations. This operation usually involves the solution of a possibly large number of high dimensional problems and the manipulation of high-dimensional structures. The required computational cost is high and therefore this operation is usually performed on a high performance system such as a computer cluster.

Online Stage

During this stage, that can be performed also on a system with a reduced computational power and storage capacity, the reduced system of equations can be solved for any new value of the input parameters. This offline-online splitting is effective in many scenarios, such as uncertainty quantification, optimization, real-time control, etc, [4, 6].

3.1 POD

In order to generate the reduced basis space, necessary for the projection of the governing equations, one can find in the literature several techniques such as the POD, the Proper Generalized Decomposition (PGD) and the Reduced Basis (RB) with a greedy sampling strategy. For more details about the different strategies, the reader may see [7, 8, 13, 21]. We apply here a POD strategy using the method of snapshots [24]. In order to assemble the snapshots matrix, the full-order model is solved for each $\mu \in \mathcal{K} = \{\mu^1, \ldots, \mu^{N_s}\} \subset \mathbb{R}^k$ where $\mathcal{K}$ is a finite dimensional training set of parameters chosen inside the parameter space $\mathcal{P}$ and k is the size of the vector μ. The number of snapshots is denoted by N_s and the number of degrees of freedom for the discrete full order solution by N_h. The snapshots matrix $\mathcal{S}$, is then given by N_s full-order snapshots:

$$\mathcal{S} = [T(\mu^1), \ldots, T(\mu^{N_s})] \in \mathbb{R}^{N_h \times N_s}. \tag{4}$$

Given a general scalar function $T : \mathcal{D} \to \mathbb{R}^d$, with a certain number of realizations $T_1, \ldots, T_{N_s}$, and denoting by $(\cdot, \cdot)_{\mathcal{D}}$ and $|| \cdot ||_{L^2(\mathcal{D})}$ the $L^2(\mathcal{D})$ inner product and norm onto the geometry $\mathcal{D}$, the POD problem consists of finding, for each value of the dimension of POD space $N_{POD} = 1, \ldots, N_s$, the scalar coefficients $a_1^1, \ldots, a_1^{N_s}, \ldots, a_{N_s}^1, \ldots, a_{N_s}^{N_s}$ and functions $\varphi_1, \ldots, \varphi_{N_s}$ that minimize the quantity:

$$E_{N_{POD}} = \sum_{i=1}^{N_s} ||T_i - \sum_{k=1}^{N_{POD}} a_i^k \varphi_k||_{L^2(\mathcal{D})}^2, \quad \forall N_{POD} = 1, \ldots, N \tag{5}$$

$$\text{with } (\varphi_i, \varphi_j)_{\mathcal{D}} = \delta_{ij}, \quad \forall i, j = 1, \ldots, N_s.$$

It can be shown [16] that the minimization problem of Eq. (5) is equivalent of solving the following eigenvalue problem:

$$C Q = Q \lambda, \quad \text{for } C_{ij} = (T_i, T_j)_{\mathcal{D}}, i, j = 1, \ldots, N_s,$$

where C is the correlation matrix obtained starting from the snapshots $\mathscr{S}$, Q is a square matrix of eigenvectors and λ is a diagonal matrix of eigenvalues.

The basis functions can then be obtained with:

$$\varphi_i = \frac{1}{N_s \lambda_{ii}^{1/2}} \sum_{j=1}^{N_s} T_j Q_{ij}. \tag{6}$$

The POD space are constructed using the aforementioned methodology resulting in the space:

$$L = [\varphi_1, \ldots, \varphi_{N^r}] \in \mathbb{R}^{N_h \times N^r}, \tag{7}$$

where $N^r < N_s$ is chosen according to the eigenvalue decay of λ, see for example [4, 21].

3.2 Main Differences with Respect to a Reference Domain Approach

We highlight here that using an embedded approach there is no need to map all the parametrized geometries to a common reference domain as usually done in the reduced order modeling community [2, 4, 20–23]. The linear and bilinear forms of Eq. (2), rewritten in a reference domain setting and in a conformal classical finite element method formulation with homogeneous Dirichlet boundary conditions, are transformed into:

$$\tilde{a}(w, T; \mu) = \tilde{\ell}(w; \mu),$$

$$\tilde{a}(w, T; \mu) = \int_{\mathscr{D}^*} \nabla w (J_T(\mu))^{-1} (J_T(\mu))^{-T} |J_T(\mu)| \nabla T dx,$$

$$\tilde{\ell}(w; \mu) = \int_{\mathscr{D}^*} |J_T(\mu)| f w dx,$$

where for a reference domain configuration $\mathscr{D}^*$, $J_T(\mu)$ and $|J_T(\mu)|$ are the Jacobian of the transformation map $\mathscr{T}_{\mathscr{M}}(\mu) : \mathscr{D}^* \to \mathscr{D}(\mu)$ and its determinants respectively. For simple geometrical parametrizations, it is possible to find an affine decomposition of the map and therefore of the differential operator ensuring a complete splitting between the offline and the online procedure, see e.g. [12]. For more complex cases such an operation becomes not trivial and therefore, in order to ensure an efficient splitting one has to rely on empirical interpolation techniques or similar methods, [3, 5, 20]. In the proposed method, even though an efficient splitting is not trivial, there is no need to rely on a transformation map.

Fig. 2 A zoom into the
embedded rectangle in order
to show the smoothing
procedure employed by the
SBM method inside the
ghost area

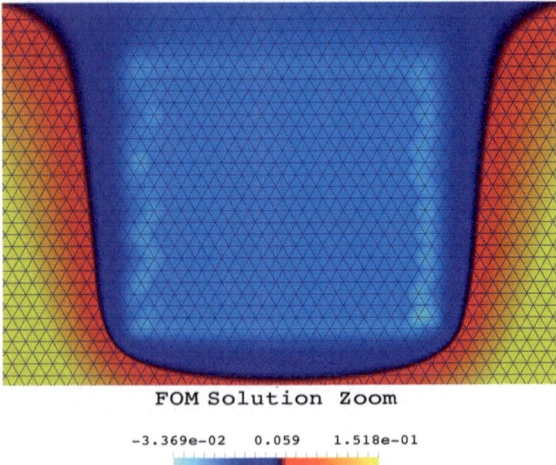

FOM Solution Zoom

-3.369e-02   0.059   1.518e-01

All the solutions are in fact referred to a common background mesh and therefore
the projection step and the reduced basis generation become straightforward. Each
snapshot however has an "out-of-interest" region which lives inside the embedded
domain and that is usually referred as "ghost area". The location of such part of
the domain depends on the parameter μ but the value assumed by the nodes inside
that area is arbitrary. The shifted boundary method used herein has the particular
advantage that the solution smoothly decreases to zero from the boundary to the
interior of the ghost area (see Fig. 2). Besides the closest points, where we have such
smooth decrease, the value inside the ghost area is set to zero. Since this choice
is arbitrary, other choices are also possible (see [14] for more details). Using such
an approach we remark that it is usually not possible to easily recover an affine
decomposition of the differential operator with respect to the geometrical parameters.
However, as highlighted in the next section, it is still possible to rely on hyper
reduction techniques, [3, 5, 27].

3.3   The Projection Stage and the Generation of the ROM

Once the POD functional space is set, the reduced field can be approximated with:

$$T^r \approx \sum_{i=1}^{N^r} a_i(\mu)\varphi_i(\boldsymbol{x}) = \boldsymbol{L}\boldsymbol{a}(\mu), \qquad (8)$$

where the reduced solution vectors $\boldsymbol{a} \in \mathbb{R}^{N^r \times 1}$ depend only on the parameter values
and the basis functions φ_i depend only on the physical space. The unknown vector
of coefficients $\boldsymbol{a}$ can then be obtained through a Galerkin projection of the full order

system of equations onto the POD reduced basis space and with the resolution of a consequent reduced algebraic system:

$$\boldsymbol{L}^T \boldsymbol{A}(\mu) \boldsymbol{L} \boldsymbol{a}(\mu) = \boldsymbol{L}^T \boldsymbol{F}(\mu), \tag{9}$$

which leads to the following algebraic reduced system:

$$\boldsymbol{A}^r(\mu) \boldsymbol{a}(\mu) = \boldsymbol{F}^r(\mu), \tag{10}$$

where $\boldsymbol{A}^r(\mu) \in \mathbb{R}^{N^r \times N^r}$, and $\boldsymbol{F}^r(\mu) \in \mathbb{R}^{N^r \times 1}$ are the reduced discretized operators and reduced forcing vector respectively. The dimension of the reduced operator, as seen also in the numerical examples, is usually much smaller than the dimension of the full order system of equations and therefore much faster to solve. We remark here that the full order discretized differential operators that appear in Eq. (3) are parameter dependent and therefore, also at the reduced order level, in order to compute the reduced differential operator, we need to assemble the full order operators. Possible ways to avoid such potentially expensive operation, relying on an affine approximation of the full order differential operator, could be to use hyper reduction techniques . In this work, since the attention is mainly devoted to the methodological development of a reduced order method in an embedded boundary setting, rather than in its efficiency, we do not rely on such hyper reduction techniques and we assemble the full order differential operators also during the online stage. Considering that the most demanding computational effort is spent during the resolution of the full order problem rather than in the assembly of the differential operators, as reported in Sect. 4, it is anyway possible to achieve a computational speedup, and the related results are reported in the next section.

4   Numerical Experiments

We consider a parameter space $\mathscr{P}$ and parameter vector $\mu \in \mathscr{P} \subset \mathbb{R}$. Let $\mathscr{D}(\mu) \subset \mathbb{R}^2$, be a bounded parametrized domain depending on μ, with boundary $\Gamma_D(\mu)$. In this Section, we report numerical results for the model problem: Find the reduced basis temperature $T(\mu) : \bar{\mathscr{D}}(\mu) \times \mathscr{P} \to \mathbb{R}$ such that in $\mathscr{P}$ we have

$$-\Delta T(\mu) = f(\mu) \quad \text{in } \mathscr{D}(\mu),$$
$$T(\mu) = g_D(\mu) \quad \text{on } \Gamma_D(\mu),$$

where $\Gamma_D(\mu)$ is the embedded boundary onto which a Dirichlet boundary condition is applied, and the imposed forces $f(\mu)=1$, $g_D(\mu) = 0$ are forcing data in $\mathscr{D}(\mu)$ and on $\Gamma_D(\mu)$, respectively. Two different geometries and parameterizations on an embedded rectangle will be examined. In the first example the y-coordinate of the

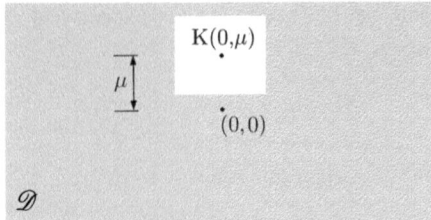

Fig. 3 Background and embedded geometry for a moving rectangle where the y-coordinate of its barycenter has been parametrized

embedded domain center is parametrized, and in the second one its aspect ratio is considered as a parameter.

4.1 Embedded Rectangle with Parameterized Center

In this first experiment the embedded domain consists of a rectangle of size 0.8×0.7 and its position inside the domain is parametrized with a geometrical parameter μ which describes the position of the rectangle embedded domain with respect to its y-center as in Fig. 3.

The horizontal coordinate of the center of the box is not parametrized and is located in the x-center of the domain. The ROM has been trained with 100 and 400 samples for $\mu \in [-0.5, 0.5]$ chosen randomly inside the parameter space. To test the accuracy of the ROM we compared its results on 50 additional samples that were not used to create the ROM and were selected randomly within the same range. The background domain size is a rectangle of size $[-2, 2] \times [-1, 1]$ discretized with mesh size $h = 0.035$, while the background mesh boundary is handled as a wall having zero temperature.

In Fig. 4, we plot the first four modes obtained with the POD procedure. In Fig. 6, we plot the full order solution, the reduced solution and the error for the scalar geometrical parametrized heat equation problem and it is possible to notice that the full and the reduced solution are qualitatively indistinguishable. To verify the behavior of the ROM and its sensitivity with respect to the number of modes in Fig. 5i we compare, for different number of modes, the average of the L^2 norm relative error for the 50 different samples used to test the ROM. The plot is reported for both the simple L^2 projection of the full order results on the POD basis functions, and for the ROM results.

Some Comments
In Table 1, for different dimensions of the reduced basis space, we report the relative error of the L^2 Galerkin projection of the snapshots onto the reduced basis space and the relative error of the ROM solution. Two different ROM solutions are examined,

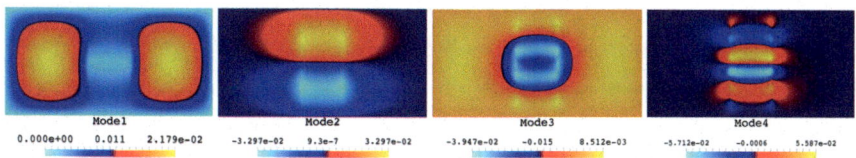

Fig. 4 The first four basis components with $\mu \in [-0.5, 0.5]$ using 100 snapshots in the offline stage

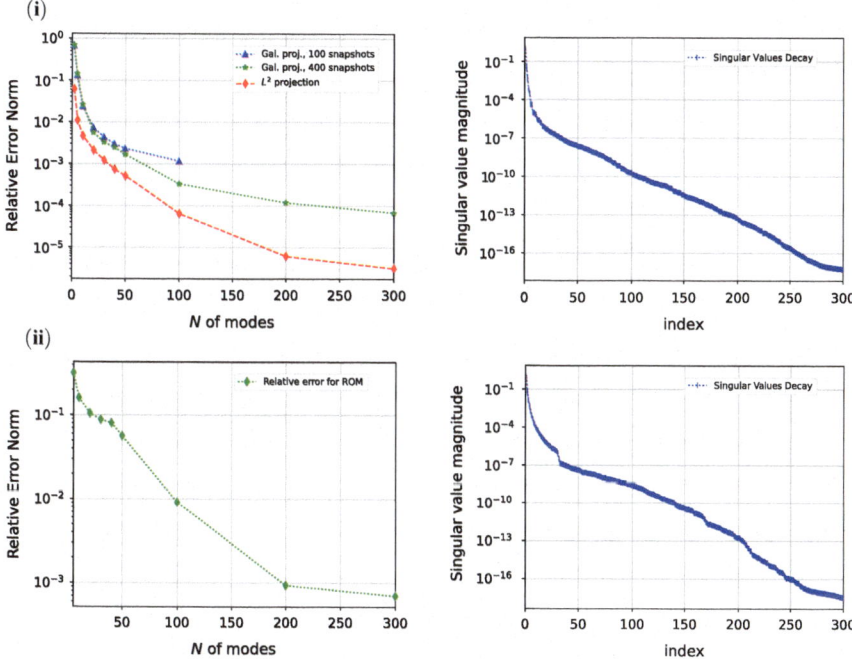

Fig. 5 Heat exchange problem results for the first (i) and second (ii) numerical experiments. On the left we plot the mean relative error for the L^2 projection of the full order solution projected onto the POD basis functions (dashed red line with square markers) and the ROM solution for various number of modes (dotted blue and green lines with triangular and star markers). The error has been computed as the mean of the error of 50 snapshots using different parameter values with respect to those used to compute the POD modes. On the right, the singular value decay of the POD procedure is visualized

using 100 and 400 snapshots during the POD procedure. The plots of Fig. 5i are generated with the ROM constructed using 100 and 400 snapshots and the ROM, as well as the L^2 projection, have been tested using 50 different parameter values not previously used to train the ROM.

In Table 2, we report the computational time comparison using different dimensions of the reduced basis space. Even for the case which employs 300 modes (the one with the largest number of modes) we still observe a good computational speedup.

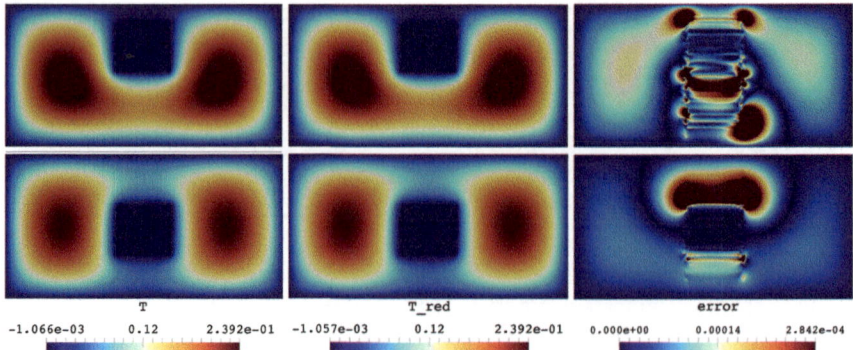

T   T_red   error

-1.066e-03   0.12   2.392e-01 -1.057e-03   0.12   2.392e-01   0.000e+00   0.00014   2.842e-04

Fig. 6 Heat exchange problem results for the first numerical example. From left to the right we report the full order solution, the reduced order solution and the absolute values of the error results respectively. The results are for two selected values of the parameter, $\mu = 0.403$ (first row) and $\mu = -0.015$ (second row)

Table 1 Relative error between the full order solution and the reduced basis solution. Results are reported for different dimensions of the reduced basis space and for fifty test samples

Num of modes	L^2 projection[b]	Galerkin projection[b]	Galerkin projection[a]
2	6.45035392e-02	7.10916700e-01	6.95126771e-01
5	1.14329338e-02	1.44949745e-01	1.36034892e-01
10	4.83332393e-03	2.64459969e-02	2.43322038e-02
20	2.19454585e-03	5.61736266e-03	7.45415783e-03
30	1.27046941e-03	3.30372025e-03	4.47413755e-03
40	7.72326410e-04	2.50189079e-03	3.08036033e-03
50	5.39532759e-04	1.69903034e-03	2.39470742e-03
100	6.79464703e-05	3.36531580e-04	1.19915352e-03
200	6.40774002e-06	1.21062274e-04	–
300	3.29000756e-06	6.94726761e-05	–

[a]100 snapshots, [b]400 snapshots

4.2 Embedded Rectangle with Parametrized Aspect Ratio

In this test problem a fixed uniform source is applied over a rectangular $\mathscr{D}$ using a parameter μ equal to the aspect ratio of the rectangle (Fig. 7); the center of $\mathscr{D}$ remains fixed within $\mathscr{T}$. The embedded domain consists of a rectangle of size $k_1 \times k_2$, for $k_1, k_2 \in \mathbb{R}$ and its size is parametrized by the parameter $\mu = \frac{k_1}{k_2}$ with the additional constraint given by $\mu k_2 = 0.2$. The ROM has been trained with 400 samples for $\mu \in [0.29, 6.67]$ chosen randomly inside the parameter space. To test the accuracy of the ROM we compared its results on 50 additional samples that were not used to create the ROM and were selected randomly within the same range. The background domain size is a square with dimensions $[-0.7, 0.7] \times [-0.7, 0.7]$ and it is discretized with mesh size $h = 0.035$.

Table 2 Execution time, savings and speed up using 400 snapshots in the online stage. The computation time includes the assembling of the full order matrices, their projection and the resolution of reduced problem. Results are reported for various dimensions of the reduced basis space

Num of modes	Execution time(s)[a]	Savings $(t_{FOM} - t_{RB})/t_{RB}$[b,c]	Speedup t_{FOM}/t_{RB}
2	4.119470×10^{-2}	96.399%	27.770
5	4.136089×10^{-2}	96.384%	27.658
10	4.168334×10^{-2}	96.356%	27.445
20	4.243647×10^{-2}	96.290%	26.957
30	4.353909×10^{-2}	96.194%	26.275
40	4.449359×10^{-2}	96.110%	25.711
50	4.494564×10^{-2}	96.071%	25.452
100	4.992923×10^{-2}	95.635%	22.912
200	6.156138×10^{-2}	94.618%	18.583
300	7.551091×10^{-2}	93.399%	15.150
FOM	1.14540×10^{0}	–	–

[a]Online stage, [b] t_{FOM} is the FOM solution time, [c] t_{RB} is the RB solution time

Fig. 7 Background and embedded geometry for a rectangle with parameter its aspect ratio. $\mu = \frac{k_1}{k_2}$ and $\mu k_2 = 0.2$

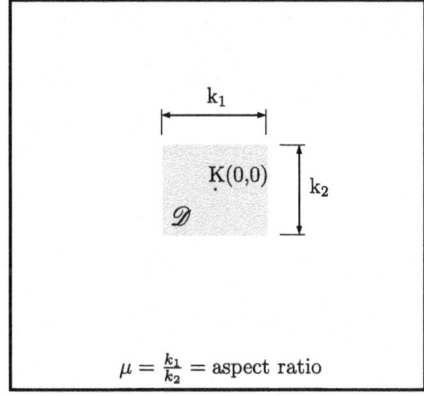

To verify the behavior of the ROM and its sensitivity with respect to the number of modes in Fig. 5ii we compare, for different number of modes, the average of the L^2 norm relative error for the 50 different samples used to test the ROM. The plot is reported for the ROM results.

Some Comments

The plots of Fig. 5ii are generated with the ROM constructed using 400 snapshots. We are pointing out here that for both experiments, we observe a discrepancy between the convergence rate of the left and right side of Fig. 5i and ii. The relative errors graph (left) shows a different convergence rate with respect to the eigenvalue decay. This happens because we compare the full order results obtained on a different training set respect to the one used to obtain the POD modes. In particular, we used the 50 different parameter values not previously used to train the ROM.

5   Conclusions and Future Perspectives

In this work we proposed a new reduced order modeling technique for parametrized geometries. We used an unfitted mesh finite element method to construct a reduced basis onto the background mesh which is independent with respect to the parameter and the parameterized geometry, applying a modified POD-Galerkin methodology. Such coupling, relying on a common background mesh permits to avoid some of the disadvantages related with a reference domain approach. The methodology has been tested on a simple geometrically parametrized heat transfer problem showing promising results. In terms of future perspectives, our interest is in testing the methodology on more complex scenarios and in particular on geometrically parametrized viscous flow problems governed by Stokes [15] and Navier-Stokes equations. Moreover our interest is also in investigating the efficiency of hyper reduction techniques to the proposed methodology in order to further increase the computational speedups and performances.

Acknowledgements We would like to thank the reviewers for their insightful comments on our work. This work is supported by the U.S. Department of Energy, Office of Science, Advanced Scientific Computing Research under Early Career Research Program Grant SC0012169, the U.S. Office of Naval Research under grant N00014-14-1-0311, ExxonMobil Upstream Research Company (Houston, TX), the European Research Council Executive Agency by means of the H2020 ERC Consolidator Grant project AROMA-CFD "Advanced Reduced Order Methods with Applications in Computational Fluid Dynamics" - GA 681447, (PI: Prof. G. Rozza), INdAM-GNCS 2018 and by project FSE European Social Fund HEaD "Higher Education and Development" SISSA operazione 1, Regione Autonoma Friuli-Venezia Giulia.

References

1. Balajewicz, M., Farhat, C.: Reduction of nonlinear embedded boundary models for problems with evolving interfaces. J. Comput. Phys. **274**, 489–504 (2014)
2. Ballarin, F., Manzoni, A., Quarteroni, A., Rozza, G.: Supremizer stabilization of POD-Galerkin approximation of parametrized steady incompressible Navier-Stokes equations. Int. J. Numer. Meth. Eng. **102**(5), 1136–1161 (2014)
3. Barrault, M., Maday, Y., Nguyen, N.: Patera: an 'empirical interpolation' method: application to efficient reduced-basis discretization of partial differential equations. Comptes Rendus Mathematique **339**(9), 667–672 (2004)
4. Benner, P., Ohlberger, M., Patera, A., Rozza, G., Urban, K.: Model reduction of parametrized systems, Vol. 17. In: MS&A Series. Springer (2017)
5. Carlberg, K., Farhat, C., Cortial, J., Amsallem, D.: The GNAT method for nonlinear model reduction: effective implementation and application to computational fluid dynamics and turbulent flows. J. Comput. Phys. **242**, 623–647 (2013)
6. Chinesta, F., Huerta, A., Rozza, G., Willcox, K.: Model order reduction. In: Encyclopedia of Computational Mechanics. Elsevier Editor (2016)
7. Chinesta, F., Ladeveze, P., Cueto, E.: A short review on model order reduction based on proper generalized decomposition. Arch. Comput. Meth. Eng. **18**(4), 395 (2011)
8. Dumon, A., Allery, C., Ammar, A.: Proper general decomposition (PGD) for the resolution of Navier-Stokes equations. J. Comput. Phys. **230**(4), 1387–1407 (2011)

9. Grepl, M., Maday, Y., Nguyen, N., Patera, A.: Efficient reduced-basis treatment of nonaffine and nonlinear partial differential equations. ESAIM: M2AN **41**(3), 575–605 (2007)
10. Grepl, M., Patera, A.: A posteriori error bounds for reduced-basis approximations of parametrized parabolic partial differential equations. ESAIM: M2AN **39**(1), 157–181 (2005)
11. Haasdonk, B., Ohlberger, M.: Reduced basis method for finite volume approximations of parametrized linear evolution equations. Math. Modell. Numer. Anal. **42**(2), 277–302 (2008)
12. Hesthaven, J., Rozza, G., Stamm, B.: Certified reduced basis methods for parametrized partial differential equations. In: SpringerBriefs in Mathematics (2016)
13. Kalashnikova, I., Barone, M.F.: On the stability and convergence of a Galerkin reduced order model (ROM) of compressible flow with solid wall and far-field boundary treatment. Int. J. Numer. Meth. Eng. **83**(10), 1345–1375 (2010)
14. Karatzas, E., Ballarin, F., Rozza, G.: Projection-based reduced order models for a cut finite element method in parametrized domains. Submitted, arXiv preprint, arXiv: 1901.03846 (2019)
15. Karatzas, E., Stabile, G., Nouveau, L., Rozza, G., Scovazzi, G.: A reduced basis approach for PDEs on parametrized geometries based on the shifted boundary finite element method and application to a Stokes flow. Comput. Meth. Appl. Mech. Eng. **347**, 568–587 (2019)
16. Kunisch, K., Volkwein, S.: Galerkin proper orthogonal decomposition methods for a general equation in fluid dynamics. SIAM J. Numer. Anal. **40**(2), 492–515 (2002)
17. Main, A., Scovazzi, G.: The shifted boundary method for embedded domain computations. Part I: Poisson and Stokes problems. J. Comput. Phys. **372**, 972–995 (2018)
18. Main, A., Scovazzi, G.: The shifted boundary method for embedded domain computations. Part II: Linear advection-diffusion and incompressible Navier-Stokes equations. J. Comput. Phys. **372**, 996–1026 (2018)
19. Quarteroni, A., Manzoni, A., Negri, F.: Reduced basis methods for partial differential equations. Springer International Publishing (2016)
20. Rozza, G.: Reduced basis methods for Stokes equations in domains with non-affine parameter dependence. Comput. Visual. Sci. **12**(1), 23–35 (2009)
21. Rozza, G., Huynh, D., Patera, A.: Reduced basis approximation and a posteriori error estimation for affinely parametrized elliptic coercive partial differential equations: application to transport and continuum mechanics. Arch. Comput. Meth. Eng. **15**(3), 229–275 (2008)
22. Rozza, G., Huynh, D.B.P., Manzoni, A.: Reduced basis approximation and a posteriori error estimation for Stokes flows in parametrized geometries: roles of the inf-sup stability constants. Numerische Mathematik **125**(1), 115–152 (2013)
23. Rozza, G., Veroy, K.: On the stability of the reduced basis method for Stokes equations in parametrized domains. Comput. Meth. Appl. Mech. Eng. **196**(7), 1244–1260 (2007)
24. Sirovich, L.: Turbulence and the dynamics of coherent structures part I: coherent structures. Q. Appl. Math. **45**(3), 561–571 (1987)
25. Song, T., Main, A., Scovazzi, G., Ricchiuto, M.: The shifted boundary method for hyperbolic systems: embedded domain computations of linear waves and shallow water flows. J. Comput. Phys. **369**, 45–79 (2018)
26. Veroy, K., Prud'homme, C., Patera, A.: Reduced-basis approximation of the viscous Burgers equation: rigorous a posteriori error bounds. Comptes Rendus Mathematique **337**(9), 619–624 (2003)
27. Xiao, D., Fang, F., Buchan, A., Pain, C., Navon, I., Du, J., Hu, G.: Non linear model reduction for the Navier-Stokes equations using residual DEIM method. J. Comput. Phys. **263**, 1–18 (2014)

POD-Based Augmented Lagrangian Method for State Constrained Heat-Convection Phenomena

Jonas Siegfried Jehle, Luca Mechelli and Stefan Volkwein

Abstract We consider an optimal boundary control problem of the heat equation with convection with bilateral control and state constraints. The goal is to keep the temperature (state) within a desired range while the norm of the boundary controls is minimized. To numerically approximate the unique solution of the problem, a first-order augmented Lagrangian method is utilized. The gradient of the augmented Lagrangian is characterized by the solution of the adjoint equation. To decrease the computational time, we apply a reduced-order approach based on proper orthogonal decomposition (POD). To gain a sufficiently accurate approximation, we propose and compare different manners to compute the snapshots and thus the POD basis. Numerical tests compare the efficiency of the proposed strategies.

Keywords Convection diffusion equation · State constraints · Augmented Lagrangian method · Proper orthogonal decomposition

1 Introduction

Nowadays, the optimal control of parabolic partial differential equations is subject of interest in many applications, e.g. in acoustics, electromagnetism and thermodynamics. In the literature, there are many contributions related to this field, like [8, 17]. In

J. S. Jehle · L. Mechelli (✉) · S. Volkwein
Department of Mathematics and Statistics, University of Konstanz,
Universitätsstraße 10, D-78457 Konstanz, Germany
e-mail: Luca.Mechelli@uni-konstanz.de

J. S. Jehle
e-mail: Jonas.Jehle@uni-konstanz.de

S. Volkwein
e-mail: Stefan.Volkwein@uni-konstanz.de

J. S. Jehle
Department of Mathematics, Shanghai Jiao Tong University, 800 Dongchuan Road,
Shanghai 200240, China

© Springer Nature Switzerland AG 2020
J. Fehr and B. Haasdonk (eds.), *IUTAM Symposium on Model Order Reduction of Coupled Systems, Stuttgart, Germany, May 22–25, 2018*, IUTAM Bookseries 36,
https://doi.org/10.1007/978-3-030-21013-7_9

particular, we are interested in the optimal control of convection-diffusion equations with bilateral control and state constraints, having energy efficient buildings in mind, i.e. the solution of the parabolic equation represents the temperature in a room and the heaters are seen as boundary controls; see [15, 16]. To solve this problem, in contrast to [15], we use the augmented Lagrangian method, which can be seen as a penalization method that allows in our case to treat the state constraints by solving iteratively many only control-constrained optimal control problems; see [4, 5, 10, 11]. This approach is different from the one in [15], where a virtual control concept is utilized (see [13]) that allows the use of a semi-smooth Newton method. For the numerical solution of the equations, we apply a Galerkin finite element approximation combined with an implicit Euler scheme in time and, in order to speed-up the computation of optimal solutions, we build a reduced-order model based on POD; cf. [7, 9]. Since there is no computable a-priori POD error estimate, we compare different techniques. Moreover, we are aware of methods to improve the quality of POD basis, e.g. TR-POD [2], OS-POD [14] and a-posteriori snapshots location techniques [1]. Methods to estimate this quality are also existing, e.g. a-posteriori error estimator in [18]. However, we do not use them in this work.

The paper is organized in the following way: in Sect. 2 we introduce the optimal control problem; in Sect. 3 we describe how we adapt the augmented Lagrangian method to our setting; in Sect. 4 we briefly introduce the POD method and comment on the three methods we use to generate the snapshots to build the POD basis; in Sect. 5 numerical tests are shown and in Sect. 6 we draw some conclusions.

2   The Optimal Control Problem

We consider an optimal control problem with time horizon $[0, T]$ with $T > 0$. Let $\Omega \subset \mathbb{R}^d, d \in \{2, 3\}$, be a bounded domain with Lipschitz-continuous boundary $\Gamma = \partial\Omega$. We suppose that $\partial\Omega$ is split into the two disjoint subsets Γ_c and Γ_o, where at least Γ_c has nonzero (Lebesgue) measure. Let us set $Q = (0, T) \times \Omega$, $\Sigma_c = (0, T) \times \Gamma_c$, and $\Sigma_o = (0, T) \times \Gamma_o$. Moreover, let $H = L^2(\Omega)$, $V = H^1(\Omega)$, $\mathcal{U} = L^2(0, T; \mathbb{R}^m)$. We recall the Hilbert space

$$W(0, T) = \{\varphi \in L^2(0, T; V) \mid \varphi_t \in L^2(0, T; V')\},$$

where V' denotes the dual space of V; cf. [6]. We consider the following economic optimal control problem:

$$\min J(z) = \frac{1}{2} \|u\|_{\mathcal{U}}^2, \quad z = (y, u) \in \mathcal{Z} := W(0, T) \times \mathcal{U} \tag{1a}$$

with $u \in \mathcal{U}$ being the boundary control for the *state equation*

$$y_t(t, x) - \Delta y(t, x) + \mathbf{v}(t, x) \cdot \nabla y(t, x) = 0 \qquad \text{a.e. in } Q,$$

$$\frac{\partial y}{\partial \mathbf{n}}(t, \mathbf{s}) + y(t, \mathbf{s}) = \sum_{i=1}^{m} u_i(t) b_i(\mathbf{s}) \quad \text{a.e. on } \Sigma_c,$$

$$\frac{\partial y}{\partial \mathbf{n}}(t, \mathbf{s}) + \gamma y(t, \mathbf{s}) = \gamma y_{\text{out}}(t) \qquad \text{a.e. on } \Sigma_o,$$

$$y(0, x) = y_o(x) \qquad \text{a.e. in } \Omega,$$

$$(1\text{b})$$

where $b_1, \ldots, b_m \in L^\infty(\Gamma_c)$ are given control shape functions and 'a.e.' stands for 'almost everywhere'. Moreover, u and y have to satisfy the bilateral inequality constraints

$$u_{ai}(t) \le u_i(t) \le u_{bi}(t), \qquad i = 1, \ldots, m \text{ a.e. in } [0, T], \qquad (1\text{c})$$

$$y_a(t, x) \le y(t, x) \le y_b(t, x) \qquad \text{a.e. in } Q. \qquad (1\text{d})$$

Assumption 1 In (1) we suppose that $\gamma \ge 0$, $y_{\text{out}} \in L^2(0, T)$, $y_o \in H$, $\mathbf{v} = (v_1, \ldots, v_d) \in L^\infty(0, T; L^\infty(\Omega; \mathbb{R}^d))$, $u_a = (u_{ai})_{1 \le i \le m}$, $u_b = (u_{bi})_{1 \le i \le m} \in \mathcal{U}$ and $y_a, y_b \in L^2(Q)$.

We introduce the time-dependent bilinear form $a(t; \cdot, \cdot) : V \times V \to \mathbb{R}$

$$a(t; \phi, \varphi) = \int_\Omega \nabla \phi \cdot \nabla \varphi + (\mathbf{v}(t) \cdot \nabla \phi) \varphi \, dx + \gamma \int_{\Gamma_o} \phi \varphi \, ds + \int_{\Gamma_c} \phi \varphi \, ds$$

for $\varphi, \phi \in V$, the time-dependent linear functional $\mathscr{F}(t) : V \to V'$

$$\langle \mathscr{F}(t), \varphi \rangle_{V', V} = \gamma y_{\text{out}}(t) \int_{\Gamma_o} \varphi \, ds \quad \text{for } \varphi \in V \text{ a.e. in } [0, T],$$

and the bounded linear operator $\mathscr{B} : \mathbb{R}^m \to V'$ defined as

$$\langle \mathscr{B}u, \varphi \rangle_{V', V} = \sum_{i=1}^{m} u_i \int_{\Gamma_c} b_i \varphi \, ds \quad \text{for all } \varphi \in V$$

and for given $u = (u_i) \in \mathbb{R}^m$. Now, a weak solution $y \in W(0, T)$ to (1b) for all $\varphi \in V$ satisfies the weak formulation

$$\frac{d}{dt} \langle y(t), \varphi \rangle_H + a(t; y(t), \varphi) = \langle \mathscr{F}(t) + \mathscr{B}(u(t)), \varphi \rangle_{V', V} \quad \text{a.e. in } (0, T] \qquad (2)$$

and $y(0) = y_o$ in H. By Assumption 1 the bilinear form $a(t; \cdot, \cdot)$ is uniformly (w.r.t. the time t) continuous and weakly coercive, it is known that (2) admits a unique solution $y \in W(0, T)$; cf. [6]. Then, we can define the bounded linear solution operator $\mathscr{S} : \mathfrak{U} \to W(0, T)$, such that $y = \mathscr{S}u$ is the unique weak solution of (1b) for $u \in \mathfrak{U}$. Now, (1) can be expressed as

$$\min \hat{J}(u) \quad \text{subject to} \quad u \in \mathcal{Z}_{\text{ad}} \tag{P}$$

where $\hat{J}(u) := J(\mathscr{S}u, u)$ and the set of admissible solutions is given by

$$\mathcal{Z}_{\text{ad}} = \left\{ u \in \mathfrak{U} \mid (\mathscr{S}u, u) \in \mathcal{Z} \text{ satisfies (1c) and (1d)} \right\}.$$

Note that (P) is a linear-quadratic, strictly convex programming problem, so that under Assumption 1 and assuming that the feasible set $\mathcal{Z}_{\text{ad}}$ is non-empty, there exists a unique optimal solution which is denoted by $\bar{z} = (\bar{y}, \bar{u})$; cf. [8, Sect. 1.5].

3 Augmented Lagrangian Method

In order to apply the *augmented Lagrangian method* to solve the optimal control problem (P), we follow [5, 11]. For $c > 0$ the augmented Lagrangian is given as

$$L_c(u, s_a, s_b, \mu_a, \mu_b) = \hat{J}(u) + \langle \mu_a, y_a - \mathscr{S}u + s_a \rangle_{L^2(Q)} + \langle \mu_b, \mathscr{S}u - y_b + s_b \rangle_{L^2(Q)}$$
$$+ \frac{c}{2} \|y_a - \mathscr{S}u + s_a\|_{L^2(Q)}^2 + \frac{c}{2} \|\mathscr{S}u - y_b + s_b\|_{L^2(Q)}^2$$

where $s_a, s_b \in L^2(Q)$ are non-negative slack variables used to transform the inequalities (1d) into equalities and $\mu_a, \mu_b \in L^2(Q)$ are the *Lagrange multipliers*. From the first-order optimality conditions

$$\frac{\partial L_c}{\partial s_a}(u, s_a, s_b, \mu_a, \mu_b)\delta_{s_a} = \langle \mu_a, \delta_{s_a} \rangle_{L^2(Q)} + c\langle y_a - \mathscr{S}u + s_a, \delta_{s_a} \rangle_{L^2(Q)} \stackrel{!}{=} 0,$$

$$\frac{\partial L_c}{\partial s_b}(u, s_a, s_b, \mu_a, \mu_b)\delta_{s_b} = \langle \mu_b, \delta_{s_b} \rangle_{L^2(Q)} + c\langle \mathscr{S}u - y_b + s_b, \delta_{s_b} \rangle_{L^2(Q)} \stackrel{!}{=} 0,$$

for all $\delta_{s_a}, \delta_{s_b} \in L^2(Q)$, we derive the two equalities

$$\mu_a + c(y_a - \mathscr{S}u + s_a) = 0 \text{ and } \mu_b + c(\mathscr{S}u - y_b + s_b) = 0 \quad \text{a.e. in } Q.$$

Hence, to ensure $s_a \geq 0$ and $s_b \geq 0$, we set

$$s_a = \max \left\{ 0, \mathscr{S}u - y_a - \frac{1}{c}\mu_a \right\} \text{ and } s_b = \max \left\{ 0, y_b - \mathscr{S}u - \frac{1}{c}\mu_b \right\}. \tag{3}$$

Then, as proved in [12], it holds

$$\hat{L}_c(u, \mu_{\mathsf{a}}, \mu_{\mathsf{b}}) := L_c(u, s_{\mathsf{a}}, s_{\mathsf{b}}, \mu_{\mathsf{a}}, \mu_{\mathsf{b}}) = \hat{J}(u) - \frac{1}{2c}(\|\mu_{\mathsf{a}}\|^2_{L^2(Q)} + \|\mu_{\mathsf{b}}\|^2_{L^2(Q)})$$
$$+ \frac{c}{2}\left\|\max\left\{0, y_{\mathsf{a}} - \mathscr{S}u + \frac{\mu_{\mathsf{a}}}{c}\right\}\right\|^2_{L^2(Q)} + \frac{c}{2}\left\|\max\left\{0, \mathscr{S}u - y_{\mathsf{b}} + \frac{\mu_{\mathsf{b}}}{c}\right\}\right\|^2_{L^2(Q)}.$$

Therefore, to find the optimal solution of (P), we can use Algorithm 1.

Algorithm 1 (Augmented Lagrangian method)

1: **Data:** Initial control $u^0 \in \mathcal{U}$, initial pair $(\mu_{\mathsf{a}}^0, \mu_{\mathsf{b}}^0) \in L^2(Q) \times L^2(Q)$, initial weight $c^0 > 0$, increment $\beta > 1$ for c^n, tolerance $\varepsilon > 0$ and $n_{\max}$ maximum number of iterations;

2: set $n = 0$ and FLAG= **true**;

3: **while** FLAG and $n < n_{\max}$ **do**

4:     For fixed $(\mu_{\mathsf{a}}^n, \mu_{\mathsf{b}}^n)$ find $u^{n+1} \in \mathcal{U}_{\mathrm{ad}} := \{u \in \mathcal{U} \mid u \text{ satisfies (1c)}\}$ that minimizes

$$\hat{L}_{c^n}(u, \mu_{\mathsf{a}}^n, \mu_{\mathsf{b}}^n) = \hat{J}(u) + \frac{c^n}{2}\left\|\max\left\{0, y_{\mathsf{a}} - \mathscr{S}u + \frac{\mu_{\mathsf{a}}^n}{c^n}\right\}\right\|^2_{L^2(Q)}$$
$$+ \frac{c^n}{2}\left\|\max\left\{0, \mathscr{S}u - y_{\mathsf{b}} + \frac{\mu_{\mathsf{b}}^n}{c^n}\right\}\right\|^2_{L^2(Q)} \underbrace{- \frac{1}{2c^n}(\|\mu_{\mathsf{a}}^n\|^2_{L^2(Q)} + \|\mu_{\mathsf{b}}^n\|^2_{L^2(Q)})}_{\text{independent of } u} \quad (\mathbf{L}^n)$$

5:     Update the Lagrange multipliers

$$\mu_{\mathsf{a}}^{n+1} = \max\{0, \mu_{\mathsf{a}}^n + c^n(y_{\mathsf{a}} - \mathscr{S}u^{n+1})\}, \quad \mu_{\mathsf{b}}^{n+1} = \max\{0, \mu_{\mathsf{b}}^n + c^n(\mathscr{S}u^{n+1} - y_{\mathsf{b}})\};$$

6:     **if** $\|u^n - u^{n+1}\|_{\mathcal{U}} \le \varepsilon$ **then**

7:         FLAG = **false**

8:     **end if**

9:     set $c^{n+1} = \beta c^n$ and $n = n + 1$;

10: **end while**

Remark 1 In future works, we want to utilize an inexact version of Algorithm 1 within a Model Predictive Control approach, as done for the primal dual active set strategy in [16]. The idea is then to compare the two different approaches in terms of approximation of the feedback control u and of the computational time, for more details we refer to [12]. Therefore, to solve the minimization problem in step 4 of Algorithm 1, we use the projected gradient method (due to the control constraints). To apply the projected gradient algorithm, we need the partial derivative with respect to u of the reduced Lagrangian $\hat{L}_c(u, \mu_{\mathsf{a}}, \mu_{\mathsf{b}})$, i.e.

$$\frac{\partial \hat{L}_c}{\partial u}(u, \mu_{\mathsf{a}}, \mu_{\mathsf{b}})u^\delta = \langle -\mathscr{B}^\star p + u, u^\delta\rangle_{\mathcal{U}} \quad \text{for } u^\delta \in \mathcal{U}$$

with p the solution of the adjoint equation

$$-\frac{d}{dt}\langle p(t), \varphi\rangle_H + a(t; \varphi, p(t)) = \langle \mathscr{G}(t; y(t), \mu_a(t), \mu_b(t)), \varphi\rangle_{V',V} \quad (4)$$

for all $\varphi \in V$ a.e. in $[0, T]$ and $p(T) = 0$ in H. Further, $\mathscr{G}(t; \cdot, \cdot, \cdot) : V \times L^2(Q) \times L^2(Q) \to V'$ is defined as

$$\langle \mathscr{G}(t; \theta, \phi, \psi), \varphi\rangle_{V',V} = c \int_\Omega \max\{0, y_a(t) - \theta + \tfrac{\phi}{c}\}\varphi - \max\{0, \theta - y_b(t) + \tfrac{\psi}{c}\}\varphi\, dx.$$

Also (4) admits a unique solution in $W(0, T)$. Therefore, we can define the bounded linear solution operator $\mathscr{A} : \mathcal{U} \times L^2(Q) \times L^2(Q) \to W(0, T)$: $p = \mathscr{A}(u, \mu_a, \mu_b)$ is the unique solution of (4) for $u \in \mathcal{U}$ and $\mu_a, \mu_b \in L^2(Q)$. $\Diamond$

4 Proper Orthogonal Decomposition

For properly chosen admissible controls $u \in \mathcal{Z}_{ad}$ and Lagrange multipliers $\mu_a, \mu_b \in L^2(Q)$, we set $y = \mathscr{S}u$ and $p = \mathscr{A}(u, \mu_a, \mu_b)$. We define the linear space

$$\mathcal{V} = \mathrm{span}\left\{ \int_0^T y(t)\omega^1(t) + p(t)\omega^2(t)\, dt \mid \omega^1, \omega^2 \in L^2(0, T)\right\} \subset V \quad (5)$$

with $d = \dim \mathcal{V} \geq 1$. The method of proper orthogonal decomposition (POD) consists in choosing an orthonormal basis $\{\psi_i\}_{i=1}^d$ in $\mathcal{V}$ such that for every $\ell \in \mathbb{N}$ with $\ell \leq d$ the mean square error between the snapshots y, p and their corresponding ℓ-th partial sum is minimized:

$$\min \int_0^T \left\| y(t) - \sum_{i=1}^\ell \langle y(t), \psi_i\rangle_V \psi_i \right\|_V^2 + \left\| p(t) - \sum_{i=1}^\ell \langle p(t), \psi_i\rangle_V \psi_i \right\|_V^2 dt \quad (6)$$

s.t. $\{\psi_i\}_{i=1}^\ell \subset V$ and $\langle \psi_i, \psi_j\rangle_V = \delta_{ij}$ for $1 \leq i, j \leq \ell$,

where δ_{ij} is the Kronecker delta.

Definition 1 A solution $\{\psi_i\}_{i=1}^\ell$ to (6) is called a POD basis of rank ℓ. We define the subspace spanned by the first ℓ POD basis functions as $V^\ell = \mathrm{span}\{\psi_1, \ldots, \psi_\ell\}$.

Using a Lagrangian framework, the solution to (6) is characterized by the following optimality conditions (cf. [7, 9]):

$$\mathcal{R}\psi = \lambda\psi, \quad (7)$$

where the operator $\mathcal{R} : V \to V$ given by

$$\mathcal{R}\psi = \int_0^T \langle y(t), \psi \rangle_V \, y(t) + \langle p(t), \psi \rangle_V \, p(t) \, \mathrm{d}t \quad \text{for } \psi \in V \qquad (8)$$

is compact, nonnegative and self-adjoint operator. Thus, there exist an orthonormal basis $\{\psi_i\}_{i \in \mathbb{N}}$ for V and an associated sequence $\{\lambda_i\}_{i \in \mathbb{N}}$ of nonnegative real numbers so that

$$\mathcal{R}\psi_i = \lambda_i \psi_i, \quad \lambda_1 \geq \cdots \geq \lambda_d > 0 \quad \text{and} \quad \lambda_i = 0, \quad \text{for } i > d. \qquad (9)$$

Moreover $V = \mathrm{span}\{\psi_i\}_{i=1}^d$. It can be also proved, see [9], that we have the following error formula for the POD basis $\{\psi_i\}_{i=1}^\ell$ of rank ℓ:

$$\int_0^T \left\| y(t) - \sum_{i=1}^\ell \langle y(t), \psi_i \rangle_V \, \psi_i \right\|_V^2 + \left\| p(t) - \sum_{i=1}^\ell \langle p(t), \psi_i \rangle_V \, \psi_i \right\|_V^2 \, \mathrm{d}t = \sum_{i=\ell+1}^d \lambda_i.$$

If a POD basis $\{\psi_i\}_{i=1}^\ell$ of rank ℓ is computed, we can derive a reduced-order model for (2): for any $u \in \mathcal{U}$ the function $y^\ell = \mathcal{S}^\ell u \in W(0, T)$ is the solution for all $\psi \in V^\ell$ of

$$\frac{\mathrm{d}}{\mathrm{d}t} \langle y^\ell(t), \psi \rangle_H + a(t; y^\ell(t), \psi) = \langle \mathcal{F}(t) + \mathcal{B}(u(t)), \psi \rangle_{V', V} \quad \text{a.e. in } (0, T].$$

For any $u \in \mathcal{U}_{\mathrm{ad}}$ the POD approximation y^ℓ for the state solution is $y^\ell = \mathcal{S}^\ell u$. Analogously a reduced-order model can be derived for the adjoint equation; see, e.g., [7]. The POD Galerkin approximation of (P) is given by

$$\min \hat{J}^\ell(u) = \min J(\mathcal{S}^\ell u, u) \quad \text{s.t.} \quad u \in \mathcal{Z}_{\mathrm{ad}}^\ell, \qquad (\hat{\mathbf{P}}^\ell)$$

where the set of admissible controls is

$$\mathcal{Z}_{\mathrm{ad}}^\ell = \{ u \in \mathcal{U} \mid u \in \mathcal{U}_{\mathrm{ad}} \text{ and } y_{\mathrm{a}}(t, x) \leq (\mathcal{S}^\ell u)(t, x) \leq y_{\mathrm{b}}(t, x) \text{ a.e. in } Q \}.$$

Remark 2 For implementing Algorithm 1 numerically, we discretize (2) and (4): for the temporal discretization, we utilize the implicit Euler method, while the spatial variable is approximated utilizing a reduced-order Galerkin approach; cf. [3, 15]. Here, we apply the POD method, where the snapshots are computed from piecewise linear finite element (FE) solutions of the state and dual variables; cf. [7, 9]. Moreover, we replace the integral over $[0, T]$ in (6) and (8) by a trapezoidal approximation; see [7]. In Sect. 5, we compare three different methods to compute the POD basis:

- (POD-M1): Use the optimal control $\bar{u}$ to compute state and adjoint snapshots.
- (POD-M2): Use a random control $\tilde{u} \in \mathcal{U}_{\mathrm{ad}}$ to compute state and adjoint snapshots.
- (POD-M3): Use a sub-optimal control $\hat{u}$, generated utilizing FE discretization for $n = 1, 2, 3$ in Algorithm 1, to compute state and adjoint snapshots. $\qquad \Diamond$

Remark 3 Due to the different magnitudes of state and adjoint snapshots, we observed that it is convenient to scale them by two different values respectively, in order to balance the optimization problem (6) involved in the basis computation. This procedure does not change the set spanned by the snapshots, but improves the POD basis approximation, as shown in Sect. 5. As mentioned, the POD method consists in minimizing the mean square error between the snapshots and their projection onto the POD subspace, moreover, once the time integral is approximated trough a trapezoidal quadrature rule, almost each term in (6) is equally weighted, but many adjoint terms can have a big magnitude for large penalty parameter c. Therefore, the first POD functions are such that the larger adjoint equation snapshots can be better approximated than the state equation ones. This fact causes the approximation problems, which can be seen in Table 2, especially for POD-M1, where in the case $c^0 = 1$ and $\beta = 4$, for example, the penalty parameter c used to generate the adjoint snapshots is equal to $4^{12} \simeq 1.7 \times 10^7$. After several tests with different scaling values (e.g. L^2 norms, mean values), we choose to report the results for the best choice we have found: we divide each state snapshot by its variance σ^2 defined as

$$\sigma^2 = \frac{1}{Nt} \sum_{i=0}^{Nt-1} (\|y(t_i)\|_H - \mu)^2, \text{ with } \mu = \frac{1}{Nt} \sum_{i=0}^{Nt-1} \|y(t_i)\|_H$$

where Nt is the numerical ending time and t_i are the numerical time steps. We do similarly for the adjoint ones. When this approach is utilized, we report each method with the name varPOD from now on. $\diamond$

5 Numerical Tests

The tests in this section are implemented on a Notebook Acer Aspire E15 E5-71G-717X, Intel ® Core™ i7-5500U 2.4 GHz (with Turbo Boost up to 3.0 GHz) and 8 GB RAM in the programming language MATLAB. For the tests, we set $T = 1$ and the unit square $\Omega = (0, 1) \times (0, 1)$ represents our room. The controls, i.e. the heaters, are placed as shown in Fig. 1 with related shape functions:

$$b_1(x) = \begin{cases} 1 \text{ for } x_1 = 0, 0 \leq x_2 \leq 0.25, \\ 0 \text{ otherwise,} \end{cases} \quad b_2(x) = \begin{cases} 1 \text{ for } x_1 = 1, 0.75 \leq x_2 \leq 1, \\ 0 \text{ otherwise,} \end{cases}$$

respectively. For the state equation, we choose the physical parameter $\gamma = 0.05$, the initial condition $y_o(x) = x_1^2 + x_2^2$ for $x = (x_1, x_2) \in \Omega$ and the outside temperature $y_{out}(t) = 2t - 1$ for all $t \in [0, T]$.
The components of the velocity field $v(t, x)$ are

$$v_1(t, x) = \begin{cases} -x_1 - x_2 \text{ for } 0 \leq t < 0.5, \\ \frac{-x_1 - x_2}{2} \text{ for } 0.5 \leq t \leq T, \end{cases} \quad v_2(t, x) = \begin{cases} \frac{x_1 + x_2}{2} \text{ for } 0 \leq t < 0.5, \\ x_1 + x_2 \text{ for } 0.5 \leq t \leq T, \end{cases}$$

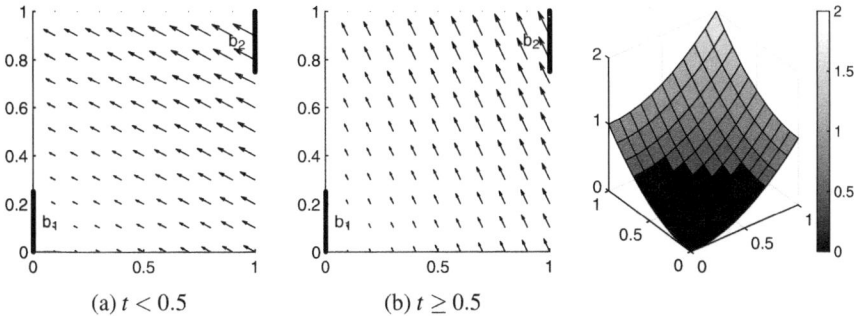

(a) $t < 0.5$ (b) $t \geq 0.5$

Fig. 1 Spatial domain Ω with the two boundary controls and the velocity fields (black arrows) at two time instances (left and middle); initial condition $y_o(x)$ (right)

for $(t, x) \in Q$. Furthermore, we take $y_a(t, x) = 3t$, $y_b(t, x) = 4$ for $(t, x) \in Q$ as the state bounds and $u_{ai}(t) = 0$, $u_{bi}(t) = 12$ for $t \in [0, T]$, $i = 1, 2$ as the control bounds. For the implicit Euler method, we use the equidistant time step $\Delta t = 0.01$. The spatial discretization is built by piecewise linear FE on a triangular mesh with $N_x = 712$ nodes. As initial guess for the control, we choose $u^0(t) = (0, 0)$ for $t \in [0, T]$. In Table 1, for different values of c^0 and β and different spatial discretizations, we report the number of iterations needed by the projected gradient method to converge in step 4 of Algorithm 1 for any iteration n. Additionally, the value of the cost functional $\hat{J}(\bar{u})$, the number of active points for the state bounds $|\mathscr{A}|$ and the computational time speed-up due to the POD approximation are shown in Table 2 together with the number of active points that differs between the FE and the POD solutions

$$\left| \mathscr{A}^{FE} \Delta \mathscr{A}^{POD} \right| = \left| \mathscr{A}^{FE} \cup \mathscr{A}^{POD} - \mathscr{A}^{FE} \cap \mathscr{A}^{POD} \right|$$

and the POD relative error for the controls defined as

$$\text{rel.err. } u = \frac{\| u^{FE} - u^{POD} \|_u}{\| u^{FE} \|_u}.$$

From these tables, one can observe that with both choices of c^0 and β the optimal control problem reaches the unique optimal solution as expected. The number of active points for the optimal state is significantly small compared to the total number of space-time grid points $N_x \times N_t = 712 \times 101 = 71912$ and it can be even reduced setting different tolerances for the augmented Lagrangian (ε) and the project gradient (ε_{pg}) methods. Note that the cost functional $\hat{J}(\bar{u})$ for POD-M3 is smaller than the FE one, but this is not surprising in this case, since what we are minimizing in Algorithm 1 is the augmented Lagrangian ($\mathbf{L}^n$), therefore a smaller contribution for $\hat{J}(\bar{u})$ is balanced by a greater violation of the constraints as can be seen from the number of active points $|\mathscr{A}|$. In Table 1, one can see that for almost all the ways

Table 1 Results for different combinations of c^0 and β (augmented Lagrangian tolerance $\varepsilon = 10^{-5}$, projected gradient tolerance $\varepsilon_{pg} = 0.2\varepsilon$, number of POD basis elements $\ell = 18$, $n_{\max} = 150$)

Discretiz.	c^0	β	Projected gradient iterations	n
FE	1	4	{4, 4, 4, 286, 3, 141, 155, 131, 308, 252, 279, 198, 2}	13
POD-M1	1	4	{4, 4, 4, 298, 3, 160, 167, 153, 476, 7, 2}	11
varPOD-M1	1	4	{4, 4, 4, 286, 3, 141, 155, 131, 308, 252, 279, 198, 2}	13
POD-M2	1	4	{4, 4, 4, 286, 3, 141, 154, 125, 304, 255, 1000, 4, 2}	13
varPOD-M2	1	4	{4, 4, 4, 286, 3, 141, 155, 131, 309, 250, 280, 202, 2}	13
POD-M3	1	4	{4, 4, 4, 286, 3, 142, 154, 111, 374, 434, 126, 2}	12
varPOD-M3	1	4	{4, 4, 4, 286, 3, 141, 155, 131, 308, 250, 275, 200, 2}	13
FE	0.1	5	{4, 4, 119, 4, 677, 3, 209, 134, 297, 249, 309, 211, 2}	13
POD-M1	0.1	5	{4, 4, 120, 4, 680, 3, 216, 139, 234, 316, 602, 184, 2}	13
varPOD-M1	0.1	5	{4, 4, 119, 4, 677, 3, 209, 134, 297, 249, 309, 210, 2}	13
POD-M2	0.1	5	{4, 4, 119, 4, 677, 3, 209, 134, 299, 249, 312, 166, 2}	13
varPOD-M2	0.1	5	{4, 4, 119, 4, 677, 3, 209, 134, 296, 249, 309, 212, 2}	13
POD-M3	0.1	5	{4, 4, 119, 4, 677, 3, 208, 131, 106, 160, 274, 6, 2}	13
varPOD-M3	0.1	5	{4, 4, 119, 4, 677, 3, 209, 134, 300, 249, 310, 209, 2}	13

Table 2 Cost functional $\hat{J}(\bar{u})$, state active points, POD errors and computational time speed-up

Discretiz.	c^0	β	$\hat{J}(\bar{u})$	$|\mathscr{A}|$	rel.err. u	$|\mathscr{A}^{\mathrm{FE}} \triangle \mathscr{A}^{\mathrm{POD}}|$	Speed-up
FE	1	4	82.12	94	–	–	–
POD-M1	1	4	91.89	590	7.78×10^{-2}	548	20.7
varPOD-M1	1	4	82.12	95	5.02×10^{-4}	1	15.2
POD-M2	1	4	82.06	101	1.58×10^{-2}	57	11.7
varPOD-M2	1	4	82.17	91	1.95×10^{-3}	3	15.1
POD-M3	1	4	79.88	175	4.77×10^{-2}	167	14.8
varPOD-M3	1	4	82.14	93	4.92×10^{-3}	1	13.8
FE	0.1	5	82.07	95	–	–	–
POD-M1	0.1	5	95.51	299	1.98×10^{-1}	270	13.4
varPOD-M1	0.1	5	82.08	95	5.31×10^{-4}	0	14.9
POD-M2	0.1	5	82.04	98	5.06×10^{-3}	3	14.6
varPOD-M2	0.1	5	82.12	93	3.20×10^{-3}	2	15.8
POD-M3	0.1	5	83.68	343	8.48×10^{-2}	304	9.6
varPOD-M3	0.1	5	82.04	96	3.42×10^{-3}	3	8.2

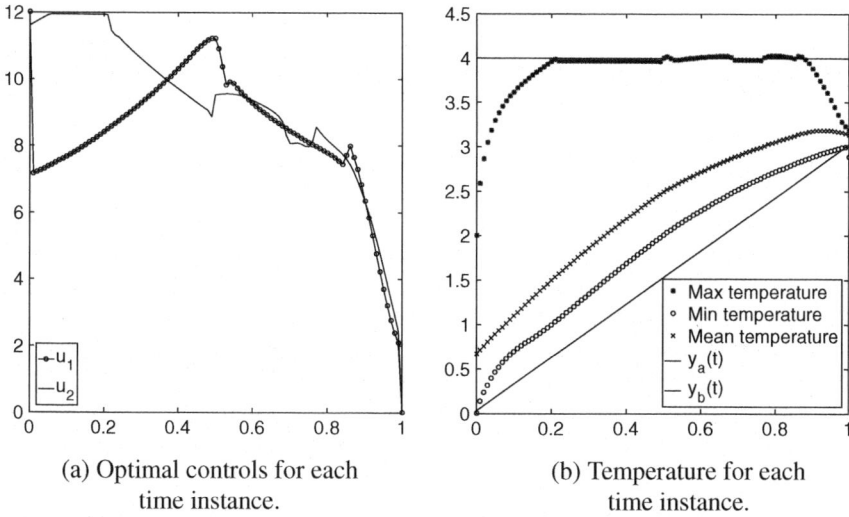

(a) Optimal controls for each
time instance.

(b) Temperature for each
time instance.

Fig. 2 Plots of optimal solution computed with FE with $c^0 = 1$ and $\beta = 4$

to generate the POD basis elements the number of iterations for Algorithm 1 is 13. Moreover, what can be observed is that to use the solution of the previous iteration as warm start helps the convergence of projected gradient for increasing values of the penalization parameter c, even if at the begin, small values of c implies really few iterations, because the effort on respecting the state constraints is weaker. Then, e.g. when $c^0 = 1$ and $\beta = 4$ at the 4th iteration of Algorithm 1 the projected gradient takes many iterations to converge, since $c^4 = 64$ already implies that the state constraints have to be way better satisfied compared to the iteration $c^3 = 16$. In Fig. 2, we can note that the optimal solution, in particular the average temperature, is almost everywhere inside the bounds and, when it is not, there are only small differences, which are related to the numerical tolerances chosen. In Table 2, it is shown also that the quality of the POD approximation is improved after rescaling the snapshots with their variance σ^2, in particular POD-M1 results can be significantly improved. As already mentioned, when the snapshots of POD-M1 are not scaled they have huge differences in magnitude, due to the fact that the adjoint variable magnitude at each time step depends directly on $c^{13} = 4^{12} \simeq 1.7 \times 10^7$ (or $c^{13} = 0.1 \times 5^{12} \simeq 2.4 \times 10^7$). Further, this improved approximation is also appearing for POD-M2 and POD-M3, even if there is less effectiveness compared to POD-M1 because the value of c for computing the snapshots is way smaller than 10^7. The speed-up is around 15 for all the methods, except for POD-M3 and varPOD-M3 in case of $c^0 = 0.1$ and $\beta = 5$ where it is even lower than 10. This can be justified from the fact that when we compute the FE suboptimal control $\hat{u}$ for $n = 3$, the projected gradient method takes 119 iterations to converge, so the idea of fixing a maximum number of finite element iterations should be considered to generate the suboptimal control $\hat{u}$ for improving the computational time speed-up, of course paying in accuracy of the approximation.

6 Conclusion

To conclude, in contrast to [15], we have applied a first-order augmented Lagrangian method to solve the optimal control problem with bilateral control and state constraints. Except for few points (compared to the total number of nodes in the space-time grid), we got the possibility to stay inside the state bounds as well. Therefore, it is meaningful to proceed by combining this method with a model predictive control approach, following the idea of [16]. Moreover, to speed-up the computational process, we have combined the previous algorithm with POD and we have compared different approaches to generate the snapshots. We have also improved the quality of the POD approximation scaling state and adjoint snapshots by different values, to balance the optimization process better which is involved in the POD basis' computation. This technique turned out to be successful: for the same number of basis elements ℓ we have that the relative errors of varPOD methods are lower than the POD relative ones without affecting the computational time speed-up.

Acknowledgements Luca Mechelli gratefully acknowledges support by the German Science Fund DFG grant Reduced-Order Methods for Nonlinear Model Predictive Control.

References

1. Alla, A., Gräßle, C., Hinze, M.: A-posteriori snapshot location for POD in optimal control of linear parabolic equations. Forthcoming article in ESAIM: M2AN. https://doi.org/10.1051/m2an/2018009
2. Arian, E., Fahl M., Sachs, E.W.: Trust-region proper orthogonal decomposition for flow control. Technical Report 2000-25, ICASE (2000)
3. Benner, P., Cohen, A., Ohlberger, M., Willcox, K.: Model Reduction and Approximation: Theory and Algorithms. SIAM, Philadelphia (2017)
4. Bergounioux, M.: Augmented Lagrangian method for distributed optimal control problems with state constraints. J. Optim. Theor. Appl. **78**(3), 493521 (1993)
5. Bertsekas, D.P.: Constrained Optimization and Lagrange Multiplier Methods. Athena Scientific, Belmont (1996)
6. Dautray, R., Lions, J.L.: Mathematical Analysis and Numerical Methods for Science and Technology, vol. 5. In: Evolution Problems I. Springer-Verlag, Berlin (2000)
7. Gubisch, M.; Volkwein, S.: Proper orthogonal decomposition for linear-quadratic optimal control. Chapter 1 in [3]
8. Hinze, M., Pinnau, R., Ulbrich, M., Ulbrich, S.: Optimization with PDE constraints. Springer-Verlag, Berlin (2009)
9. Holmes, P., Lumley, J.L., Berkooz, G., Rowley, C.W.: Turbulence, coherent structures, dynamical systems and symmetry, 2nd edn. In: Cambridge Monographs on Mechanics. Cambridge University Press, Cambridge (2012)
10. Ito, K., Kunisch, K.: Augmented Lagrangian methods for nonsmooth, convex optimization in Hilbert spaces. Nonlinear Anal.: Theory Meth. Appl. **41**, 591–616 (2000)
11. Ito, K., Kunisch, K.: Lagrange multiplier approach to variational problems and applications. In: Advances in design and control. SIAM, Philadelphia (2008)
12. Jehle, J.S.: First-order augmented Lagrangian methods for state constraint problems. Master's thesis (2018). http://nbn-resolving.de/urn:nbn:de:bsz:352-2-1q60suvytbn4t1

13. Krumbiegel, K., Rösch, A.: A virtual control concept for state constrained optimal control problems. Comput. Optim. Appl. **43**, 213–233 (2009)
14. Kunisch, K., Volkwein, S.: Proper orthogonal decomposition for optimality systems. ESAIM: M2AN 42, 1–23 (2008)
15. Mechelli, L., Volkwein, S.: POD-based economic optimal control of heat-convection phenomena. In Falcone, M., Ferretti, R., Grüne, L., McEneaney, W.M. (eds.) Numerical Methods for Optimal Control Problems, pp. 63–87. Springer International Publishing, Cham (2018)
16. Mechelli, L., Volkwein, S.: POD-based economic model predictive control for heat-convection phenomena. Lect. Notes Comput. Sci. Eng. **126**, 663–670 (2019)
17. Tröltzsch, F.: Optimal control of partial differential equations. In: Theory, Methods and Applications. American Mathematical Society (2010)
18. Tröltzsch, F., Volkwein, S.: POD a-posteriori error estimates for linear-quadratic optimal control problems. Comput. Optim. Appl. **44**, 83–115 (2009)

Coupling of Incompressible Free-Surface Flow, Acoustic Fluid and Flexible Structure Via a Modal Basis

Florian Toth and Manfred Kaltenbacher

Abstract Free surface flow of incompressible liquids interacts with compressible gases and flexible structures in applications like liquid tanks or air chamber supported floating platforms. After briefly describing the physical modeling of the coupled system, we suggest a procedure for model order reduction based on the modal bases of the un-coupled domains. To again couple the different physical domains, one needs to assemble the coupling conditions on the interfaces between gas, liquid, and structure domains in terms of the modal coordinates. We demonstrate the effectiveness of the model order reduction by applying it to a geometrically simple but strongly coupled tank system, for which we compute the sloshing modes.

Keywords Free surface waves · Acoustics · Flexible structure · Modal coupling

1 Introduction

Air chamber supported floating platforms, or liquid tanks like liquefied natural gas (LNG) containers are typical example applications in which the interaction of a fluid in liquid and gaseous phase and a structure must be considered. In air chamber supported floating platforms, large pockets of compressed air provide buoyancy and can significantly reduce the wave induced loading on the platform [19]. For partially filled liquid tanks the dynamics of the fluid with free surface (sloshing) must be known to allow for save operation [10]. Model order reduction is an attractive option to reduce the computational effort in computational models. Especially for parameter dependent problems, methods based on proper orthogonal decomposition (POD) are frequently applied, e.g. for modelling arterial blood flow [1], parametric eigenvalue

F. Toth (✉) · M. Kaltenbacher
Institute of Mechanics and Mechatronics, TU Wien, Getreidemarkt 9, Vienna, Austria
e-mail: florian.toth@tuwien.ac.at

M. Kaltenbacher
e-mail: manfred.kaltenbacher@tuwien.ac.at

© Springer Nature Switzerland AG 2020   141
J. Fehr and B. Haasdonk (eds.), *IUTAM Symposium on Model Order Reduction of Coupled Systems, Stuttgart, Germany, May 22–25, 2018*, IUTAM Bookseries 36,
https://doi.org/10.1007/978-3-030-21013-7_10

problems of vibrating structures [8], or the treatment of the free-surface problems in osmotic cell swelling [14].

While model order reduction has already been applied to various coupled multi-field problems [3], the case of free-surface-flow, acoustic fluid and flexible structure has not been explored yet. A semi analytic procedure based on the acoustic modes in an rectangular air-camber was proposed already by Newman to model wave effects on floating platforms [17]. A coupling approach for flexible structure and air chambers modeled by generalized (modal) coordinates and water waves modeled by the boundary element method was proposed shortly afterwards [13]. There exist multiple examples of model order reduction techniques for coupled vibro-acoustical systems [4, 7, 16, 21] most building on the ideas of component mode synthesis [5] and related methods from structural dynamics [2, 9]. We will follow a similar procedure to include liquid sloshing dynamics originating from gravity waves on an interface between fluids of different densities.

2 Physical Modeling

We consider a system consisting of three domains: flexible structure Ω_s, liquid Ω_w and gas Ω_a as depicted in Fig. 1. For the flexible structure we assume a linear elastic material, i.e. Hook's law $\sigma = C\epsilon$ where σ, ϵ, and C denote the stress, strain and stiffness tensors, respectively, and small displacements, i.e. the linearized strain displacement relationship $\epsilon = \left(\nabla u + (\nabla u)^T\right)/2$, where u denotes the displacement vector. The liquid is assumed as inviscid and incompressible, which is a reasonable assumption e.g. for water or LNG. However, in the gaseous domain we take compressibility into account, which is necessary e.g. for (compressed) air. Additionally, we assume that both fluid domains are in equilibrium and only consider small perturbations to this equilibrium. This yields Poisson's equation for the dynamic pressure in the liquid domain, and the acoustic wave equation in the gaseous domain, as detailed in [22]. The equations for the dynamic quantities then read

Fig. 1 Sketch of the physical model domains: incompressible liquid (blue), compressible gas (red) and flexible structure (gray), with their respective coupling interfaces

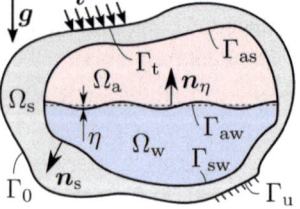

$$\nabla \cdot \nabla p_{\mathrm{w}} = 0 \quad \text{in } \Omega_{\mathrm{w}}, \tag{1a}$$

$$\frac{\partial^2 p_{\mathrm{a}}}{\partial t^2} - c_{\mathrm{a}}^2 \, \nabla \cdot \nabla p_{\mathrm{a}} = 0 \quad \text{in } \Omega_{\mathrm{a}}, \tag{1b}$$

$$\rho_{\mathrm{s}} \frac{\partial^2 \boldsymbol{u}}{\partial t^2} - \frac{1}{2} \nabla \cdot \boldsymbol{C} \left(\nabla \boldsymbol{u} + (\nabla \boldsymbol{u})^T \right) = \boldsymbol{0} \quad \text{in } \Omega_{\mathrm{s}}, \tag{1c}$$

where c_{a} denotes the speed of sound in the gaseous domain, p_i denotes the dynamic pressure variation in the fluid domains, and ρ_i denotes the density, where the index $i = $ a,w,s is used to refer to gaseous (air), liquid (water) and structure domains, respectively.

A non-standard coupling condition concerns the interface between liquid and gas phase, Γ_{aw}. Here, one needs to enforce both the dynamic condition, i.e. equality of the pressure, and the kinematic condition, i.e. particle velocities are equal. The derivation of this coupling condition [22] introduced a variable describing the position of the interface, η, as usually done in the modeling of water waves [15], which, however can subsequently be removed. Care also has to be taken at the liquid-solid interface: Here the contribution of buoyancy must be taken into account in the dynamic coupling condition. Together with the usual conditions for acoustic-structural coupling on Γ_{as} (which relate structural stress and acoustic pressure as well as structure displacement and acoustic particle velocity, see e.g. [11]) we obtain the following set of relations

$$g\rho_{\mathrm{w}} \frac{\partial p_i}{\partial \boldsymbol{n}_\eta} = \rho_i \frac{\partial^2}{\partial t^2} (p_{\mathrm{a}} - p_{\mathrm{w}}) \quad \text{on } \Gamma_{\mathrm{aw}}, \tag{2a}$$

$$\frac{\partial p_i}{\partial \boldsymbol{n}_{\mathrm{s}}} = -\rho_i \frac{\partial^2 \boldsymbol{u}}{\partial t^2} \cdot \boldsymbol{n}_{\mathrm{s}} \quad \text{on } \Gamma_{is}, \tag{2b}$$

$$\boldsymbol{\sigma} \cdot \boldsymbol{n}_{\mathrm{s}} = -p_{\mathrm{a}} \boldsymbol{n}_{\mathrm{s}} \quad \text{on } \Gamma_{\mathrm{as}}, \tag{2c}$$

$$\boldsymbol{\sigma} \cdot \boldsymbol{n}_{\mathrm{s}} = -(\rho_{\mathrm{w}} \boldsymbol{g} \cdot \boldsymbol{u} + p_{\mathrm{w}}) \boldsymbol{n}_{\mathrm{s}} \quad \text{on } \Gamma_{\mathrm{ws}}, \tag{2d}$$

$$\boldsymbol{\sigma} \cdot \boldsymbol{n}_{\mathrm{s}} = \boldsymbol{t} \quad \text{on } \Gamma_{\mathrm{t}}, \tag{2e}$$

$$\boldsymbol{\sigma} \cdot \boldsymbol{n}_{\mathrm{s}} = \boldsymbol{0} \quad \text{on } \Gamma_0. \tag{2f}$$

Again, the index i is used to denote air or water domain. Thus, Eqs. (2a) and (2b) hold for both domains. The normal vector $\boldsymbol{n}_\eta$ of the coupling surface between liquid and gas, Γ_{aw}, is defined as pointing from the higher density liquid to the lower density gas, i.e. from Ω_{w} to Ω_{a}, oriented like $-\boldsymbol{g}$. The acceleration of gravity is denoted $\boldsymbol{g}$, and g is used for its magnitude. The normal vector of the structure $\boldsymbol{n}_{\mathrm{s}}$ points from the fluid into the structure. We impose a surface traction $\boldsymbol{t}$ to the structure via an inhomogeneous Neumann condition on Γ_{t}. The natural stress-free condition is the homogeneous version applied at Γ_0.

3 Finite Element Formulation

We introduce appropriate test functions γ, φ, and δ for gas pressure, liquid pressure and displacement vector, respectively, multiply our coupled system of PDEs by these test functions and integrate over the whole computational domain. Integration by parts,[1] and incorporation of the boundary conditions, yields the variational formulation: Find $p_a \in H^1(\Omega_a)$, $p_w \in H^1(\Omega_w)$, and $\boldsymbol{u} \in (H^1(\Omega_s))^2$ such that

$$
\int_{\Omega_a} \nabla\gamma \cdot \nabla p_a dx + \int_{\Omega_a} \frac{1}{c_a^2}\gamma\frac{\partial^2 p_a}{\partial t^2}dx + \int_{\Gamma_{aw}} \frac{\rho_a}{g\rho_w}\gamma\frac{\partial^2 p_a}{\partial t^2}ds
$$
$$
- \int_{\Gamma_{aw}} \frac{\rho_a}{g\rho_w}\gamma\frac{\partial^2 p_w}{\partial t^2}ds + \int_{\Gamma_{as}} \rho_a\gamma\frac{\partial^2 \boldsymbol{u}}{\partial t^2}\cdot \boldsymbol{n}_s ds = 0 , \qquad (3a)
$$

$$
\int_{\Omega_w} \nabla\varphi \cdot \nabla p_w dx + \int_{\Gamma_{aw}} \frac{1}{g}\varphi\frac{\partial^2 p_w}{\partial t^2}ds
$$
$$
- \int_{\Gamma_{aw}} \frac{1}{g}\varphi\frac{\partial^2 p_a}{\partial t^2}ds + \int_{\Gamma_{sw}} \rho_w\varphi\frac{\partial^2 \boldsymbol{u}}{\partial t^2}\cdot \boldsymbol{n}_s ds = 0 , \qquad (3b)
$$

$$
\frac{1}{2}\int_{\Omega_s} \nabla\boldsymbol{\delta} : \boldsymbol{C}\left(\nabla\boldsymbol{u} + \nabla\boldsymbol{u}^T\right) dx - \int_{\Gamma_{sw}} \rho_w\boldsymbol{g}\cdot\boldsymbol{u}\boldsymbol{n}_s \cdot \boldsymbol{\delta} ds + \int_{\Omega_s} \rho_s\boldsymbol{\delta}\cdot\frac{\partial^2 \boldsymbol{u}}{\partial t^2}dx
$$
$$
- \int_{\Gamma_{as}} p_a\boldsymbol{n}_s \cdot \boldsymbol{\delta} ds - \int_{\Gamma_{sw}} p_w\boldsymbol{n}_s \cdot \boldsymbol{\delta} ds = \int_{\Gamma_t} \boldsymbol{\delta}\cdot\boldsymbol{t} ds , \qquad (3c)
$$

for all test functions $\gamma \in H^1(\Omega_a)$, $\varphi \in H^1(\Omega_w)$, and $\boldsymbol{\delta} \in (H^1(\Omega_s))^2$.

After the standard spatial discretization by finite elements and collecting the nodal degrees of freedom in the vectors $\boldsymbol{q}_i$ we arrive at the following system of ordinary differential equations

$$
\begin{bmatrix} \boldsymbol{M}_a & \boldsymbol{M}_{aw} & \boldsymbol{M}_{as} \\ \boldsymbol{M}_{wa} & \boldsymbol{M}_w & \boldsymbol{M}_{ws} \\ \boldsymbol{0} & \boldsymbol{0} & \boldsymbol{M}_s \end{bmatrix}\begin{bmatrix} \ddot{\boldsymbol{q}}_a \\ \ddot{\boldsymbol{q}}_w \\ \ddot{\boldsymbol{q}}_s \end{bmatrix} + \begin{bmatrix} \boldsymbol{K}_a & \boldsymbol{0} & \boldsymbol{0} \\ \boldsymbol{0} & \boldsymbol{K}_w & \boldsymbol{0} \\ \boldsymbol{K}_{sa} & \boldsymbol{K}_{sw} & \boldsymbol{K}_s \end{bmatrix}\begin{bmatrix} \boldsymbol{q}_a \\ \boldsymbol{q}_w \\ \boldsymbol{q}_s \end{bmatrix} = \begin{bmatrix} \boldsymbol{0} \\ \boldsymbol{0} \\ \boldsymbol{f}_s \end{bmatrix} , \qquad (4)
$$

where a dot over a variable denotes the derivative with respect to time. Note, that the mass matrix associated with the liquid domain, $\boldsymbol{M}_w$, originates solely from the surface integral over the coupling surface Γ_{aw} in Eq. (3b). Neglecting any forcing

---

[1] Using the relation $\nabla \cdot (\boldsymbol{\delta}\cdot\boldsymbol{\sigma}) = (\nabla \cdot \boldsymbol{\sigma})\cdot\boldsymbol{\delta} + \boldsymbol{\sigma} : (\nabla\boldsymbol{\delta})$, where ':' denotes the double dot product, i.e. the sum of the products of conjugated tensor elements.

on the structure f_s and using a harmonic ansatz for the solution one directly arrives at the coupled eigenvalue problem. This can be solved by any eigenvalue solver capable of handling real-valued, un-symmetric, generalized eigenvalue problems, e.g. the FEAST solver [12, 20] which was used here.

4 Reduced Order Model

In order to derive a reduced order model, we decompose the system in its physical sub-domains liquid, gas and structure. We neglect all couplings and, thus, arrive at three un-coupled eigenvalue problems

$$K_i v_i = \omega^2 M_i v_i \,, \tag{5}$$

which can be solved to obtain a set of m_i eigenvectors $v_{i,k}$ and corresponding eigenfrequencies for each domain. To reduce the number of degrees of freedom we now introduce modal coordinates $\phi_{i,k}$ in each domain and approximate the solution fields in the usual form

$$q_i(t) = \sum_{k=1}^{m_i} v_{i,k} \phi_{i,k}(t) = V_i \phi_i(t). \tag{6}$$

To obtain a coupled system in the chosen modal basis we proceed as follows: We assemble the coupling terms in Eq. (3) for unit modal coordinates individually. The linear forms to be assembled are equivalent to known right-hand-side forces, e.g. for the coupling between structure and gas domain we need to assemble

$$f_{as,k} = \int_{\Gamma_{as}} \rho_a \gamma u_k \cdot n_s ds \,, \tag{7a}$$

$$f_{sa,k} = - \int_{\Gamma_{as}} \delta \cdot n_s p_{a,k} ds \,, \tag{7b}$$

where the displacements u_k and pressures $p_{a,k}$ are known. They are obtained from the fields defined by the k-th un-coupled basis vector, i.e. from $v_{s,k}$ for the structure and $v_{a,k}$ for the gas domain, respectively. By this procedure we obtain m_i modal force vectors containing n_i elements, where n_i denotes the number of degrees of freedom in domain i. We are now able to write the variational formulation, Eq. (3), in terms of the modal coordinates, e.g. for the air domain

$$M_a V_a \ddot{\phi}_a + F_{aw} \ddot{\phi}_w + F_{as} \ddot{\phi}_s + K_a V_a \phi_a = 0 \,, \tag{8}$$

where the forcing vectors for unit modal structure displacement and unit modal liquid pressure have been collected in the matrices $F_{as} = [f_{as,1}, \ldots, f_{as,M_s}]$ and

F_{aw}, respectively. Finally, we project each of the equations into its modal sub-space to arrive at the coupled system

$$
\begin{bmatrix} V_a^T M_a V_a & V_a^T F_{aw} & V_a^T F_{as} \\ V_w^T F_{wa} & V_w^T M_w V_w & V_w^T F_{ws} \\ 0 & 0 & V_s^T M_s V_s \end{bmatrix} \begin{bmatrix} \ddot{\phi}_a \\ \ddot{\phi}_w \\ \ddot{\phi}_s \end{bmatrix} +
$$
$$
\begin{bmatrix} V_a^T K_a V_a & 0 & 0 \\ 0 & V_w^T K_w V_w & 0 \\ V_s^T F_{sa} & V_s^T F_{sw} & V_s^T K_s V_s \end{bmatrix} \begin{bmatrix} \phi_a \\ \phi_w \\ \phi_s \end{bmatrix} = \begin{bmatrix} 0 \\ 0 \\ V_s^T f_s \end{bmatrix}. \tag{9}
$$

Again, the coupled system can be written as a generalized eigenvalue problem by neglecting right-hand-side forcing and the use of an harmonic ansatz.

5 Application Example

As a simple example we consider a 2D model of a square box with a wall thickness of 0.1 m, which encloses a square fluid domain with a side length of 1 m. The square domain is filled up to a level of 0.7 m by liquid, and the top 0.3 m are filled by gas. For the 2D model we assume a plane strain state in the structure domain. The material is assumed linear elastic and isotropic with a Young's modulus of 5 kPa, a Poisson's number of 0.25 and a density of 5 kg/m³. The incompressible fluid has a density of 4 kg/m³ and the compressible gas has a density of 1 kg/m³ and a speed of sound of 1 m/s. The acceleration of gravity is assumed pointing downwards with a magnitude of 9.81 m/s². As boundary condition we constrain the displacement on the bottom side of the structure (in both directions).

The eigenvalue analyses of the individual domains yield the modal basis for the reduction of the system. We use the m_i first modes of the acoustic, water and structural domain (see Fig. 2 and Table 1 for the corresponding natural frequencies). We select

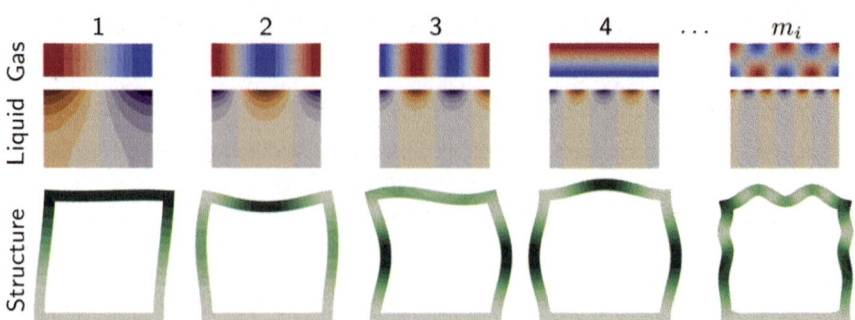

Fig. 2 Modal basis used for the model order reduction, showing the first 4 and the last used mode

10 acoustic modes, $m_a = 10$, 6 water wave modes $m_w = 6$ and 15 structural modes $m_s = 15$.

Thus, we arrive at a reduced coupled system with 21 degrees of freedom, as compared to 961 in the full model. The choice of the reduced basis requires some engineering experience, and was done based of the following considerations: (a) one should only use physical modes, i.e. ones that are appropriately represented by the discretised FE model, and (b) one must use a sufficient number of modes not to over-constrain the system, as each neglected mode can be seen as a linear constraint. In the current case the limit for (a) is posed by the water domain where the free surfaces waves must be properly resolved, which is the case approximately until mode 6 at about 14 Hz. The acoustic modes in the air domain are selected in a similar frequency range, i.e. up to mode 10 at about 17 Hz. Finally, the structural modes are selected up to mode 15 in order to allow a sufficient deformation basis to honour criterion (b).

We solve the eigenvalue problem in the modal sub-space, from Eq. (9), and compare the results to the solution of the full model, from Eq. (4).

Observing the relative error of the natural frequencies of the reduced model with respect to the full model we obtain good agreement. The relative error is generally low with a maximum relative error lower than 6% for the first 10 modes given in Table 1. Figure 3 shows a representative selection of mode shapes for the coupled system (modes 1, 2, 4, 5, and 7). Visually, we obtain excellent agreement for most of the modes (mind the different signs for some). For some modes (5 and 7) slight discrepancies can be recognized between the mode shapes of reduced and full model. A quantitative measure of the mode shape correlation is given by the modal assurance criterion (MAC) [6, 18] given in Table 1 and visually in Fig. 4 (left). The reason for the discrepancies in the mode shapes most likely is an in-sufficient modal basis in the

Table 1 Obtained natural frequencies ω_i in rad/s for the individual domains and for the coupled system, as well as the relative error of the frequencies from reduced model with respect to the full model, and the modal assurance criterion (MAC) between the modes of reduced and full model

Mode	Gas	Liquid	Structure	Coupled (full)	Coupled (reduced)	Relative error	MAC
1	3.145	5.490	2.706	2.123	2.152	−0.01361	0.931
2	6.309	7.882	10.187	3.205	3.175	+0.00939	0.986
3	9.512	9.705	16.931	5.724	5.694	+0.00513	0.998
4	10.592	11.288	18.535	6.124	6.039	+0.01422	0.999
5	11.049	12.740	32.335	7.346	6.947	+0.05733	0.688
6	12.329	14.117	39.542	8.079	7.916	+0.02055	0.977
7	12.774		41.478	9.094	8.668	+0.04908	0.319
8	14.236		43.505	9.901	9.588	+0.03264	0.944
9	16.114		56.202	10.585	10.195	+0.03827	0.449
10	16.594		71.600	10.923	10.642	+0.02640	0.740

Fig. 3 Representative mode shapes of the coupled system computed using the reduced and full model, respectively

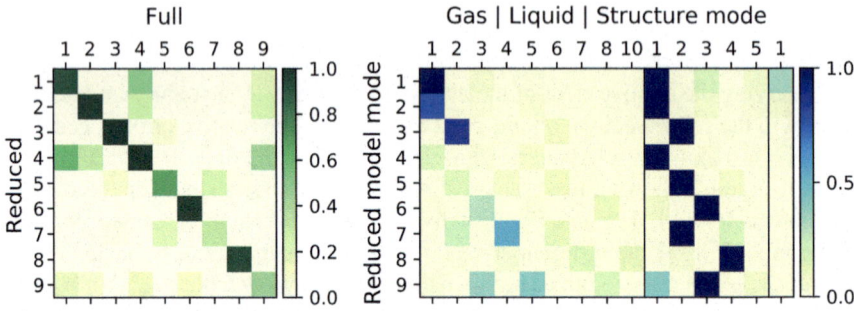

Fig. 4 MAC between the mode shapes of the reduced and full model (left) and relative contribution of the individual basis modes to the mode shapes of the reduced coupled system (right)

liquid domain. As the un-coupled liquid modes show decaying pressure values with increasing distance from the free surface, they are not able to account for structural motions originating from the lower regions. As the liquid is incompressible, any motion of the structure-liquid interface must have an impact on the free surface. This impact might be included by augmenting the modal basis by some modes for the static liquid pressure change due to interface deformation. Finally, the strength of coupling between the different domains is depicted by visualizing the relative contribution of each basis vector to the modes of the coupled system in Fig. 4 (right). For the first coupled mode we see significant contributions of acoustic mode 1, liquid mode 1 and structural mode 1. For the coupled mode 3, the main contribution comes from liquid mode 2 and acoustic mode 2. For the coupled mode 7 one sees an interaction between liquid mode 2 and acoustic modes 4, 2 and 6. Another indication for the strong coupling is that many of the un-coupled natural frequencies do not appear again as a natural frequency of the coupled system.

6   Conclusion

The suggested procedure works well, even for the chosen example case exhibiting strong coupling between the different physical domains. The number of unknowns in the reduced order model could be reduced by an order of magnitude from 961 nodal degrees of freedom to 21 modal coordinates, while maintaining acceptable accuracy. This was achieved simply by choosing a set of un-coupled modes as a modal basis. A coupled system in this basis is obtained by assembling the coupling terms for the basis vectors. The computational effort for the suggested procedure can be expected to be significantly lower than for the treatment of the full model. Furthermore, it might be a feasible option in cases where the assembly of the full system is too expensive, but the sub-systems can still be handled. As long as the used modes from the sub-domains are able to represent the coupled modes, accurate results can be expected. While the procedure has only been applied to the computation of the coupled eigenvalue problem, it has been formulated in a fully general manner, and could, thus, also be applied to solve reduced coupled problems in the time and frequency domain. Of course, time and frequency domain problems are only accurately represented by the reduced model in the frequency range covered by the reduced basis. While the extension of the proposed procedure to non-linear problems seems challenging, the application to other linear multi-physics problems should work in a straightforward way.

References

1. Ballarin, F., Rozza, G., Maday, Y.: Model reduction of parametrized systems. In: MS&A Modeling, Simulation and Applications, chap. Reduced-Order Semi-Implicit Schemes for Fluid-Structure Interaction Problems, pp. 149–167. Springer, Cham (2017)
2. Bampton, M.C.C., Craig, R.R.J.: Coupling of substructures for dynamic analyses. AIAA J. 6(7), 1313–1319 (1968)
3. Benner, P., Feng, L.: Model order reduction for coupled problems. Technical report Max Planck Institute for Dynamics of Complex Technical Systems (2015)
4. Bennighof, J.: Vibroacoustic frequency sweep analysis using automated multi-level substructuring. In: 40th Structures, Structural Dynamics, and Materials Conference and Exhibit (1999)
5. Craig, R.J.: Coupling of substructures for dynamic analyses—an overview. In: 41st Structures, Structural Dynamics, and Materials Conference and Exhibit (2000)
6. Ewins, D.J.: Modal Testing, 2 edn. Wiley (2009)
7. Herrmann, J., Maess, M., Gaul, L.: Substructuring including interface reduction for the efficient vibro-acoustic simulation of fluid-filled piping systems. Mech. Syst. Signal Process. 24(1), 153–163 (2010)
8. Horger, T., Wohlmuth, B., Dickopf, T.: Simultaneous reduced basis approximation of parameterized elliptic eigenvalue problems. ESAIM: Math. Model. Numer. Anal. 51(2), 443–465 (2017)
9. Hurty, W.C.: Dynamic analysis of structural systems using component modes. AIAA J. 3(4), 678–685 (1965)
10. Ibrahim, R.A.: Liquid Sloshing Dynamics. Cambridge University Press (2016)
11. Kaltenbacher, M.: Numerical Simulation of Mechatronic Sensors and Actuators: Finite Elements for Computational Multiphysics, 3rd edn. Springer (2015). ISBN: 978-3-642-40169-5

12. Kestyn, J., Polizzi, E., Tang, P.T.P.: Feast eigensolver for non-hermitian problems, arXiv (2015)
13. Lee, C.H., Newman, J.N.: Wave effects on large floating structures with air cushions. Marine Struct. **13**(4–5), 315–330 (2000)
14. Lehrenfeld, C., Rave, S.: Mass conservative reduced order modeling of a free boundary osmotic cell swelling problem, arXive (2018)
15. Lighthill, J.: Waves in Fluids. Cambridge University Press (2001)
16. Maess, M., Gaul, L.: Substructuring and model reduction of pipe components interacting with acoustic fluids. Mech. Syst. Signal Process. **20**(1), 45–64 (2006)
17. Newman, J.N.: Diffraction of water waves by an air chamber, In: 15th International Workshop on Water Waves and Floating Bodies (2000)
18. Pastor, M., Binda, M., Harčarik, T.: Modal assurance criterion. Proc. Eng. **48**, 543–548 (2012)
19. Pinkster, J.A., Meevers Scholte, E.J.A.: The behaviour of a large air-supported mob at sea. Marine Struct. **14**(1–2), 163–179 (2001)
20. Polizzi, E.: Density-matrix-based algorithm for solving eigenvalue problems. Phys. Rev. B 79(11) (2009)
21. Puri, R.S., Morrey, D., Bell, A.J., Durodola, J.F., Rudnyi, E.B., Korvink, J.G.: Reduced order fully coupled structural–acoustic analysis via implicit moment matching. Appl. Math. Modell. **33**(11), 4097–4119 (2009)
22. Toth, F., Kaltenbacher, M.: Fully coupled linear modelling of incompressible free-surface flow, compressible air and flexible structures. Int. J. Numer. Meth. Eng. **107**(11), 947–969 (2016)

Model Order Reduction of Coupled, Parameterized Elastic Bodies for Shape Optimization

Benjamin Fröhlich, Florian Geiger, Jan Gade, Manfred Bischoff
and Peter Eberhard

Abstract In this contribution, coupled, parameterized second order systems are considered where the coupled, parameterized system is derived from the assembly of several parameterized component models. Two approaches for the Parametric Model Order Reduction of such coupled systems are presented and compared in a reduced order shape optimization example. In the first approach, the coupled, parameterized system is derived by coupling the parameterized, full order component models. Then, Parametric Model Order Reduction is executed for the coupled system. In the second approach, the parameterized, component models are first reduced independently of their actual mounting situation. Afterwards, the parameterized, reduced order component models are coupled to derive the parameterized, reduced order system model. It is shown that the first approach yields smaller parameterized, reduced order system models. However, the second approach allows to reuse and to recombine the parameterized, reduced order component models arbitrarily. It therefore introduces more flexibility in the modeling process, enabling for example a toolbox based optimization with parameterized, reduced order models.

Keywords Parametric model order reduction · Coupled systems · Shape optimization · Moment matching

B. Fröhlich · P. Eberhard (✉)
Institute of Engineering and Computational Mechanics, University of Stuttgart,
Pfaffenwaldring 9, 70569 Stuttgart, Germany
e-mail: peter.eberhard@itm.uni-stuttgart.de

B. Fröhlich
e-mail: benjamin.froehlich@itm.uni-stuttgart.de

F. Geiger · J. Gade · M. Bischoff
Institute for Structural Mechanics, University of Stuttgart, Pfaffenwaldring 7,
70569 Stuttgart, Germany
e-mail: geiger@ibb.uni-stuttgart.de

J. Gade
e-mail: gade@ibb.uni-stuttgart.de

M. Bischoff
e-mail: bischoff@ibb.uni-stuttgart.de

© Springer Nature Switzerland AG 2020
J. Fehr and B. Haasdonk (eds.), *IUTAM Symposium on Model Order Reduction
of Coupled Systems, Stuttgart, Germany, May 22–25, 2018*, IUTAM Bookseries 36,
https://doi.org/10.1007/978-3-030-21013-7_11

1 Introduction

The finite element method is a well-established method for modeling elastic structures. For fine spatial discretizations, the corresponding systems of second order differential equations become very large making evaluations time consuming. Model Order Reduction (MOR) is a powerful tool to decrease this high numerical effort. However, models of sophisticated systems are often created in a modular fashion. This means that the underlying *component* models are modeled separately and coupled afterwards to derive the overall *system model*. This has several advantages as it simplifies the modeling process and it allows to exchange single components to conduct variant studies. A straightforward approach for the MOR of such coupled systems is to couple the full order component models and to conduct the MOR for the system model. A second approach is to reduce the components separately and to couple the reduced order component models to derive the reduced order system model as presented in [7, 10] for first order systems and applied in [6, 11] to second order and parameterized second order systems.

The novel contribution of this paper is that geometrically parameterized component models are considered. Coupling these geometrically parameterized component models allows to derive parameterized system models in versatile design variations. The two MOR approaches for coupled systems are extended to a Parametric Model Order Reduction (PMOR) for coupled, geometrically parameterized systems. Both approaches are well suited for reduced order modeling of coupled systems with a parameterized shape. A comparison of the approaches is done in a reduced order shape optimization example.

The paper is structured as follows. The next section reviews the theoretical background of the paper. Two PMOR methods for the reduction of coupled, parameterized systems are illustrated in Sect. 3. Both approaches are applied to a shape optimization example in Sect. 4. The conclusion can be found in Sect. 5.

2 Theoretical Background

In this contribution, the linear finite element method (FEM) is used to model elastic structures resulting in a set of second order differential equations

$$\mathbf{M}(\mathbf{p})\ddot{\mathbf{q}} + \mathbf{K}(\mathbf{p})\mathbf{q} = \mathbf{f}_{\text{ext}}. \tag{1}$$

The nodal displacements are gathered in $\mathbf{q} \in \mathbb{R}^N$ and the external forces are described by $\mathbf{f}_{\text{ext}} \in \mathbb{R}^N$. The matrices $\mathbf{M}(\mathbf{p}) \in \mathbb{R}^{N \times N}$ and $\mathbf{K}(\mathbf{p}) \in \mathbb{R}^{N \times N}$ describe the mass and the stiffness matrix. Furthermore, parameterized FEM models are considered meaning that both the mass and the stiffness matrix are parameterized with a parameter vector $\mathbf{p} \in \mathbb{R}^d$ to describe e.g., a parameterized geometry or parameterized material properties. Rewriting the external forces $\mathbf{f}_{\text{ext}}$ as a product of the input matrix

$\mathbf{B} \in \mathbb{R}^{N \times k}$ and system inputs $\mathbf{u} \in \mathbb{R}^k$ and adding an output equation results in a second order, linear, parameter-variant system

$$\mathbf{M}(\mathbf{p})\ddot{\mathbf{q}} + \mathbf{K}(\mathbf{p})\mathbf{q} = \mathbf{B}\mathbf{u} \tag{2}$$
$$\mathbf{y} = \mathbf{C}\mathbf{q}$$

where the output matrix $\mathbf{C} \in \mathbb{R}^{l \times N}$ extracts the outputs of interest $\mathbf{y} \in \mathbb{R}^l$ from the nodal displacements. The transfer behavior of system (2) can be equivalently described in Laplace domain with the parameterized transfer function

$$\mathbf{H}(s, \mathbf{p}) = \mathbf{C} \left(s^2 \mathbf{M}(\mathbf{p}) + \mathbf{K}(\mathbf{p}) \right)^{-1} \mathbf{B}, \tag{3}$$

where $s \in \mathbb{C}$ denotes the Laplace variable.

It is the idea of linear model order reduction by projection to approximate the solution of Eq. (2) as $\mathbf{q} \approx \mathbf{V}\bar{\mathbf{q}}$. This means that the full order solution $\mathbf{q}$ is approximated in a lower dimensional subspace $\mathscr{V}$ spanned by the columns of the projection matrix $\mathbf{V} \in \mathbb{R}^{N \times n}$ with $n \ll N$. Using $\mathbf{V}$ to perform a Galerkin projection yields the parameterized, reduced order system

$$\bar{\mathbf{M}}(\mathbf{p})\ddot{\bar{\mathbf{q}}} + \bar{\mathbf{K}}(\mathbf{p})\bar{\mathbf{q}} = \bar{\mathbf{B}}\mathbf{u} \tag{4}$$
$$\bar{\mathbf{y}} = \bar{\mathbf{C}}\bar{\mathbf{q}}$$

with the reduced system matrices

$$\bar{\mathbf{M}}(\mathbf{p}) = \mathbf{V}^{\mathrm{T}}\mathbf{M}(\mathbf{p})\mathbf{V}, \quad \bar{\mathbf{K}}(\mathbf{p}) = \mathbf{V}^{\mathrm{T}}\mathbf{K}(\mathbf{p})\mathbf{V}, \quad \bar{\mathbf{B}} = \mathbf{V}^{\mathrm{T}}\mathbf{B}, \quad \bar{\mathbf{C}} = \mathbf{C}\mathbf{V}. \tag{5}$$

In this contribution, interpolatory projection methods or moment-matching methods, respectively, are used to determine the subspace $\mathscr{V}$. These methods were successfully applied to the reduction of linear, time-invariant systems, see, e.g., [1]. Later, so called multi-moment matching methods extended these methods for parameterized systems [3]. A numerically stable framework for parameterized systems is presented in [2]. The idea is to construct a parameterized, reduced order model with the parameterized, reduced order transfer function

$$\bar{\mathbf{H}}(s, \mathbf{p}) = \bar{\mathbf{C}} \left(s^2 \bar{\mathbf{M}}(\mathbf{p}) + \bar{\mathbf{K}}(\mathbf{p}) \right)^{-1} \bar{\mathbf{B}} \tag{6}$$

interpolating the original transfer function $\mathbf{H}(s, \mathbf{p})$ at selected frequency expansion points $\hat{\sigma}$ and parameter expansion points $\hat{\mathbf{p}}$. According to [2], choosing a base as

$$\mathrm{span}(\mathbf{V}) = \mathrm{span}([\hat{\sigma}^2 \mathbf{M}(\hat{\mathbf{p}}) + \mathbf{K}(\hat{\mathbf{p}})]^{-1}\mathbf{B}) \tag{7}$$

ensures that

$$\mathbf{H}(s, \mathbf{p}) = \bar{\mathbf{H}}(s, \mathbf{p}) \quad \text{for} \quad s = \hat{\sigma}, \quad \mathbf{p} = \hat{\mathbf{p}}. \tag{8}$$

Additionally, if the inputs and outputs are collocated, meaning that $\mathbf{C} = \mathbf{B}^T$, the gradients of the transfer function with respect to the Laplace variable and the parameters are matched too, such that

$$\frac{\partial \mathbf{H}(s, \mathbf{p})}{\partial s} = \frac{\partial \bar{\mathbf{H}}(s, \mathbf{p})}{\partial s} \quad \text{and} \quad \frac{\partial \mathbf{H}(s, \mathbf{p})}{\partial \mathbf{p}} = \frac{\partial \bar{\mathbf{H}}(s, \mathbf{p})}{\partial \mathbf{p}} \quad \text{for} \quad s = \widehat{\sigma}, \quad \mathbf{p} = \widehat{\mathbf{p}}, \quad (9)$$

see [2]. It is also possible to construct a parameterized, reduced order model whose transfer function interpolates the full order transfer function at multiple frequency and parameter expansion points. For that, projection matrices for different frequency and parameter expansion points have to be derived with Eq. (7) and gathered in a concatenated projection matrix.

3 PMOR for Coupled, Parameterized Systems

This section addresses the construction of projection matrices for coupled, parameterized systems. Two approaches for the PMOR of coupled, parameterized systems will be presented. For the sake of simplicity, all derivations will be shown for a coupled system consisting of two component models. However, both methods are also applicable when coupling an arbitrary number of components.

3.1 Governing Equations

First, the coupling of the parameterized component models to derive the parameterized system model is described. Quantities belonging to component 1 or component 2, therefore describing quantities on *component level* will be indicated with the subscript '1' or '2'. Quantities belonging to the coupled system and therewith describing quantities on *system level* will be marked with the subscript 'S'.

The system level equation of motion for two arbitrary components is given by

$$\underbrace{\begin{bmatrix} \mathbf{M}_1(\mathbf{p}_1) & \mathbf{0} \\ \mathbf{0} & \mathbf{M}_2(\mathbf{p}_2) \end{bmatrix}}_{=: \, \mathbf{M}_S(\mathbf{p}_S)} \underbrace{\begin{bmatrix} \ddot{\mathbf{q}}_1 \\ \ddot{\mathbf{q}}_2 \end{bmatrix}}_{=: \, \ddot{\mathbf{q}}_S} + \underbrace{\begin{bmatrix} \mathbf{K}_1(\mathbf{p}_1) + \mathbf{J}_1^T \mathbf{K}_c \mathbf{J}_1 & -\mathbf{J}_1^T \mathbf{K}_c \mathbf{J}_2 \\ -\mathbf{J}_2^T \mathbf{K}_c \mathbf{J}_1 & \mathbf{K}_2(\mathbf{p}_2) + \mathbf{J}_2^T \mathbf{K}_c \mathbf{J}_2 \end{bmatrix}}_{=: \, \mathbf{K}_S(\mathbf{p}_S)} \underbrace{\begin{bmatrix} \mathbf{q}_1 \\ \mathbf{q}_2 \end{bmatrix}}_{=: \, \mathbf{q}_S} \quad (10)$$

$$= \underbrace{\begin{bmatrix} \mathbf{B}_1 & \mathbf{0} \\ \mathbf{0} & \mathbf{B}_2 \end{bmatrix}}_{=: \, \mathbf{B}_S} \underbrace{\begin{bmatrix} \mathbf{u}_1 \\ \mathbf{u}_2 \end{bmatrix}}_{=: \, \mathbf{u}_S}.$$

Here, the parameterized mass and stiffness matrices $\mathbf{M}_1(\mathbf{p}_1), \mathbf{M}_2(\mathbf{p}_2), \mathbf{K}_1(\mathbf{p}_1), \mathbf{K}_2(\mathbf{p}_2)$ of component 1 and component 2 are ordered block diagonal. The system level parameter vector $\mathbf{p}_S = [\mathbf{p}_1, \ \mathbf{p}_2]$ is concatenated by component level parameter vectors. The two components are coupled via stiffnesses $\mathbf{K}_c$ where the distribution matrices $\mathbf{J}_1$ and $\mathbf{J}_2$ extract the coupling degrees of freedom from the nodal displacements and distribute the coupling forces on the structure.

3.2 Reduction on System Level

One approach is to couple the parameterized, full order component models to derive the parameterized full order system model as in Eq. (10). Then, interpolatory methods as described in Sect. 2 can be applied. The system level projection matrix

$$\text{span}(\mathbf{V}_S) = \text{span}([\widehat{\sigma}^2 \mathbf{M}_S(\widehat{\mathbf{p}}_S) + \mathbf{K}_S(\widehat{\mathbf{p}}_S)]^{-1} \mathbf{B}_S) \tag{11}$$

ensures that the parameterized, full order transfer function of the coupled system and the parameterized, reduced order transfer function (and its derivatives for $\mathbf{C}_S = \mathbf{B}_S^T$) match at the frequency and parameter expansion points $\widehat{\sigma}$ and $\widehat{\mathbf{p}}_S$ according to Eqs. (8) and (9).

3.3 Reduction on Component Level

A second approach is to reduce the parameterized component models and to couple the parameterized, reduced order components. For this, only the upper half of the coupled equation of motion (10), which is given by

$$\mathbf{M}_1(\mathbf{p}_1)\ddot{\mathbf{q}}_1 + \mathbf{K}_1(\mathbf{p}_1)\mathbf{q}_1 = \mathbf{B}_1\mathbf{u}_1 + \mathbf{J}_1^T \underbrace{(\mathbf{K}_c\mathbf{J}_2\mathbf{q}_2 - \mathbf{K}_c\mathbf{J}_1\mathbf{q}_1)}_{=: \ \mathbf{u}_c} \tag{12}$$

is considered. In a first step, the coupling forces $\mathbf{u}_c$, which depend on the nodal displacements $\mathbf{q}_1$ and $\mathbf{q}_2$, are regarded as additional system inputs with the additional input matrix $\mathbf{J}_1^T$. Therefore, a projection matrix $\mathbf{V}_1$ for the component level expansion points $\widehat{\sigma}$ and $\widehat{\mathbf{p}}_1$ of component 1 with the augmented input matrix is given by

$$\text{span}(\mathbf{V}_1) = \text{span}(\underbrace{[\widehat{\sigma}^2 \mathbf{M}_1(\widehat{\mathbf{p}}_1) + \mathbf{K}_1(\widehat{\mathbf{p}}_1)]^{-1}}_{=: \ \widehat{\mathbf{G}}_1} [\mathbf{B}_1, \ \mathbf{J}_1^T]) = \text{span}([\widehat{\mathbf{G}}_1\mathbf{B}_1 \ \widehat{\mathbf{G}}_1\mathbf{J}_1^T]). \tag{13}$$

The projection matrix $\mathbf{V}_2$ for component 2 can be derived analogously. Arranging $\mathbf{V}_1$ and $\mathbf{V}_2$ block diagonal yields the separate base reduction (SBR) projection matrix

$$\text{span}(\mathbf{V}_{\text{SBR}}) = \text{span}(\text{blkdiag}(\mathbf{V}_1, \ \mathbf{V}_2)) = \text{span}(\begin{bmatrix} \widehat{\mathbf{G}}_1\mathbf{B}_1 & \widehat{\mathbf{G}}_1\mathbf{J}_1^{\mathsf{T}} & \mathbf{0} & \mathbf{0} \\ \mathbf{0} & \mathbf{0} & \widehat{\mathbf{G}}_2\mathbf{B}_2 & \widehat{\mathbf{G}}_2\mathbf{J}_2^{\mathsf{T}} \end{bmatrix}), \quad (14)$$

(compare to [7, 10]).

3.4 Comparison of Both Approaches

It will be shown in the following that reducing the coupled, parameterized system (10) with $\mathbf{V}_{\text{SBR}}$ yields a meaningful coupled, parameterized, reduced order system model. The following results were first shown in [7, 10] for first order systems. Due to the limited scope of this paper only the general idea is presented. For a more elaborate derivation for second order systems and parameterized second order systems the reader is referred to [6, 11].

First, the system level projection matrix $\mathbf{V}_S$ calculated for the system level expansion points $\widehat{\sigma}$ and $\widehat{\mathbf{p}}_S = [\widehat{\mathbf{p}}_1, \ \widehat{\mathbf{p}}_2]$ from Eq. (11) is rewritten. This rewriting of $\mathbf{V}_S$ is not necessary for an actual application of the system level approach. It is only necessary to enable an insightful comparison of the system level and the component level approach. For rewriting, the Sherman–Morrison-Woodbury formula, [12], is used with

$$\left(\mathbf{L} + \mathbf{UPN}^{\mathsf{T}}\right)^{-1} = \mathbf{L}^{-1} - \mathbf{L}^{-1}\mathbf{U}\left(\mathbf{P}^{-1} + \mathbf{N}^{\mathsf{T}}\mathbf{L}^{-1}\mathbf{U}\right)^{-1}\mathbf{N}^{\mathsf{T}}\mathbf{L}^{-1}, \quad (15)$$

$$\mathbf{L} = \begin{bmatrix} \widehat{\sigma}^2\mathbf{M}_1(\widehat{\mathbf{p}}_1) + \mathbf{K}_1(\widehat{\mathbf{p}}_1) & \mathbf{0} \\ \mathbf{0} & \widehat{\sigma}^2\mathbf{M}_2(\widehat{\mathbf{p}}_2) + \mathbf{K}_2(\widehat{\mathbf{p}}_2) \end{bmatrix} = \begin{bmatrix} \widehat{\mathbf{G}}_1^{-1} & \mathbf{0} \\ \mathbf{0} & \widehat{\mathbf{G}}_2^{-1} \end{bmatrix}, \quad (16)$$

$$\mathbf{UPN}^{\mathsf{T}} = \begin{bmatrix} \mathbf{J}_1^{\mathsf{T}}\mathbf{K}_c\mathbf{J}_1 & -\mathbf{J}_1^{\mathsf{T}}\mathbf{K}_c\mathbf{J}_2 \\ -\mathbf{J}_2^{\mathsf{T}}\mathbf{K}_c\mathbf{J}_1 & \mathbf{J}_2^{\mathsf{T}}\mathbf{K}_c\mathbf{J}_2 \end{bmatrix} = \begin{bmatrix} \mathbf{0} & \mathbf{J}_1^{\mathsf{T}} \\ \mathbf{J}_2^{\mathsf{T}} & \mathbf{0} \end{bmatrix}\begin{bmatrix} \mathbf{K}_c & \mathbf{0} \\ \mathbf{0} & \mathbf{K}_c \end{bmatrix}\begin{bmatrix} -\mathbf{J}_1 & \mathbf{J}_2 \\ \mathbf{J}_1 & -\mathbf{J}_2 \end{bmatrix}. \quad (17)$$

Plugging Eqs. (16) and (17) into Eq. (11) yields

$$\mathbf{V}_S = \underbrace{\begin{bmatrix} \widehat{\mathbf{G}}_1\mathbf{B}_1 & \mathbf{0} \\ \mathbf{0} & \widehat{\mathbf{G}}_2\mathbf{B}_2 \end{bmatrix}}_{=: \ \Delta_B} + \underbrace{\begin{bmatrix} \mathbf{0} & \widehat{\mathbf{G}}_1\mathbf{J}_1^{\mathsf{T}} \\ \widehat{\mathbf{G}}_2\mathbf{J}_2^{\mathsf{T}} & \mathbf{0} \end{bmatrix}}_{=: \ \Delta_J}\mathbf{T}\underbrace{\begin{bmatrix} \mathbf{B}_1 & \mathbf{0} \\ \mathbf{0} & \mathbf{B}_2 \end{bmatrix}}_{=: \ \mathbf{B}_S}. \quad (18)$$

with $\mathbf{T} = \mathbf{T}(\mathbf{K}_c, \widehat{\mathbf{G}}_1, \mathbf{J}_1, \widehat{\mathbf{G}}_2, \mathbf{J}_2)$, see [6].

First, it can be seen that $\mathbf{TB}_S$ does not change the span of Δ_J, meaning that $\text{span}(\Delta_J\mathbf{TB}_S) = \text{span}(\Delta_J)$. Second, it can be seen that the expressions $\widehat{\mathbf{G}}_1\mathbf{B}_1$, $\widehat{\mathbf{G}}_2\mathbf{B}_2$, $\widehat{\mathbf{G}}_1\mathbf{J}_1^{\mathsf{T}}$ and $\widehat{\mathbf{G}}_2\mathbf{J}_2^{\mathsf{T}}$ in the rewritten projection matrix $\mathbf{V}_S$ from Eq. (18) also appear in $\mathbf{V}_{\text{SBR}}$ from Eq. (14). Herewith, it is possible to find a linear combination of the columns of $\mathbf{V}_{\mathrm{SBR}}$ spanning the same space as $\mathbf{V}_{\mathrm{S}}$ meaning that $\mathrm{span}(\mathbf{V}_{\mathrm{S}}) \subset \mathrm{span}(\mathbf{V}_{\mathrm{SBR}})$.

As the most important result, this means that the component level projection matrices $\mathbf{V}_1$ for $\widehat{\sigma}$ and and $\widehat{\mathbf{p}}_1$ for component 1 and $\mathbf{V}_2$ for $\widehat{\sigma}$ and and $\widehat{\mathbf{p}}_2$ for component 2 can be computed independently of each other. However, reducing the coupled, parameterized system from Eq. (10) with $\mathbf{V}_{\mathrm{SBR}} = \mathrm{blockdiag}(\mathbf{V}_1, \mathbf{V}_2)$ ensures to obtain a coupled, parameterized, reduced order system model whose transfer function interpolates the transfer function of the coupled, parameterized, full order system model at $\widehat{\sigma}$ and $\widehat{\mathbf{p}}_S = [\widehat{\mathbf{p}}_1, \widehat{\mathbf{p}}_2]$. As in the non-coupled case it is also possible to derive concatenated component level projection matrices $\mathbf{V}_1$ and $\mathbf{V}_2$ for multiple expansion points in order to improve the approximation quality of the coupled, parameterized, reduced order system model.

The component level approach shows two major advantages. First, the PMOR can be conducted for the parameterized component models and, therefore, for smaller full order models. Second, no information about the coupled system (as the mounting situation or the coupling stiffness $\mathbf{K}_c$) are necessary during the PMOR of the component models. Only the degrees of freedom at which coupling forces might act have to be provided in form of the distribution matrices $\mathbf{J}_1^{\mathrm{T}}$ and $\mathbf{J}_2^{\mathrm{T}}$. It allows therefore to derive parameterized, reduced order component models and to couple them afterwards arbitrarily. However, from Eq. (14) it can be seen that the number of columns of $\mathbf{V}_{\mathrm{SBR}}$ directly depends on the number of columns of the distribution matrices $\mathbf{J}_1^{\mathrm{T}}$ and $\mathbf{J}_2^{\mathrm{T}}$ and therewith on the number of interface degrees of freedom. Therefore, a disadvantage of the component level approach compared to the system level approach is that it yields rather large reduced order models if many coupling degrees of freedom are considered.

4 Application to Shape Optimization in a Coupled Setup

In the following, both approaches are compared in a shape optimization example in a coupled setup. To this end, two geometrically parameterized, component models are generated. They are shown in Fig. 1 both for a reference shape (blue) and for an illustrative shape (yellow).

Geometrically parameterized, solid finite elements as presented in [4] are used to derive the parameterized system matrices for the parameterized shape resulting in full order models with 8 685 degrees of freedom for both components. The curvature of component 1 is parameterized by a quadratic Bézier curve with three parameters. One additional parameter is used to control the height of the component. Component 2 is parameterized with two parameters controlling height and width. The interfaces are modeled with standard Rigid Body Elements (RBE2), see e.g., [9], introducing a rigid connection between the coupling nodes (black) with 6 degrees of freedom each and the interfaces at the boundaries. The usage of RBE2 elements allows an easy coupling of interfaces that do not share coinciding nodes. Additionally, the number of interface degrees of freedom can be reduced. However, this comes at the price of

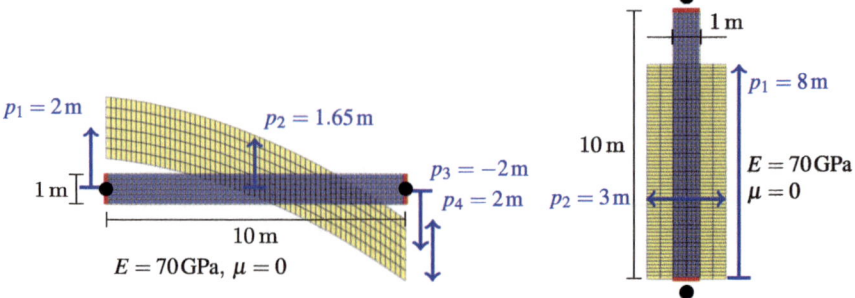

Fig. 1 Component 1 and component 2 in initial shape (blue) and in an changed shape (yellow) with coupling nodes (black), interfaces (red) and material properties

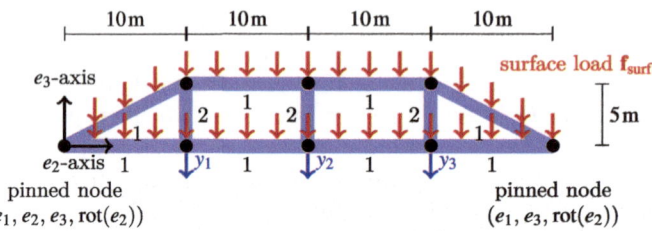

Fig. 2 Coupled system model with numbers of component models, coupling nodes and pinned nodes (black), surface load (red) and displacements y_1 to y_3 (blue)

a stiffer interface and a locally nonrepresentative stress distribution (compare e.g., [5]).

The coupled system is shown in Fig. 2. It can be seen that component 1 is used 8 times in different geometric parameterizations and component 2 is used 3 times resulting in a coupled, parameterized system model with 95 535 degrees of freedom. A detailed view of three aligning components is shown in Fig. 3. It shows that the three RBE2 master nodes coincide. However, they are not coupled by an algebraic constraint but with interface stiffnesses according to Sect. 3.1. This means that every RBE2 master node keeps its six independent degrees of freedom. The interface stiffness is chosen five orders of magnitude larger compared to component models in order to approximate a rigid connection between the coinciding RBE2 master nodes. The structure is subjected to a surface load $\mathbf{f}_{\text{surf}}$ acting on the horizontal components.

The objective is to adapt the shape of the components (and therefore the shape of the coupled system) in order to minimize the displacements y_1 to y_3. The optimization problem reads

$$\underset{\mathbf{p}_S}{\text{minimize}} \quad J = \left(\mathbf{C}_S \mathbf{K}_S(\mathbf{p}_S)^{-1} \mathbf{f}_{\text{surf}}\right)^2$$

$$\text{with} \quad \mathscr{P} := \{\mathbf{p}_S \in \mathbb{R}^{38} \mid m(\mathbf{p}_S) \le m(\mathbf{p}_{S0}) \mid \mathbf{Ab} = \mathbf{0} \mid \mathbf{p}_{Sl} \le \mathbf{p}_S \le \mathbf{p}_{Su}\}. \tag{19}$$

Fig. 3 Detailed view of three aligning components with three coinciding, but independent, RBE2 master nodes (black) which are connected by stiffnesses (not shown)

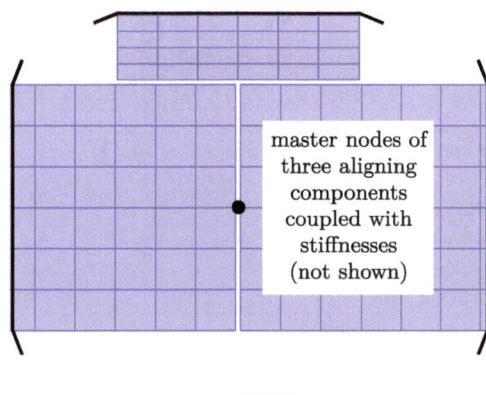

master nodes of three aligning components coupled with stiffnesses (not shown)

Fig. 4 Shape of the optimized, coupled system model

The equality constraint $\mathbf{Ab} = \mathbf{0}$ ensures that the lower, horizontal components remain straight and that the components only adapt their shape in a geometrically compliant way. Additionally, the total mass of the coupled structure $m(\mathbf{p}_S)$ must not exceed the initial mass $m(\mathbf{p}_{S_0})$. For the underlying system description of the coupled system, the input vector $\mathbf{B}_S \in \mathbb{R}^{95\,535}$, and the transposed output vector $\mathbf{C}_S^T \in \mathbb{R}^{95\,535}$ extracting the sum of the displacements are concatenated. Therewith, the special case of collocated inputs and outputs is created artificially to yield a parameterized, reduced order system with the properties of Eq. (9). The optimization problem is now solved with the coupled, parameterized, full order system model with the `matlabR2015b` SQP solver `fmincon`. The optimized shape of the coupled system is shown in Fig. 4 and it can be seen that the optimal shape is similar to the shape of a tied-arch bridge. The computation time for solving Eq. (19) and calculating the gradients during the optimization is 105.0 s, see Table 1.

In the following, the optimization problem is also solved with coupled, parameterized, reduced order system models where both the system level approach from Sect. 3.2 and the component level approach from Sect. 3.3 are used to conduct the PMOR. First, the reduction settings for both approaches are explained which is followed by a comparison and a discussion. The reduction settings and the results are later summarized in Table 1. Note that the focus of this contribution is only about highlighting the general advantages and disadvantages of the two PMOR approaches for coupled systems and not about the selection of expansion points or about interface reduction.

In the first approach, the system level approach, the parameterized, full order component models are coupled with stiffnesses at the RBE2 master nodes according to Sect. 3.1. The reduction is then conducted with the coupled, parameterized, full order system model meaning that the PMOR is done for a model with 95 535 degrees

Table 1 Reduction settings and computation time in online and offline phase

	Full order model	PMOR on system level	PMOR on component level	
			Component 1	Component 2
Order of full model	95 535	95 535	8 685	8 685
Parameter expansion points	–	80	16	4
System inputs and outputs	2	2	13	12
Order of reduced components	–	–	104	34
Order of coupled, reduced system	–	152	934	
Computation time reduction	–	269.2 s	10.8 s	
Computation time optimization	105.0 s	0.1 s	1.2 s	

of freedom. The columns of the system level projection matrix $\mathbf{V}_S$ are derived with Eq. (11) where all frequency expansion points are chosen as $\widehat{\sigma} = 0$ as a static problem is investigated. The expansion points are chosen randomly from the feasible design space $\mathbf{p}_S \in \mathbb{P}$. In this example, 80 parameter expansion points yields a model that is accurate enough. So, the order of the reduced system becomes $80 \cdot 2 - 4 = 156$, as 4 columns of $\mathbf{V}_S$ are truncated due to rank deficiency. The computation time for calculating the projection matrix is 269.2 s. The computation time for solving Eq. (19) and calculating the gradients during the optimization is 0.1 s. Note that choosing random parameters as expansion points is a rather brute force approach for prototyping. More practical approaches could involve e.g., Latin Hypercube sampling, see [8], or an approach in which the base $\mathbf{V}_S$ is enriched in an iterative, error controlled process, compare [4].

In the second approach, the component level approach, one projection matrix for each component model has to be derived. Afterwards, the parameterized, reduced order components are coupled with stiffnesses at the RBE2 master nodes.

The stiffness matrices $\mathbf{K}_1(\mathbf{p}_1)$ and $\mathbf{K}_2(\mathbf{p}_2)$ of the unconstrained component models are just positive semi-definite but not positive definite meaning that rigid body deformations are left. Therefore, the frequency expansion points are chosen as $\widehat{\sigma} = 2\pi \mathrm{j} 5 Hz$. The parameter expansion points for the component models are distributed as regular grids within the feasible design spaces of the component models. For both models two grid points per dimension are selected as this already delivers (in this example) accurate enough reduced order component models. Both components have twelve coupling inputs where component 1 has one additional system input to consider the surface load $\mathbf{f}_{surf}$. After truncating 104 columns due to rank deficiency the reduced order of component 1 is $4^2 \cdot 13 - 104 = 104$. Here, a relatively large number of columns is truncated. The reason is that the geometry parameterization for component 1 introduced in Fig. 1 allows one rigid body movement design variation. However, such a design variation will not change the transfer behavior of the elastic component and will therefore also only add rank deficient columns to $\mathbf{V}_1$. The reduced order of component 2 is $2^2 \cdot 12 - 14 = 34$ after truncating 14 rank deficient columns. The computation time for deriving the projection matrices for both components is 10.8 s.

Using component 1 eight times and component 2 three times yields the coupled, parameterized reduced order system model with $8 \cdot 104 + 3 \cdot 34 = 934$ degrees of freedom. The computation time for solving Eq. (19) and calculating the gradients during the optimization is 1.2 s.

Both approaches are successfully applied in a shape optimization example. In the system level approach, PMOR has to be conducted for a single system level model with 95 535 degrees of freedom. In contrast, in the component level approach, two parameterized component models have to be reduced. However, the reduction can be performed with much smaller models of an order of only 8 685 for both component 1 and component 2. This means that the component level approach requires much less effort during the offline phase. In this case, the reduction on component level is about 25 times faster than the reduction on system level. The drawback is that the component level approach requires to take the coupling degrees of freedom as additional system inputs into account. This leads to larger reduced order component models and, therefore, to larger reduced order system models compared to the system level approach. Consequently, the optimization problem is solved 16 times slower with reduced order model obtained by the component level approach compared to the reduced order model obtained by the system level approach. Still, the computation times for both approaches are faster compared to the full order solution. However, the larger reduced order model when using the component level approach might become an issue in dynamical problems.

Though, the component level approach shows two additional major advantages. First, the PMOR of component models is much simpler to conduct as the PMOR for system models as the component models comprise less parameters. Therefore, the user has more physical insight into the problem and it is a much simpler task to select expansion points or to conduct an error analysis. Second, the component level approach allows to derive a data base of parameterized, reduced order component models and to combine them arbitrarily. It is not necessary to redo the PMOR if the modular setup is changed and allows therefore a reduced order toolbox based modeling or shape or topology optimization.

5 Conclusion and Outlook

Two approaches for Parametric Model Order Reduction of coupled, parameterized systems are presented. Both approaches are appropriate for the reduction of parameterized, coupled systems and were applied in a reduced order shape optimization example. The system level approach showed better performance in the online phase as it allows to derive smaller reduced order system models. The component level approach required much less effort for obtaining the reduced order component models. A further advantage of this approach is that the parameterized, reduced order component models can be reused and recombined arbitrarily enabling a reduced order, toolbox based shape or topology optimization. A drawback of the component level approach is that the coupled, parameterized reduced order system models become larger compared to the system level approach.

A promising topic for future investigations is the application of reduction strategies to systems with a large number of system inputs and outputs or a better selection of expansion points.

Acknowledgements The authors gratefully thank the German Research Foundation (DFG) for the support of this research work within the collaborative research centre SFB/CRC 1244, "Adaptive Skins and Structures for the Built Environment of Tomorrow" with the projects B01 and A04.

References

1. Bai, Z., Su, Y.: Dimension reduction of large-scale second-order dynamical systems via a second-order Arnoldi method. SIAM J. Sci. Comput. **16**, 1692–1709 (2005)
2. Baur, U., Beattie, C., Benner, P., Gugercin, S.: Interpolatory projection methods for parameterized model reduction. SIAM J. Sci. Comput. **33**(5), 2489–2518 (2011)
3. Feng, L.H., Rudnyi, E.B., Korvink, J.G.: Preserving the film coefficient as a parameter in the compact thermal model for fast electrothermal simulation. IEEE Trans. Comput.-Aided Des. Integr. Circuits Syst. **24**(12), 1838–1847 (2005)
4. Fröhlich, B., Gade, J., Geiger, F., Bischoff, M., Eberhard, P.: Geometric element parameterization and parametric model order reduction in finite element based shape optimization. Comput. Mech. **63**(5), 853–868 (2019). Springer, Berlin, Heidelberg
5. Heirmann, G.H.K., Desmet, W.: Interface reduction of flexible bodies for efficient modeling of body flexibility in multibody dynamics. Multibody Syst. Dyn. **24**(2), 219–234 (2010)
6. Holzwarth, P., Eberhard, P.: Interpolation and truncation model reduction techniques in coupled elastic multibody systems. In: Proceedings of the ECCOMAS Thematic Conference on Multibody Dynamics, pp. 366–378 (2015)
7. Lutowska, A.: Model order reduction for coupled systems using low-rank approximations. Doctoral Thesis, Eindhoven University of Technology, Netherlands (2012)
8. McKay, M.D., Beckman, R.J., Conover, W.J.: Comparison of three methods for selecting values of input variables in the analysis of output from a computer code. Technometrics **21**(2), 239–245 (1979)
9. Nastran: Reference Manual. MSC Software Corporation, Newport Beach (2016)
10. Schilders, W.H.A., Lutowska A.: A novel approach to model order reduction for coupled multiphysics problems. In: Quarteroni, A., Rozza, G., (ed.), Reduced Order Methods for Modeling and Computational Reduction, pp. 1–49. Springer, Berlin (2014)

11. Walker, N., Fröhlich, B., Eberhard, P.: Model order reduction for parameter dependent substructured systems using Krylov subspaces. In: Proceedings of the 9th Vienna International Conference on Mathematical Modelling, Vienna (2018)
12. Woodbury, M.A.: In: Kuntzmann, J. (ed.) Inverting Modified Matrices. No. 42 in Statistical Research Group Memorandum Reports. Princeton University, Princeton (1950)

A Novel Penalty-Based Reduced Order Modelling Method for Dynamic Analysis of Joint Structures

Jie Yuan, Loic Salles, Chian Wong and Sophoclis Patsias

Abstract This work proposes a new reduced order modelling method to improve the computational efficiency for the dynamic simulation of a jointed structures with localized contact friction non-linearities. We reformulate the traditional equation of motion for a joint structure by linearising the non-linear system on the contact interface and augmenting the linearised system by introducing an internal non-linear penalty variable. The internal variable is used to compensate the possible non-linear effects from the contact interface. Three types of reduced basis are selected for the Galerkin projection, namely, the vibration modes (VMs) of the linearised system, static modes (SMs) and also the trial vector derivatives (TVDs) vectors. Using these reduced basis, it would allow the size of the internal variable to change correspondingly with the number of active non-linear DOFs. The size of the new reduced order model therefore can be automatically updated depending on the contact condition during the simulations. This would reduce significantly the model size when most of the contact nodes are in a stuck condition, which is actually often the case when a jointed structure vibrates. A case study using a 2D joint beam model is carried out to demonstrate the concept of the proposed method. The initial results from this case study is then compared to the state of the art reduced order modeling.

Keywords Nonlinear Reduced Order Modelling · Structural dynamics · Mechanical joints · Component mode synthesis · Contact friction

J. Yuan (✉) · L. Salles
Department of Mechanical Engineering, Vibration University Technology Centre,
Imperial College London, London, UK
e-mail: Jie.Yuan@imperial.ac.uk

L. Salles
e-mail: L.Salles@imperial.ac.uk

C. Wong · S. Patsias
Rolls Royce plc, Derby, UK
e-mail: Chian.Wong@Rolls-Royce.com

S. Patsias
e-mail: Sophoclis.Patsias@Rolls-Royce.com

© Springer Nature Switzerland AG 2020   165
J. Fehr and B. Haasdonk (eds.), *IUTAM Symposium on Model Order Reduction
of Coupled Systems, Stuttgart, Germany, May 22–25, 2018*, IUTAM Bookseries 36,
https://doi.org/10.1007/978-3-030-21013-7_12

1 Introduction

Jointed structures have been widely used in gas turbine engine to transfer the loading from one component to the other. The joints between the substructures such as the shrouds, underplatform damper and dovetail joint in gas turbine engines are also regarded as the primary damping sources for energy dissipation. However, they also significantly complicate the dynamic behaviour of an assembly by the change of stability, the jump phenomenon and energy localization [4]. It is therefore very important to understand and improve the dynamics of the joints for a improved design. The use of finite element (FE) method for analysing such dynamical systems is however often impeded by the unacceptable computational expense due to the tremendous size of the model and strong inherit contact friction nonlinearities. The harmonic balanced method (HBM) provides a very efficient approach to obtain the steady state dynamic behaviour of such jointed structures comparing to the time integration method [4]. However, HBM would expand the size of the orginal system by multiplying the chosen number of harmonic coefficients [6]. One of the viable approaches to take these nonlinearities into account is to reduce the model size by several orders of magnitudes by employing reduced order modelling (ROM) techniques [4, 6, 12].

Component mode synthesis (CMS) techniques have been extensively used for model order reduction for linear and localised nonlinear dynamic systems where the physical nonlinear DOFs on the interface are retained as unknowns [8, 13]. A CMS-Hybrid approach based on free interface modes and flexible residual have been successfully applied to the Imperial VUTC in-house FORSE solver [6]. Another effective CMS approach is on the use of Craig-Bampton (CB) method [2]. A review of CMS based ROM techniques for the applications to the linear vibration and localized nonlinear vibration can be referred to [3, 12]. The main drawback of these two approaches is the size of the reduced model is proportional to the number of DOFs involved in nonlinearities. It could become extremely large when nonlinear interface regions are intensive and densely meshed [10, 12]. In terms of modeling contact friction on the interface, a node-to-node modeling approach has been widely used [1] and also experimentally validated for turbine underplatform damper at Imperial VUTC [5]. The contact friction conditions can be described as in stuck, slip and gap states. More details on the contact friction modelling would be introduced in Sect. 2. Figure 1 shows the forced frequency response of a turbine blade and also the average contact conditions of the interface nodes during the non-linear dynamic analysis. An interesting observation is that most of the contact nodes are actually in a stuck condition under vibrational loads. When the contact interface is in a stuck condition, the coupling between two contact interfaces can be represented as the linear springs. An inspiration from this observation is that we can linearise the non-linear system using linear springs, and compensate the non-linear effect from those contact nodes in a slip or gap condition with a internal penalty variable.

This paper aims to investigate this novel penalty-based ROM approach in order to further reduce the model size comparing to those reduced models using classical CMS methods. The paper is organized as follows: The reformulated equation of

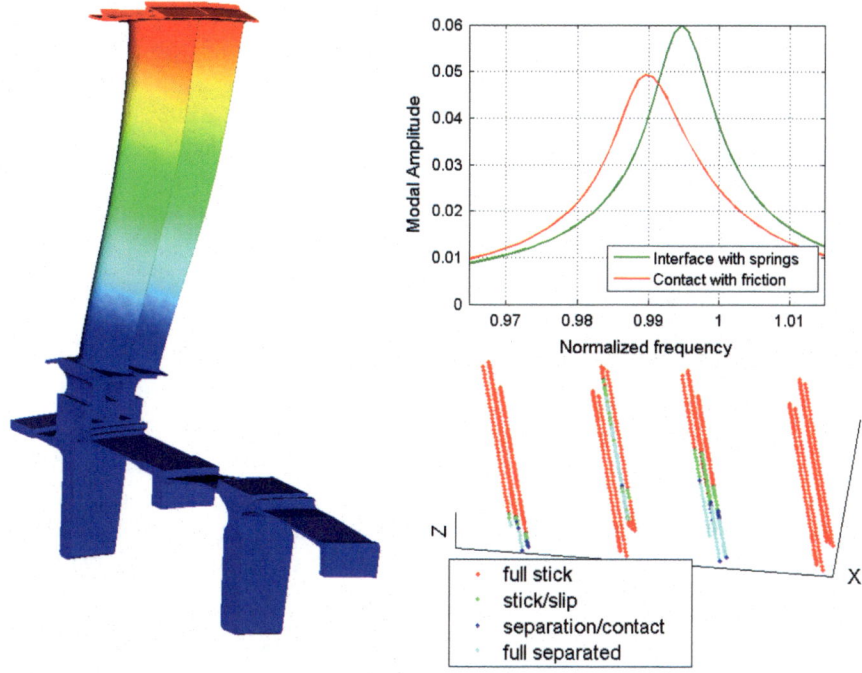

Fig. 1 An example of the contact conditions of a turbine during the vibrations

motion (EOM) for this penalty approach will be firstly presented; it is followed by the presentation of reduced basis for Galerkin projection; we will then elaborate how this method can be coupled with the harmonic balanced method; the performance of this approach will be then demonstrated using a 2D jointed beam case study.

2 Formulation

2.1 Equation of Motion

A dynamic system consisting of two connected substructures with localized contact friction interfaces is considered as an example. The partial differential governing equation of such a system without taking into account of damping matrix is:

$$\mathbf{M}\ddot{u} + \mathbf{K}u = F - F_{nl}(u, \ddot{u}) \tag{1}$$

It is assumed the size of each substructures is N and the size of contact DOFs for each substructure is M. Using the finite element modelling method, the assembled system with two substructures can be expressed as:

$$\begin{bmatrix} \mathbf{M_1} & \mathbf{0} \\ \mathbf{0} & \mathbf{M_2} \end{bmatrix} \begin{bmatrix} \ddot{u}_1 \\ \ddot{u}_2 \end{bmatrix} + \begin{bmatrix} \mathbf{K_1} & \mathbf{0} \\ \mathbf{0} & \mathbf{K_2} \end{bmatrix} \begin{bmatrix} u_1 \\ u_2 \end{bmatrix} = \begin{bmatrix} F_1 \\ F_2 \end{bmatrix} - (\begin{bmatrix} \mathbf{B_1^T} \\ \mathbf{B_2^T} \end{bmatrix} F^T(u_1, u_2) + \begin{bmatrix} \mathbf{B_1^N} \\ \mathbf{B_2^N} \end{bmatrix} F^N(u_1, u_2))$$

$$(2)$$

where $\mathbf{M_1}, \mathbf{M_2}, \mathbf{K_1}, \mathbf{K_2}$ are the mass and stiffness matrix of two substructures with the same dimensions of $N x N$; F^T and F^N are the $2M x 1$ non-linear contact friction force vector, which are in a function of the relative displacement of the contact interface in a joint structure. $\mathbf{B^T}$ and $\mathbf{B^N}$ are the $2N x 2M$ boolean matrix related to the boundary DOFs in tangential and normal direction separately.

2.2 Contact Friction Modelling

A node-to-node approach with Jenkin model is used to model the contact friction phenomenon in a joint, which includes stuck, slip and gap states. These contact friction states are dependent on the preloading levels as well as the amplitude of relative movements on the interface. The Jenkin model has been widely used and also validated with experiments [1]. The formulation for a 3D contact node pair using this Jenkin model can be written as:

$$F_x^T = \begin{cases} k_t(\Delta x - \Delta x_c) + N_x^0, \text{ stuck} \\ \mu F_z^N \sin(\theta), \text{ slip} \\ 0, \text{ gap} \end{cases} \tag{3}$$

$$F_y^T = \begin{cases} k_t(\Delta y - \Delta y_c) + N_y^0, \text{ stuck} \\ \mu F_z^N \cos(\theta), \text{ slip} \\ 0, \text{ gap} \end{cases} \tag{4}$$

$$F_z^N = \begin{cases} k_n \Delta z + N_z^0, \text{ stuck} \\ k_n \Delta z + N_z^0, \text{ slip} \\ 0, \text{ gap} \end{cases} \tag{5}$$

where Δx, Δy and Δz are the time-dependent tangential and normal relative displacement of a contact node pair; Δx_c, Δy_c are internal variables representing the tangential position of the slider, which also needs to update at each time step; F_x^T, F_y^T are the tangential force in x and y direction; F_z^N is the normal force in z direction; θ is the angle between the tangential force component in x and y direction, which is determined by the predicted tangential force:

$$\theta = arctan(\frac{k_t(\Delta x - \Delta x_c) + N_x^0}{k_t(\Delta y - \Delta y_c) + N_y^0}) \tag{6}$$

The stuck condition occurs when the predicted tangential force $\sqrt{F_x^{T\,2} + F_y^{T\,2}}$ is less than the critical slipping force μF_z^N. The contact force would behave linearly and there would be no energy dissipation. The slip condition occurs when the predicted tangential friction force is larger than the critical value μF_z^N. The tangential contact force would behave non-linearly with a value of μF_z^N and the energy dissipation would happen then. The gap condition would happen when predicted normal force F_z^N is less than zero, and all the contact force would be zero then.

3 Penalty-Based Approach

3.1 EOM Modification

For the proposed penalty-based approach, the original EOM in Eq. 2 is linearised on the interface using contact stiffness by assuming all of the contact nodes are in a stuck condition. The modified linearised EOM of the system can be expressed as:

$$
\begin{bmatrix} \mathbf{M}_1 & \mathbf{0} \\ \mathbf{0} & \mathbf{M}_2 \end{bmatrix} \begin{bmatrix} \ddot{u}_1 \\ \ddot{u}_2 \end{bmatrix} + \begin{bmatrix} \mathbf{K}_1 + \mathbf{K}_{\text{Joint}} & -\mathbf{K}_{\text{Joint}} \\ -\mathbf{K}_{\text{Joint}} & \mathbf{K}_2 + \mathbf{K}_{\text{Joint}} \end{bmatrix} \begin{bmatrix} u_1 \\ u_2 \end{bmatrix} = \begin{bmatrix} F_1 \\ F_2 \end{bmatrix} \tag{7}
$$

where $\mathbf{K}_{\text{Joint}}$ is the $N x N$ stiffness matrix containing the local stiffness matrix associated to the joint DOFs. The linearised stiffness matrix in Eq. 7 is denoted as $\mathbf{K}_{\text{linearized}}$.

When any contact nodes are in a slip or gap condition, an internal variable Δp would be needed to augment linearised EOM. The dimension of Δp is $M x 1$, which is the half number of the total joint DOFs. The internal variable would become zero when a contact node are in a stuck condition. The expression of the internal variable can be formulated as:

$$
\Delta p = \begin{cases} 0, \text{ stuck} \\ \mathbf{K}_{\text{Joint}}^{\text{nl,nl}\,-1}(F_{nl}(\Delta u) - \mathbf{K}_{\text{Joint}}^{\text{nl}} \Delta u), \text{ slip} \\ \mathbf{K}_{\text{Joint}}^{\text{nl,nl}\,-1}(F_{nl}(\Delta u) - \mathbf{K}_{\text{Joint}}^{\text{nl}} \Delta u), \text{ gap} \end{cases} \tag{8}
$$

where $\mathbf{K}_{\text{Joint}}^{\text{nl,nl}}$ is the $M x M$ joint stiffness matrix associated to non-linear internal variable; $\mathbf{K}_{\text{Joint}}^{\text{nl}}$ is the $N x M$ joint matrix relating to the DOFs in each substructure. Δu is the assembly of relative displacement (Δx, Δy, Δz) of all the contact pairs in joint interfaces. By integrating the internal penalty variable, the modified EOM can be further augmented:

$$
\begin{bmatrix} \mathbf{M}_1 & \mathbf{0} & \mathbf{0} \\ \mathbf{0} & \mathbf{M}_2 & \mathbf{0} \\ \mathbf{0} & \mathbf{0} & \mathbf{0} \end{bmatrix} \begin{bmatrix} \ddot{u}_1 \\ \ddot{u}_2 \\ \Delta \ddot{p} \end{bmatrix} + \begin{bmatrix} \mathbf{K}_1 + \mathbf{K}_{\text{Joint}} & -\mathbf{K}_{\text{Joint}} & \mathbf{K}_{\text{Joint}}^{\text{nl}} \\ -\mathbf{K}_{\text{Joint}} & \mathbf{K}_2 + \mathbf{K}_{\text{Joint}} & -\mathbf{K}_{\text{Joint}}^{\text{nl}} \\ \mathbf{K}_{\text{Joint}}^{\text{nl}\prime} & -\mathbf{K}_{\text{Joint}}^{\text{nl}\prime} & \mathbf{K}_{\text{Joint}}^{\text{nl,nl}} \end{bmatrix} \begin{bmatrix} u_1 \\ u_2 \\ \Delta p \end{bmatrix} = \begin{bmatrix} F_1 \\ F_2 \\ F_{nl}(u_1, u_2) \end{bmatrix} \tag{9}
$$

Here, $\mathbf{M_{New}}$, $\mathbf{K_{New}}$ are used to denote as new assembled mass and stiffness matrix. It is worth noting that the zero part of Δp associated to the contact nodes in a stuck condition can be further eliminated in EOM. The following section would detail how the ROM formulates to enable such an automatic updating.

3.2   Reduced Basis

Galerkin projection is used to reduce the size of a physical model by transforming it into a subspace. The solution of the system can be expressed as a linear combination of vectors spanning the subspace. The selection of the reduced basis is crucial in determining the accuracy and computational efficiency of a reduced system. More about Galerkin projection can be referred to [7, 12]. The reduced basis for the proposed penalty approach contain three parts, namely the vibrational modes of the corresponding linearised system, constrain (static) modes and also the modal derivatives vectors (Trial Vector Derivatives). The vibration modes can be obtained by solving the eigenvalue problem of the linearised system Eq. 2:

$$- \omega^2 \begin{bmatrix} \mathbf{M_1} & 0 \\ 0 & \mathbf{M_2} \end{bmatrix} [\phi] + \begin{bmatrix} \mathbf{K_1} + \mathbf{K_{Joint}} & -\mathbf{K_{Joint}} \\ -\mathbf{K_{Joint}} & \mathbf{K_2} + \mathbf{K_{Joint}} \end{bmatrix} [\phi] = [0] \qquad (10)$$

Like CMS techniques, the static modes are used to approximate the high frequency response on the contact interface, which can be obtained by applying unit displacement vectors on the DOFs related to internal penalty variable.

$$\begin{bmatrix} \mathbf{K_1} + \mathbf{K_{Joint}} & -\mathbf{K_{Joint}} & -\mathbf{K_{Joint}^{nl}} \\ -\mathbf{K_{Joint}} & \mathbf{K_2} + \mathbf{K_{Joint}} & \mathbf{K_{Joint}^{nl}} \\ -\mathbf{K_{Joint}^{nl'}} & \mathbf{K_{Joint}^{nl'}} & \mathbf{K_{Joint}^{nl,nl}} \end{bmatrix} \begin{bmatrix} \varphi \\ \mathbf{I} \end{bmatrix} = \begin{bmatrix} 0 \\ \mathbf{R} \end{bmatrix} \qquad (11)$$

where $\mathbf{I}$ is the $M \times M$ identity matrix; φ is the $2N \times M$ matrix including all the static modes; $\mathbf{R}$ are the $M \times M$ reaction force matrix.

TVDs are used to calibrate the linear reduced basis in order to consider the effect of the non-linearities from the contact friction. It is particularly usefully when all contact nodes are in a gap condition. The detailed formulation to calculate the TVDs can be referred to [11]. The first-order modal derivatives of the linear reduced basis can be calculated as follows:

$$\varphi_{i,j} = \mathbf{K}_{linearised}^{-1} \frac{\partial \mathbf{K}_{linearised}}{\partial \mathbf{q_j}} \varphi_j, \ \phi_{i,j} = \mathbf{K}_{linearised}^{-1} \frac{\partial \mathbf{K}_{linearised}}{\partial \mathbf{q_j}} \phi_j \qquad (12)$$

Where $\varphi_{i,j}$ and $\phi_{i,j}$ are the TVDs from the linear vibration modes ϕ_j and constrain modes φ_j. The number of TVD vectors is equal to the squared number of the linear reduced basis. The proper orthogonal decomposition is then used to reduced the size of TVDs.

The transformation of the physical system to the modal domain can be shown as follows:

$$
\begin{bmatrix} u_1 \\ u_2 \\ \Delta p \end{bmatrix} = \begin{bmatrix} \phi\ \phi_{\mathbf{T}}\ \varphi \\ \mathbf{0}\ \ \mathbf{0}\ \ \mathbf{I} \end{bmatrix} \begin{bmatrix} \eta_1 \\ \eta_2 \\ \Delta p \end{bmatrix} = \begin{bmatrix} \phi\ \phi_{\mathbf{T}}\ \varphi \\ \mathbf{0}\ \ \mathbf{0}\ \ \mathbf{I} \end{bmatrix} \begin{bmatrix} \mathbf{I}\ \mathbf{0}\ \mathbf{0} \\ \mathbf{0}\ \mathbf{I}\ \mathbf{0} \\ \mathbf{0}\ \mathbf{0}\ \mathbf{B} \end{bmatrix} \begin{bmatrix} \eta_1 \\ \eta_2 \\ \Delta p_R \end{bmatrix} \tag{13}
$$

$$
\mathbf{T} = \begin{bmatrix} \phi\ \phi_{\mathbf{T}}\ \varphi \\ \mathbf{0}\ \ \mathbf{0}\ \ \mathbf{I} \end{bmatrix} \begin{bmatrix} \mathbf{I}\ \mathbf{0}\ \mathbf{0} \\ \mathbf{0}\ \mathbf{I}\ \mathbf{0} \\ \mathbf{0}\ \mathbf{0}\ \mathbf{B} \end{bmatrix}, \ \mathbf{M_R} = \mathbf{T}'\mathbf{M}_{New}\mathbf{T}, \ \mathbf{K_R} = \mathbf{T}'\mathbf{K}_{New}\mathbf{T} \tag{14}
$$

where η_1 and η_2 are modal participation factors for vibration models and TVDs; Δp_R is the non-zero part of Δp; $\mathbf{B}$ is the index matrix to extract the non-zero part of Δp, which would be updated during the simulation depending on the contact conditions; the size of transformation matrix $\mathbf{T}$ would be also updated accordingly. As a result, the reduced mass and stiffness system $\mathbf{K_R}$ and $\mathbf{M_R}$ would be adaptively changing to reduce the computational time.

4 Harmonic Balanced Method with Continuation Techniques

Harmonic balanced method is used for solving the Eq. 7. The idea of this method is to represent the steady state non-linear dynamic response using truncated Fourier series with n harmonic series:

$$
u(t) = \tilde{u}_0 + \sum_{i=1}^{n} (\tilde{u}_i^c \cos m_i \omega t + \tilde{u}_i^s \sin m_i \omega t) \tag{15}
$$

where $\tilde{u}_i^{c,s}$ are cosine and sine harmonic coefficients; ω is the principal vibration frequency; $\tilde{u}_0$ is the zero harmonic response. The Newton-Raphson method, in coupling with the alternating frequency-time (AFT) method, is used to solve these nonlinear equations. The AFT technique is used to transform the frequency-domain solution to the time domain for non-contact force calculation, and transforme non-linear contact force back to frequency domain. More details about this part can be referred to [4]. Figure 2 illustrates the implementation process about how the contact friction model,iterative Newton-Raphson solver work with reduced order modeling in HBM. The physical nonlinear DOFs u_b on the contact interface is firstly expanded from modal subspace by $\mathbf{T}$ before employing the AFT procedure. After AFT procedure, the nonlinear force F_{nl} is then projected back to the modal subspace. The continuation techniques are then used to obtain the forced frequency response. More details about continuation techniques can be referred to [9].

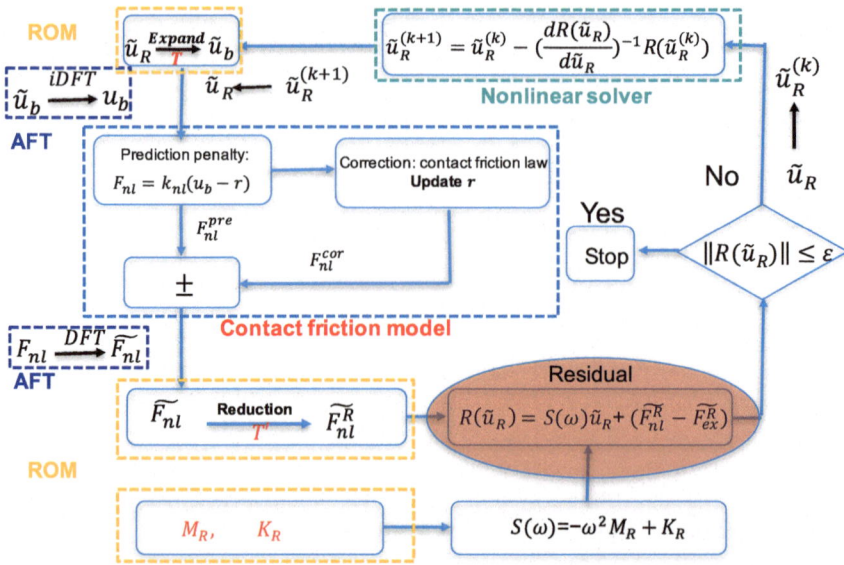

Fig. 2 An illustration of how Newton-Raphson solver works with AFT scheme and ROM methods

The automatic size updating depending on the contact condition is achieved as follows. When evaluating the continuous forced frequency response, the contact condition of all interface nodes would need to be assessed using the previous converged solution. This would help to generate the index matrix **B**. The size of system can be then reduced for the new simulation. If using the continuation techniques with HBM, the size updating would be performed at the predictor stage. The size of the system would be kept same in the corrector stage with the iterative solver. The implementation of the size updating with HBM is still on-going research.

5   Case Study

Figure 3 shows a jointed beam model with linear springs connecting the two equivalent beam substructures. The length of each beam is 0.3 m. The width and height of the cross section is 25 mm and 6 mm respectively. They are modelled by using the Euler-Bernoulli beam elements, where each node has three DOFs (u_x, u_y, r_z). The beams are made of steel with a nominal density of 7850 kg/m^3 and Young's modulus of 2.1e11 Nm^{-2}. The tangential stiffness of the springs in the joint is 1e4 N/m while normal contact stiffness is 5e6 N/m and bending stiffness of 8e6 Nm/rad.

Figure 4 shows the first nine natural frequencies (NFs) and modes of this linearised jointed beam system. These nine modes all belong to the bending modes. Due to the

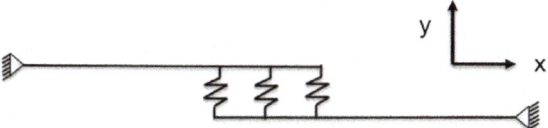

Fig. 3 A 2D FE model of a jointed beam with contact non-linear springs

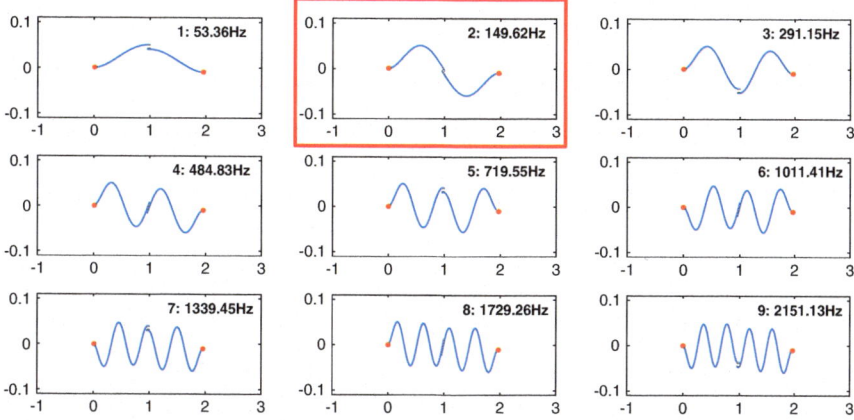

Fig. 4 Modes of a linearised jointed beam

large value of stiffness used in the joint, local elastic modes in the joint do not appear. The forced frequency response of the second bending mode would be studied.

The proposed method and also other reference methods are applied to the linearised joint beam model at first. The reference methods include CB, Rubin, Dual Craig-Bampton, MacNeal, Joint interface method and TVD methods. The formulation of these methods can be referred to [12]. The idea here is to compare the quality of these methods when all the contact nodes are in a stuck condition. Figure 5 shows the comparison of the natural frequency (NF) errors between the proposed method and other reference methods. Except for the penalty method, all other methods have the same number of normal modes, namely 20. In terms of the static modes, CB, Rubin and MacNeal methods have the same number as the non-linear DOFs while DCB and JIM methods have only half number of these non-linear DOFs. The static modes with TVD method is independent of the non-linear DOFs. In this case, the size of penalty method would be equal to the number of VMs in the linearised structure, because the number of static modes is zero due to the stuck condition. Figure 5 shows the proposed method achieves the best accuracy with the smallest reduction ratio. RR stands for the reduction ratio, which is the ratio between the size of a ROM to the size of a full system.

For linear analysis, it is well known that the pre-processing effort would be the main challenge when using the CMS techniques, because the inversion of matrix during the dynamic analysis would be only needed once for each frequency. For non-

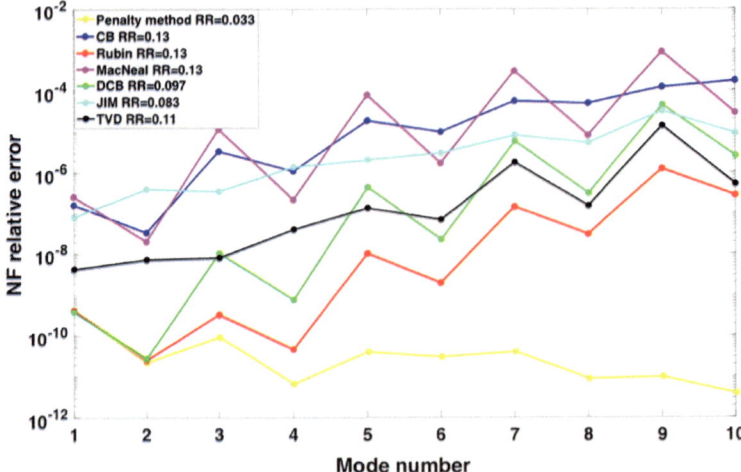

Fig. 5 NF relative errors of the jointed beams between different ROM methods (RR:reduction ratio)

linear analysis, the reduced size of system would be also a challenge especially when the assemble structure contains intensive contact interfaces. It is because the iterative solver would be needed to solve the reduced system until the solution converges. Furthermore, using the HBM, the final size of system would be expanded by the number of harmonics. Based on the authors' previous simulation, the computational time would be cubic relation to the reduced number of DOFs [5]. Therefore, the RR would be particularly important to non-linear vibration analysis. In terms of the off-line cost, the proposed method has the same computational time for pre-processing when comparing to the classical CMS techniques.

For nonlinear analysis, the linear springs on the contact interface are replaced by using Jenkin contact friction model. Figure 5 shows the comparison of the frequency response functions (FRFs) between the proposed method and Rubin, CB methods. The structure is excited in the middle of the first substructure in the y direction. The two structure would be separated if the excitation level was large, which would activate the soften effect of the in-phase and out-of-phase bending modes. For Rubin and CB methods, the number of nominal modes is 10 while the number of static modes is equal to the number of non-linear DOFs in the joint. Figure 6 shows the peak of the out-of-phase resonance shafts on the left but the amplitude of the response remain unchanged. It means that the jointed structure experiences the separation in resonance frequency region. CB and Rubin have the same FRFs as that from the full solution. Using 10 vibrational modes and 3 static modes, the penalty approach leads to the noticeable errors. When increasing the number of vibration modes to 20, the accuracy with the proposed method improves but one still can observe the clear discrepancy. This is because the introduced non-linear force on the contact interface would affect the linear reduced basis and cause mode interaction between them

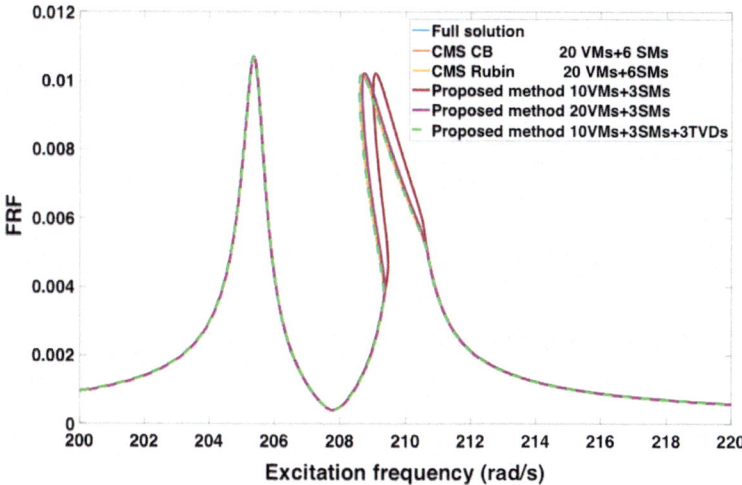

Fig. 6 Forced frequency response comparisons between CB,Rubin and Penalty ROM methods

[11]. This mean the linear reduced basis need to be calibrated in order to accurately represent the dynamics of coupled systems. The TVD method is one of the effective approaches for the calibration when this coupling between linear vibration modes is significant [11]. Three TVDs vectors are then added into the reduced basis to assess how this would improve the accuracy of the propose method. The results show that the proposed method with TVDs can obtain the same FRF as that from the full solution. Comparing to the CB and Rubin methods, the size of this reduced jointed structure model using the proposed method can be reduced by 40% even near the resonance frequency region.

6 Conclusions and Future Work

A novel penalty-based reduced order modelling approach has been proposed for dynamic analysis of a jointed structure with localised contact friction non-linearities. The formulation of the proposed approach has been presented where the contact friction is modelled by Jenkin model. We also showed how the proposed method can be effectively integrated with the harmonic balanced method, AFT and non-linear solver. A case study using jointed beam has been carried out to demonstrate the proposed method. The result obtained from the penalty approach was compared with the full solution and also classical Rubin and CB methods. The initial results show the method can effectively capture the FRFs of the non-linear dynamic system. TVDs are particularly useful to improve the accuracy when the contact interface is largely in a non-linear condition. Comparing to the classical CMS methods, the proposed method can reduce the size of ROM further when most of contact nodes are in a stuck condition.

The main objective of this paper was to present the formulation of this new reduced order modelling approach for a joint structure, and demonstrate the concept with a simple case study. Further developments would be needed to couple the proposed method with continuation techniques for HBM. This would enable the automatic size updating during the forced frequency response simulations. Also, high fidelity models are needed to further test and validate the proposed method and effects of TVDs on the dynamics of joint structures.

Acknowledgements The authors would like to acknowledge the support of Rolls-Royce plc for this research through the Vibration University Technology Centre (UTC) at the Imperial College London, UK. Special acknowledgement goes also to GEMiniDS WP3—Innovate UK Project 113088, which is jointly supported by Innovate UK and Rolls-Royce plc.

References

1. Bograd, S., Reuss, P., Schmidt, A., Gaul, L., Mayer, M.: Modeling the dynamics of mechanical joints. Mech. Syst. Signal Process. **25**(8), 2801–2826 (2011)
2. Craig, R., Bampton, M.: Coupling of substructures for dynamic analyses. AIAA J. **6**(7), 1313–1319 (1968)
3. Gruber, F.M., Rixen, D.J.: Evaluation of substructure reduction techniques with fixed and free interfaces. Strojniški vestnik-J. Mech. Eng. **62**(7–8), 452–462 (2016)
4. Krack, M., Salles, L., Thouverez, F.: Vibration prediction of bladed disks coupled by friction joints. Arch. Comput. Meth. Eng. **24**(3), 589–636 (2017)
5. Pesaresi, L., Salles, L., Jones, A., Green, J., Schwingshackl, C.: Modelling the nonlinear behaviour of an underplatform damper test rig for turbine applications. Mech. Syst. Signal Process. **85**, 662–679 (2017)
6. Petrov, E.: A high-accuracy model reduction for analysis of nonlinear vibrations in structures with contact interfaces. J. Eng. Gas Turbines Power **133**(10), 102503 (2011)
7. Pinnau, R.: Model reduction via proper orthogonal decomposition. In: Model Order Reduction: Theory, Research Aspects and Applications, pp. 95–109. Springer (2008)
8. Rubin, S.: Improved component-mode representation for structural dynamic analysis. AIAA J. **13**(8), 995–1006 (1975)
9. Sarrouy, E., Sinou, J.-J.: Non-linear periodic and quasi-periodic vibrations in mechanical systems-on the use of the harmonic balance methods. In: Advances in Vibration Analysis Research, InTech (2011)
10. Witteveen, W., Irschik, H.: Efficient mode based computational approach for jointed structures: joint interface modes. AIAA J. **47**(1), 252–263 (2009)
11. Witteveen, W., Pichler, F.: Efficient model order reduction for the dynamics of nonlinear multilayer sheet structures with trial vector derivatives. Shock Vibr. **2014** (2014)
12. Yuan, J., El-Haddad, F., Salles, L., Wong, C.: Numerical assessment of reduced order modeling techniques for dynamic analysis of jointed structures with contact nonlinearities. J. Eng. Gas Turbines Power **141**(3), 031027 (2019)
13. Yuan, J., Scarpa, F., Allegri, G., Titurus, B., Patsias, S., Rajasekaran, R.: Efficient computational techniques for mistuning analysis of bladed discs: a review. Mech. Syst. Signal Process. **87**, 71–90 (2017)

POD-DEIM Model Order Reduction for the Monodomain Reaction-Diffusion Sub-Model of the Neuro-Muscular System

Nehzat Emamy, Pascal Litty, Thomas Klotz, Miriam Mehl
and Oliver Röhrle

Abstract We apply POD-DEIM model order reduction to a 0D/1D model used to simulate the propagation of action potentials through the myocardium or along skeletal muscle fibers. This corresponding system of ODEs (reaction) and PDEs (diffusion) is called the monodomain equation. 0D sets of ODEs describing the ionic currents flowing across the cell membrane are coupled along muscle fibers through a 1D diffusion process for the transmembrane potential. Due to the strong coupling of the transmembrane potential and other state variables describing the behavior of the membrane, a total reduction strategy including all degrees of freedom turns out to be more efficient than a reduction of only the transmembrane potential. The total reduction approach is four orders of magnitude more accurate than partial reduction and shows a faster convergence in the number of POD modes with respect to the mesh refinement. A speedup of 2.7 is achieved for a 1D mesh with 320 nodes. Considering the DEIM approximation in combination with the total reduction, the nonlinear functions corresponding to the ionic state variables are also approximated in addition to the nonlinear ionic current in the monodomain equation. We observe that the same number of DEIM interpolation points as the number of POD modes is the optimal choice regarding stability, accuracy and runtime for the current POD-DEIM approach.

N. Emamy (✉) · P. Litty · M. Mehl
Institute for Parallel and Distributed Systems, University of Stuttgart, Universitätsstr. 38,
70569 Stuttgart, Germany
e-mail: nehzat.emamy@ipvs.uni-stuttgart.de

M. Mehl
e-mail: miriam.mehl@ipvs.uni-stuttgart.de

T. Klotz · O. Röhrle
Institute of Applied Mechanics (CE), SimTech Research Group on Continuum
Biomechanics and Mechanobiology, University of Stuttgart, Pfaffenwaldring 5a,
70569 Stuttgart, Germany
e-mail: thomas.klotz@mechbau.uni-stuttgart.de

O. Röhrle
e-mail: roehrle@simtech.uni-stuttgart.de

© Springer Nature Switzerland AG 2020
J. Fehr and B. Haasdonk (eds.), *IUTAM Symposium on Model Order Reduction
of Coupled Systems, Stuttgart, Germany, May 22–25, 2018*, IUTAM Bookseries 36,
https://doi.org/10.1007/978-3-030-21013-7_13

Keywords Model order reduction · Proper orthogonal decomposition (POD) ·
Discrete empirical interpolation method (DEIM) · Neuro-muscular system ·
Reaction-diffusion equation

1 Introduction

The monodomain model is widely used in the field of computational electrophysiology to simulate the propagation of action potentials through the myocardium [1] or
along skeletal muscle fibers in detailed multi-scale models of skeletal muscle tissue
[2, 3]. Such models can be used to systematically investigate the complex physiological behavior of the neuromuscular system in a controlled environment (e.g. see
[4–7]). However, even though the monodomain model is an approximation of the
more complex bidomain model [8, 9], i.e., assuming that the intracellular and extracellular conductivities are proportional, it is computationally challenging, since the
underling physics requires a fine discretization in space and time (see e.g. [10, 11]
and the references therein).

To reduce the computational costs, model order reduction techniques can be
applied. In literature, for spiking neurons, model order reduction is applied in [12],
where a proper orthogonal decomposition (POD) is used together with the proposed
discrete empirical interpolation method (DEIM) to approximate the nonlinear ionic
current resulting from the Hodgkin and Huxley model [13]. For parameter identification in cardiac electrophysiology, the POD approach is used to build a reduced basis
for the extracellular and transmembrane potentials [14]. The bidomain equations are
considered and the phenomenological two-variable model of Mitchell and Schaeffer
[15] is employed to describe the ionic current in a simplified way. The POD-DEIM
approach is used in [16] to estimate the cardiac conductivities and in [17] to study
electromyographic signals for skeletal muscles. In [16], the monodomain equation
is reduced and the ionic model of Rogers-McCulloch [18] with one gating variable
is considered to control the depolarization and repolarization phases of the cardiac
action potential. In [17], the bidomain equations are reduced, where a characteristic
shape for the action potential is considered.

In our study, we consider the monodomain equation together with the biophys-
ically motivated model of Hodgkin and Huxley, where three gating variables are
considered for the activation/inactivation of potassium and sodium ion channels.
The usual reduction approach as used in the above mentioned studies, is to consider
snapshots of the transmembrane potential or nonlinear ionic current separately to
build the POD bases (partial reduction). As the transmembrane potential and ionic
state variables are strongly coupled, we focus on a reduction strategy to reduce
them all together in the current study (total reduction). We study total and partial
POD-based reduction strategies with respect to the spatial convergence, accuracy
and runtime. As the total reduction strategy appears to be superior, we consider it for
applying the POD-DEIM approach. For this purpose, some minor modification of
the discrete system of equations are required. As a test case for our studies mentioned

above, we consider a 1 cm muscle fiber, which is stimulated once in the middle. To generalize our results, we further consider frequent stimulations of the fiber. We train our model with frequencies between 50 and 100 Hz. The trained and untrained data show a weak sensitivity with respect to frequency, which is negligible provided that sufficient POD modes are considered for building the basis.

The current study is organized as follows. In Sect. 2, we show the monodomain equation and its discrete version, respectively. In Sect. 3, we introduce the partial and total reduction strategies as well as the POD-DEIM approach considering the total reduction. Numerical experiments comparing partial and total reduction and showing accuracy and runtime results for POD-Galerkin and POD-DEIM are presented in Sect. 4. In this section, we also use the total reduction strategy to build the reduced basis considering frequent stimulations from the nervous system and assess their accuracy for different 'untrained' stimulation frequencies.

2 Propagation of Action Potentials Along Muscle Fibers

Considering skeletal muscles, the propagation of action potentials along a muscle fiber can be approximated by the 1D monodomain equation (e.g. see [3, 4]), providing a good trade-off between numerical complexity and the possibility to include biophysical details like the behavior of the muscle fiber membrane, resulting in a reaction-diffusion equation:

$$\frac{\partial V_{\mathrm{m}}}{\partial t} = \frac{1}{A_{\mathrm{m}} C_{\mathrm{m}}} \left(\frac{\partial}{\partial x} \left(\sigma_{\mathrm{eff}} \frac{\partial V_{\mathrm{m}}}{\partial x} \right) - A_{\mathrm{m}} I_{\mathrm{ion}} \left(\mathbf{y}, V_{\mathrm{m}}, I_{\mathrm{stm}} \right) \right) \text{ in } \Gamma, \quad (1a)$$

$$\frac{\partial \mathbf{y}}{\partial t} = G \left(\mathbf{y}, V_{\mathrm{m}} \right) \text{ in } \Gamma. \quad (1b)$$

Therein V_{m} is the transmembrane potential, I_{stm} is an externally applied current representing a stimulus from the nervous system, I_{ion} is the ionic current flowing through the ion channels and -pumps in the membrane, C_{m} is the capacitance of the muscle fiber membrane (sarcolemma), A_{m} is the fiber's surface to volume ratio, σ_{eff} is the effective conductivity and x denotes the spatial coordinate along the fiber. Further, $\mathbf{y}$ summarizes additional state variables, e.g. the states of different ion channels (cf. [13]) and G summarizes the right-hand side of all nonlinear ODEs associated with the state variables $\mathbf{y}$. Here, the ionic currents across the muscle fiber membrane, i.e. the non-linear reaction terms, are simulated with the model of Hodgkin and Huxley [13], where $\mathbf{y} \in \mathbb{R}^3$ represent the three gating variables related to activation/inactivation of potassium and sodium ion channels. We assume that no charges can leave the fiber domain, leading to homogeneous Neumann boundary conditions for the transmembrane potential V_{m} at both ends of a 1 cm long fiber, which is considered as our test case. Further, the initial values for the transmembrane potential V_{m} and $\mathbf{y}$ are set

according to the Hodgkin and Huxley model. The stimulation current is applied at
the middle of the fiber.

Discrete System. For the time integration of the monodomain equation (1a), we
apply the Godunov splitting scheme, where we integrate the reaction and diffusion
terms in two separate steps. The reaction term is integrated together with (1b) using
the explicit Euler scheme. For the diffusion term we use the implicit Euler scheme.

$$
\begin{bmatrix} V_m^* \\ \mathbf{y}^* \end{bmatrix} = \begin{bmatrix} V_m^{(t)} \\ \mathbf{y}^{(t)} \end{bmatrix} + dt \begin{bmatrix} -\frac{1}{C_m} I_{\text{ion}}\left(V_m^{(t)}, \mathbf{y}^{(t)}, I_{\text{stm}}^{(t)}\right) \\ G\left(V_m^{(t)}, \mathbf{y}^{(t)}\right) \end{bmatrix}, \tag{2a}
$$

$$
\begin{bmatrix} V_m^{(t+1)} \\ \mathbf{y}^{(t+1)} \end{bmatrix} = \begin{bmatrix} V_m^* \\ \mathbf{y}^* \end{bmatrix} + dt \begin{bmatrix} \frac{1}{A_m C_m} \frac{\partial}{\partial x}\left(\sigma_{\text{eff}} \frac{\partial V_m^{(t+1)}}{\partial x}\right) \\ 0 \end{bmatrix} \quad \text{in } \Gamma \tag{2b}
$$

V_m^* denotes an intermediate transmembrane potential for the time integration. After
the spatial discretization using the FEM method, we have the following discrete
system of equations, which define our full order model,

$$
\begin{bmatrix} \mathbf{v}_m^* \\ \mathbf{y}^* \end{bmatrix} = \begin{bmatrix} \mathbf{v}_m^{(t)} \\ \mathbf{y}^{(t)} \end{bmatrix} + \begin{bmatrix} \mathbf{F}(\mathbf{v}_m^{(t)}, \mathbf{y}^{(t)}) + \mathbf{I}_{\text{stm}}^{(t)} \\ \mathbf{G}(\mathbf{v}_m^{(t)}, \mathbf{y}^{(t)}) \end{bmatrix}, \tag{3a}
$$

$$
\begin{bmatrix} \mathbf{v}_m^{(t+1)} \\ \mathbf{y}^{(t+1)} \end{bmatrix} = \begin{bmatrix} \mathbf{v}_m^* \\ \mathbf{y}^* \end{bmatrix} + \begin{bmatrix} \mathbf{A}\mathbf{v}_m^{(t+1)} \\ 0 \end{bmatrix}, \tag{3b}
$$

where $\mathbf{v}_m \in \mathbb{R}^n$ and $\mathbf{y} \in \mathbb{R}^{3n}$ are the spatially discrete representations of V_m and $\mathbf{y}$
on a spatial mesh with n nodes. The spatially discretized nonlinear part of I_{ion}/C_m
is represented by $\mathbf{F} : \mathbb{R}^{4n} \mapsto \mathbb{R}^n$. The discrete vector representing the stimulation
current as an input term is $\mathbf{I}_{\text{stm}} \in \mathbb{R}^n$, which is adjustable linearly with the number
of nodes in the mesh. The discrete diffusion operator acting on $\mathbf{v}_m$ is $\mathbf{A} \in \mathbb{R}^{n \times n}$ and
$\mathbf{G} : \mathbb{R}^{4n} \mapsto \mathbb{R}^{3n}$ is the discretized version of G.

3   Reduction Strategy and POD-Galerkin/POD-DEIM

POD-Galerkin. Reduction strategies are based on collected snapshots of the process
under consideration. In this work, we use n_s snapshots in time (denoted by superscripts for the respective variables)[1] and accordingly establish a snapshot matrix $\mathbf{S}$
in which each column represents spatially discrete state variables on the used mesh
with n nodes. We consider two strategies with respect to these snapshots. First, for a

---

[1]Snapshots may be subsequent time steps or only a selection of time steps. In general, snapshots
may be based on parameters independent of time. Therefore, we do not use parenthesis for the
superscripts denoting the snapshots as we do for time step data.

partial reduction, we choose the snapshots of the transmembrane potential. Second, for a total reduction, we consider the snapshots of the transmembrane potential and the three state variables:

$$\mathbf{S}_{\text{part}} = \begin{bmatrix} v_{\text{m}1}^1 & v_{\text{m}1}^2 & \cdots & v_{\text{m}1}^{n_s} \\ \vdots & \vdots & \cdots & \vdots \\ v_{\text{m}n}^1 & v_{\text{m}n}^2 & \cdots & v_{\text{m}n}^{n_s} \end{bmatrix} \in \mathbb{R}^{n \times n_s} \qquad \mathbf{S}_{\text{total}} = \begin{bmatrix} v_{\text{m}1}^1 & v_{\text{m}1}^2 & \cdots & v_{\text{m}1}^{n_s} \\ y_{11}^1 & y_{11}^2 & \cdots & y_{11}^{n_s} \\ y_{21}^1 & y_{21}^2 & \cdots & y_{21}^{n_s} \\ y_{31}^1 & y_{31}^2 & \cdots & y_{31}^{n_s} \\ \vdots & \vdots & \cdots & \vdots \\ v_{\text{m}n}^1 & v_{\text{m}n}^2 & \cdots & v_{\text{m}n}^{n_s} \\ y_{1n}^1 & y_{1n}^2 & \cdots & y_{1n}^{n_s} \\ y_{2n}^1 & y_{2n}^2 & \cdots & y_{2n}^{n_s} \\ y_{3n}^1 & y_{3n}^2 & \cdots & y_{3n}^{n_s} \end{bmatrix} \in \mathbb{R}^{4n \times n_s}.$$

For model order reduction, singular value decomposition (SVD) is applied on the snapshot (sample of trajectory) matrix $\mathbf{S} = [\mathbf{s}_1, ..., \mathbf{s}_{n_s}]$,

$$\mathbf{S} = \mathbf{V}\mathbf{\Sigma}\mathbf{W}^{\text{T}} \text{ with } \mathbf{V} \in \mathbb{R}^{p \times r}, \mathbf{\Sigma} \in \mathbb{R}^{r \times r}, \mathbf{W} \in \mathbb{R}^{p \times r},$$

where $p = n$ for the partial reduction approach and $p = 4n$ for the total reduction, r is the rank of $\mathbf{S}$, $r \leq \min(p, n_s)$. We follow the notation in [19], where $\mathbf{\Sigma} = \text{diag}(\sigma_1, ..., \sigma_r)$ with $\sigma_1 \geq \sigma_2 \geq ... \geq \sigma_r > 0$. The matrix $\mathbf{V}$ contains the left singular vectors $\mathbf{v}_i$ of $\mathbf{S}$ corresponding to the singular value σ_i. Reduction approaches truncate this SVD to parts associated to the largest singular values $\sigma_1, \sigma_2, ..., \sigma_k$, $k < r$, yielding

$$\mathbf{S} \approx \mathbf{V}_k\mathbf{\Sigma}_k\mathbf{W}_k^{\text{T}} \text{ with } \mathbf{V}_k \in \mathbb{R}^{p \times k}, \mathbf{\Sigma}_k \in \mathbb{R}^{k \times k}, \mathbf{W}_k \in \mathbb{R}^{p \times k}.$$

We use the KerMor[2] [20] library and consider the Galerkin projection of the system (3) above. Using the partial reduction, we have the following reduced system to solve for the reduced transmembrane potential $\tilde{\mathbf{v}}_{\text{m}}$ and the (unreduced) state vector $\mathbf{y}$:

$$\begin{bmatrix} \tilde{\mathbf{v}}_{\text{m}}^* \\ \mathbf{y}^* \end{bmatrix} = \begin{bmatrix} \tilde{\mathbf{v}}_{\text{m}}^{(t)} \\ \mathbf{y}^{(t)} \end{bmatrix} + \begin{bmatrix} \mathbf{V}_k^{\text{T}}\mathbf{F}(\mathbf{v}_{\text{m}}^{(t)}, \mathbf{y}^{(t)}) + \mathbf{V}_k^{\text{T}}\mathbf{I}_{\text{stm}}^{(t)} \\ \mathbf{G}(\mathbf{v}_{\text{m}}^{(t)}, \mathbf{y}^{(t)}) \end{bmatrix}, \tag{4a}$$

$$\begin{bmatrix} \tilde{\mathbf{v}}_{\text{m}}^{(t+1)} \\ \mathbf{y}^{(t+1)} \end{bmatrix} = \begin{bmatrix} \tilde{\mathbf{v}}_{\text{m}}^* \\ \mathbf{y}^* \end{bmatrix} + \begin{bmatrix} \mathbf{V}_k^{\text{T}}\mathbf{A}\mathbf{V}_k\tilde{\mathbf{v}}_{\text{m}}^{(t+1)} \\ 0 \end{bmatrix}, \tag{4b}$$

where we recover $\mathbf{v}_{\text{m}}$ from $\mathbf{v}_{\text{m}} = \mathbf{V}_k\tilde{\mathbf{v}}_{\text{m}}$. Considering the total reduction, we get the system to be solved for the fully reduced state $\tilde{\mathbf{z}}$:

---

[2]http://www.ians.uni-stuttgart.de/MoRePaS/software/kermor/.

$$\tilde{z}^* = \tilde{z}^{(t)} + V_k^T FG(v_m^{(t)}, y^{(t)}) + V_k^T I_{enh}^{(t)}, \tag{5a}$$

$$\tilde{z}^{(t+1)} = \tilde{z}^* + V_k^T A_{enh} V_k \tilde{z}^{(t+1)}, \tag{5b}$$

where $FG(v_m, y)$, I_{enh}, and A_{enh} as well as the full state recovery are defined by

$$
FG(v_m, y) := \begin{bmatrix} F_1(v_m, y) \\ G_1(v_m, y) \\ F_2(v_m, y) \\ G_2(v_m, y) \\ \vdots \\ F_n(v_m, y) \\ G_n(v_m, y) \end{bmatrix}, \ I_{enh} := \begin{bmatrix} I_{stm,1} \\ 1 \\ I_{stm,2} \\ 1 \\ \vdots \\ I_{stm,n} \\ 1 \end{bmatrix}, \ A_{enh} := \begin{bmatrix} A_1 \\ 1 \\ A_2 \\ 1 \\ \vdots \\ A_n \\ 1 \end{bmatrix}, \ V_k \tilde{z} = \begin{bmatrix} v_{m,1} \\ y_1 \\ v_{m,2} \\ y_2 \\ \vdots \\ v_{m,n} \\ y_n \end{bmatrix}.
$$

The reduced matrix $V_k^T A V_k$ or $V_k^T A_{enh} V_k$ is precomputed in an offline phase, thus dramatically reducing the computational cost in the actual execution of time steps (online phase) for the linear component of the system. The linear systems of equations in (4b) and (5b) are solved using a direct solver, which suffices for our case of 1D problem.

POD-DEIM. Besides the total reduction strategy in equation (5a), we apply the discrete empirical interpolation method (DEIM) [19] to the nonlinear functions F and G:

$$\tilde{z}^* = \tilde{z}^{(t)} + V_k^T U_m (P_m^T U_m)^{-1} P_m^T FG(v_m^{(t)}, y^{(t)}) + V_k^T I_{enh}^{(n)}, \tag{6a}$$

$$\tilde{z}^{(t+1)} = \tilde{z}^* + V_k^T A_{enh} V_k \tilde{z}^{(t+1)}, \tag{6b}$$

Here, we reduce the full state representation $FG(v_m^{(t)}, y^{(t)})$ to a representation $U_m c$ in the subspace defined by the columns of $U \in \mathbb{R}^{4n \times m}$ found by applying a truncated SVD to a snapshot matrix of FG, only. The coefficients $c \in \mathbb{R}^m$ are found by solving the interpolation problem

$$P_m^T U_m c = P_m^T FG(v_m^{(t)}, y^{(t)})$$

for m projection points defined in $P_m = [e_{\wp_1}, ..., e_{\wp_m}]$ with selected unit vectors $e_{\wp_i}$, $i = 1, \ldots, m$, where the indices $\{\wp_1, ..., \wp_m\}$ are found through the DEIM algorithm with respect to the basis $\{u_1, ..., u_m\}$. The DEIM algorithm selects an index in each iteration corresponding to the maximum residual such that the growth of an error bound is limited. Applying the above, the computational complexity of the nonlinear functions is independent of n (spatial dimension) as FG is only evaluated at the m interpolation points. Moreover, the matrix vector multiplication of size

Table 1 Comparison of the singular values using the partial and total reduction strategies for different meshes. k/r ratios show the number of singular values greater than 10^{-5} relative to the rank of snapshot matrix

Partial reduction							Total reduction					
n	10	20	40	80	160	320	10	20	40	80	160	320
r	10	20	40	80	160	320	40	80	160	320	640	1280
k	8	18	38	78	120	124	32	72	121	135	157	160
k/r	0.8	0.9	0.95	0.98	0.75	0.39	0.8	0.9	0.76	0.42	0.245	0.16

$k \times 4n$ times $4n$ in equation (5a) is replaced by $k \times m$ times m in (6a) as the matrix $\mathbf{V}_k^{\mathrm{T}} \mathbf{U}_m (\mathbf{P}_m^{\mathrm{T}} \mathbf{U}_m)^{-1} \in \mathbb{R}^{k \times m}$ is precomputed.

4 Numerical Results

Evaluation of the Truncation Potential for Snapshot SVD. In the following, we provide numerical tests based on snapshots gathered every time step (5×10^{-4} ms) for 10 ms ($n_s = 20000$) for meshes with $n = 10, 20, 40, 80, 160$ and 320 nodes. The singular values are shown in Fig. 1 for the two reduction strategies. There are jumps in the singular values, at which one can set a threshold for the reduction. A convergence with respect to mesh refinement could be observed for both strategies. We consider a threshold of 10^{-5} and show the number of singular values, which are greater than this threshold in Table 1. The number of the POD modes k appears to converge by the mesh refinement for both strategies. Small ratios k/r indicate that the reduction is effective, specially for fine meshes. Using the total reduction, a faster convergence in k and much smaller ratios k/r are achieved in comparison to the partial reduction.
Partial versus Total Reduction Error and Runtime Analysis. In Fig. 2, we compare the two strategies partial and total reduction in terms of the mean relative L2-norm errors of the transmembrane potential versus runtime using a mesh with 80 nodes. Each point on the plot resembles additional 5 modes for the partial reduction and 20 modes for the total reduction yielding comparable percentages of the total number of modes ($k/r = 5/80 = 20/320$). The relative L2-norm errors are computed with respect to the full order model (3) and averaged over the whole computational time. Using the total reduction, more accurate solutions are achieved in smaller runtimes rather than using the partial reduction. For the same runtime of approximately 24 s, the error is four orders of magnitude smaller for the total reduction. The last point in the partial reduction strategy shows a jump in the error, which means that considering all modes, the error with respect to the full order model is negligibly small as expected.
Runtimes and speedups. In Fig. 3, we show the errors of $\mathbf{v}_m$ versus runtime and speedup for different meshes using the total reduction. Considering a threshold of 5×10^{-5} for the errors, speedups of 1.5, 1.8 and 2.7 are achieved for meshes with

$n = 80, 160$ and 320 nodes. As expected, the speedups become considerable for finer meshes. However, spatial convergence in terms of the number of detected modes (as visible in Fig. 1/Table 1) should be considered which renders too fine meshes inefficient as no additional information is captured.

POD-DEIM Error and Runtime Analysis. We show the errors of the transmembrane potential versus runtime for the POD-DEIM approach in Figs. 4, 5 and 6. A mesh with 80 nodes is used. As we use the total reduction strategy, the maximum number of POD modes is equal to $r = 4n = 320$, the rank of the snapshot matrix. Ratios $m/r = 0.4, ..., 1$ are the relative number of interpolation points with respect to the rank of the snapshot matrix. Each data series m/r starts by considering $k = 20$ POD modes. The number of modes is increased by 20 between data points. In Fig. 4, we observe that the errors are reduced by increasing the number of POD modes up to the point where the number of interpolation points becomes smaller than the number of POD modes, $m < k$. If $m/r < 0.6$, the solution is not stable. If more than 60 percent of the interpolation points are used ($m/r > 0.6$), the error is dominated by the error of the DEIM approximation and not reduced further by adding more POD modes. A smaller number of interpolation points rather than the POD modes would

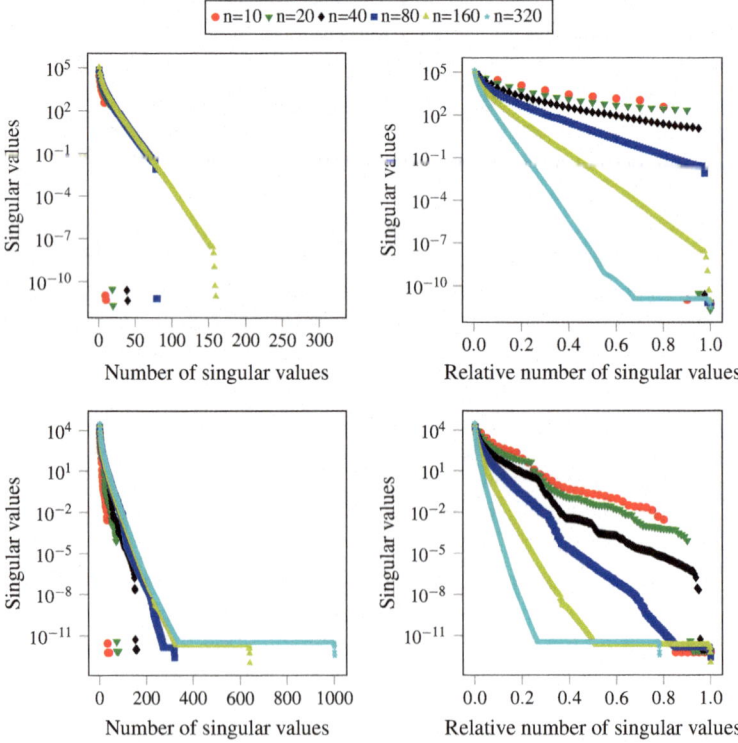

Fig. 1 Singular value decomposition of the snapshot matrix. Partial reduction on the snapshots of the transmembrane potential (top). Total reduction on the snapshots of the transmembrane potential and three ion-channel state variables (bottom)

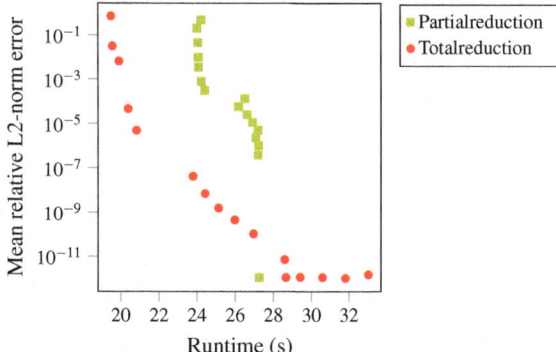

Fig. 2 Comparison of the partial and total reduction strategies. Errors of the transmembrane potential with respect to the full order model are shown. A mesh with 80 nodes is employed. Each point on the plot resembles additional 5 and 20 (equivalent) modes in the partial and total reduction strategies, respectively

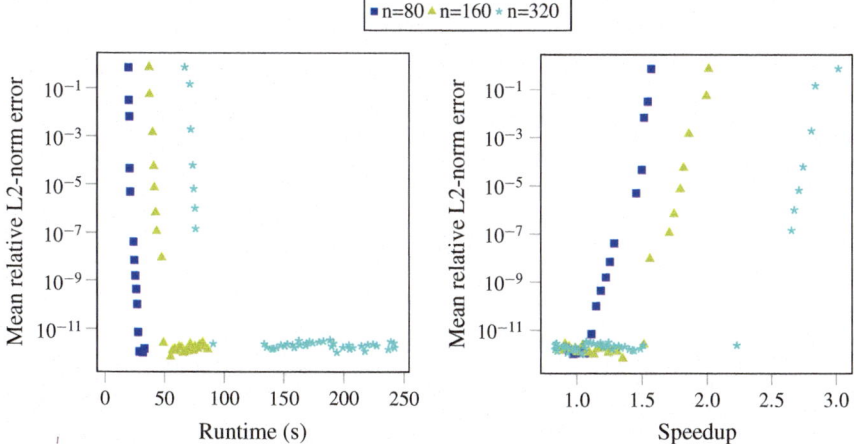

Fig. 3 Errors of the transmembrane potential versus runtime and speedup with respect to the full order model. The POD approach with the total reduction strategy is applied. Starting from 20 modes, we increase the number of modes between points on the plot by 20

mean missing parts of the DEIM basis for the nonlinear functions, which are already present in the POD basis of the solution. In Fig. 5, we observe that for a given level of accuracy, the number of the DEIM interpolation points should be chosen equal to or greater than the number of POD modes, $m \geq k$. The minimum runtime is achieved for $m = k$. In Fig. 6, we show our optimal choice $m = k$, which is stable and provides the minimum runtime. This is in agreement with [12], where they mention the equal number of POD and DEIM modes as the best choice.

Generalization of the Reduced Model. To assess the generalizability of our reduced order model, we generate the snapshot matrix $\mathbf{S}_{\text{total}}$ with full model simulations with

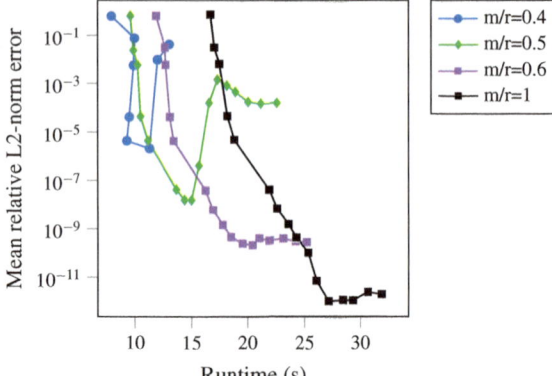

Fig. 4 POD-DEIM approach for model order reduction. Errors of the transmembrane potential with respect to the full order model are shown. The errors are reduced by increasing the number of POD modes k up to the point where $m < k$. Ratios m/r show the number of DEIM interpolation points relative to the rank of the snapshot matrix. Each data series m/r starts with $k = 20$ POD modes, we increase the number of modes by 20 between points

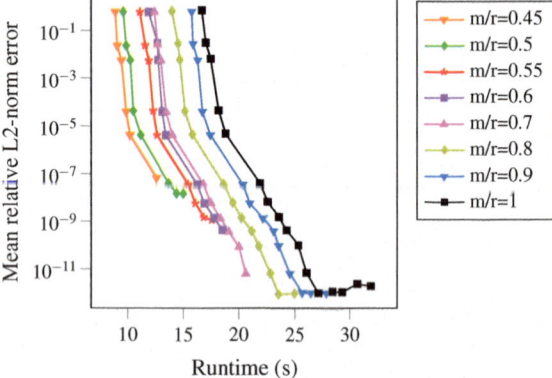

Fig. 5 POD-DEIM approach for the case that the number of DEIM interpolation points are greater than or equal to the number of POD modes, $m \geq k$. Errors of the transmembrane potential with respect to the full order model are shown. Ratios m/r show number of DEIM interpolation points relative to the rank of the snapshot matrix. Each data series m/r starts with $k = 20$ POD modes, we increase the number of modes by 20 between points

frequencies of the stimulation current between 50 and 100 Hz. We simulate 200 ms with time step of 0.005 ms. The snapshots are gathered from each time step on a mesh with 80 nodes. The POD approach with the total reduction strategy is then evaluated for 'untrained' testing frequencies between 20 and 100 Hz.

Errors of the transmembrane potential with respect to the full order model are shown in Fig. 7. The errors show a weak dependence on the frequency of the stimulation for both the training and testing frequencies. Slightly higher errors are observed for frequencies between 70 and 90 Hz, an issue which can be overcome by using

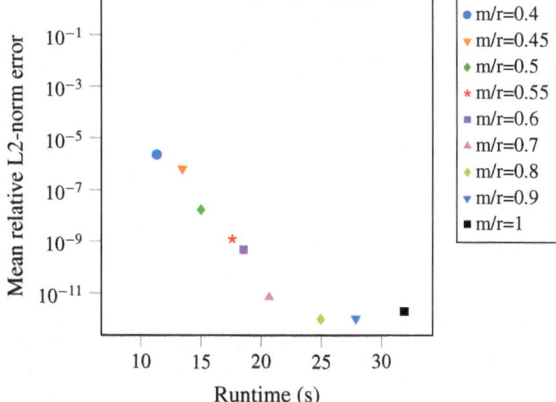

Fig. 6 POD-DEIM approach for the case that the number of DEIM interpolation points and the number of POD modes are the same, $m = k$. Errors of the transmembrane potential with respect to the full order model are shown. The points on the plot are the points with minimum runtime for each series m/r from Fig. 5. The ratios m/r show number of DEIM interpolation points relative to the rank of the snapshot matrix

more POD modes. Therefore, the obtained POD basis with adequate number of modes could be used for different frequencies of stimulation without a considerable deviation. If the time interval between two stimuli is greater than the refractory period of the membrane in the Hodgkin-Huxley model, which is about 20 ms (see eg. [21]), the monodomain model in combination with the Hodgkin-Huxley model behaves like a time invariant system. Therefore, for this special case a single stimulation should basically contain all information about the system's dynamics in the physiological stimulation conditions and therefore be sufficient as the training data. This is in agreement with our training strategy, where we did not require to train for frequencies below 50 Hz. Probably, this will not hold anymore when using a membrane model which also includes fatigue effects, e.g. the Shorten model [22].

5 Conclusion and Outlook

Our numerical results show that POD-DEIM model order reduction yields a considerable reduction of runtime for the simulation of the so-called monodomain equations, i.e., action potential propagation along muscle fibers. We observed that, due to the strong coupling between the transmembrane potential and ionic state variables, a total reduction of all state variables simultaneously gives the best results. Comparing to the usual partial reduction, where the transmembrane potential is reduced separately, we consider additionally the three ionic state variables resulting from the Hodgkin and Huxley model [13] to build one total snapshot matrix. The SVD decomposition of the snapshot matrices in both partial and total reduction strategies show jumps

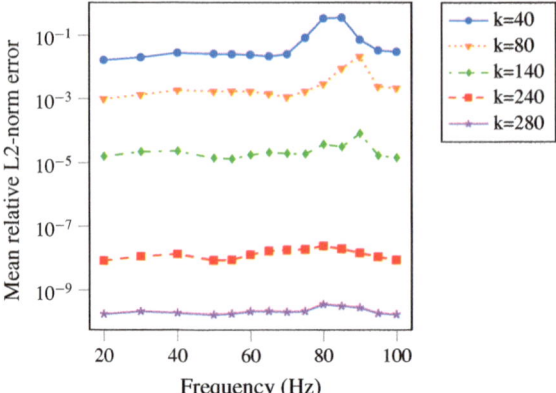

Fig. 7 POD approach considering different frequencies of the stimulation. Errors of the transmembrane potential with respect to the full order model are shown. The POD basis is extracted using the total reduction of the snapshots taken from the full order model simulations with frequencies 50, 60, 70, 80, 90 and 100 Hz

in the singular values for coarse meshes. Using finer meshes, the jumps disappear. This leads to the conclusion, that intermediate modes are not yet resolved on the coarse meshes. For all meshes, however, only a certain number of singular values is of considerable size. This number does not further increase with mesh resolution after all modes of interest are resolved. Reference [16] present a similar reduction approach that we, however, can not directly compare our results with because their method would be only equivalent to our partial reduction. However, it seems that the slow decays, which they mention for the singular values are due to the very fine mesh used for gathering the snapshots. Comparing our total and partial reduction strategies, more significant modes of the solution are retrieved on the same mesh using the total reduction. As an example, for a mesh with 80 nodes, 135 singular values are above a threshold of 10^{-5} using the total reduction, which are about 42 percent of the total singular values. The singular values above the threshold are 78 using the partial reduction, which are equivalent to 98 percent of the total singular values. The effectiveness of the total reduction is approved further by comparing the mean relative L2-norm errors of the transmembrane potential for both strategies on the same mesh above. Using 30 modes for the partial reduction and 120 modes for the total reduction, the error of the total reduction is four orders of magnitude smaller with the same runtime. Adopting the total reduction strategy, speedups are computed for different meshes with respect to the full order model. A speedup of 1.5 is achieved with the mesh with 80 nodes. A higher speedup of 2.7 is achieved for a finer mesh with 320 nodes. Further mesh refinement is not performed as we have shown the mesh convergence for the problem. In order to efficiently evaluate the nonlinear ionic current in the monodomain equation, we use the DEIM approximation for the nonlinear functions of the ODEs resulting from the Hodgkin and Huxley model, which are coupled with the monodomain equation. Varying the number of

DEIM interpolation points for a fixed number of POD modes could considerably reduce the runtime for a preserved accuracy as expected. However, if the number of interpolation points is smaller than the number of POD modes, the solution is either unstable (when using less than 60 percent of the interpolation points) or the errors are dominated by the DEIM approximation error. This could be interpreted as a hint that the nonlinear terms actually are decisive for the overall number of relevant modes. An equal number of DEIM points and POD modes appears to be the optimal choice, which is stable and accurate with a minimal runtime. Applying the current approach to study the effect of the stimulation frequency shows a weak dependence on the frequency of stimulation, which is acceptable employing the Hodgkin and Huxley model. Our overall conclusion is that the potential for model order reduction lies mainly in reduction of the reaction term of the monodomain equation. This should be verified by considering more complicated models for chemical reactions such as the Shorten model [22], which contains more than 50 state variables and includes fatigue effects. Application of model order reduction to deforming fibers is still an open question, which would be interesting for the following studies.

Acknowledgements This research was funded by the Baden-Württemberg Stiftung as part of the DiHu project of the High Performance Computing II program and the Cluster of Excellence for Simulation Technology (EXC 310/1).

References

1. Pullan, A.J., Buist, M.L., Cheng, L.K.: Mathematically Modelling the Electrical Activity of the Heart: From Cell to Body Surface and Back Again. World Scientific Publishing Company, Singapore (2005)
2. Röhrle, O., Davidson, J.B., Pullan, A.J.: A physiologically based, multi-scale model of skeletal muscle structure and function. Front. Physiol. **3**, (2012)
3. Heidlauf, T., Röhrle, O.: Modeling the Chemoelectromechanical Behavior of Skeletal Muscle Using the Parallel Open-Source Software Library OpenCMISS. Computational and Mathematical Methods in Medicine **1–14** (2013)
4. Heidlauf, T., Röhrle, O.: A multiscale chemo-electro-mechanical skeletal muscle model to analyze muscle contraction and force generation for different muscle fiber arrangements. Front. Physiol. **5**(498), 1–14 (2014)
5. Mordhorst, M., Heidlauf, T., Röhrle, O.: Predicting electromyographic signals under realistic conditions using a multiscale chemo-electro-mechanical finite element model. Interface Focus **5**(2), 1–11 (2015)
6. Heidlauf, T., Klotz, T., Rode, C., et al.: A multi-scale continuum model of skeletal muscle mechanics predicting force enhancement based on actin-titin interaction. Biomech. Model. Mechanobiol. **15**(6), 1423–1437 (2016)
7. Heidlauf, T., Klotz, T., Rode, C., Siebert, T., Röhrle, O.: A continuum-mechanical skeletal muscle model including actin-titin interaction predicts stable contractions on the descending limb of the force-length relation. PLOS Comput. Biol. **13**(10), 1–25, 10 (2017)
8. Miller, W.T., Geselowitz, D.B.: Simulation studies of the electrocardiogram. i. the normal heart. Circ. Res. **43**(2), 301–315 (1978)
9. Tung, L.: A bi-domain model for describing ischemic myocardial DC potentials. PhD thesis, Massachusetts Institute of Technology (1978)

10. Bradley, C.P., Emamy, N., Ertl, T., et al.: Enabling detailed, biophysics-based skeletal muscle models on HPC systems. Front. Physiol. **9** (2018)
11. Clayton, R.H., Bernus, O., Cherry, E.M., et al.: Models of cardiac tissue electrophysiology: Progress, challenges and open questions. Prog. Biophys. Mol. Biol. **104**(1–3), 22–48 (2011)
12. Kellems, A.R., Chaturantabut, S., Sorensen, D.C., Cox, S.J.: Morphologically accurate reduced order modeling of spiking neurons. J. Comput. Neurosci. **28**(3), 477–494 (2010)
13. Hodgkin, A.L., Huxley, A.F.: A quantitative description of membrane current and its application to conduction and excitation in nerve. J. Physiol. **117**(4), 500–544 (1952)
14. Boulakia, M., Schenone, E., Gerbeau, J-F.: Reduced-order modeling for cardiac electrophysiology. Application to parameter identification. Int. J. Numer. Methods Biomed. Eng. **28**(6–7), 727–744 (2012)
15. Mitchell, C.C., Schaeffer, D.G.: A two-current model for the dynamics of cardiac membrane. Bull. Math. Biol. **65**(5), 767–793 (2003)
16. Yang, H., Veneziani, A.: Efficient estimation of cardiac conductivities via POD-DEIM model order reduction. Appl. Numer. Math. **115**, 180–199 (2017)
17. Mordhorst, M., Strecker, T., Wirtz, D., Heidlauf, T., Röhrle, O.: POD-DEIM reduction of computational EMG models. J. Comput. Sci. **19**, 86–96 (2017)
18. Rogers, J.M., Mc Culloch, A.D.: A collocation-Galerkin finite element model of cardiac action potential propagation. IEEE Trans. Biomed. Eng. **41**(8), 743–757 (1994)
19. Chaturantabut, S., Sorensen, D.C.: Nonlinear model reduction via discrete empirical interpolation. SIAM J. Sci. Comput. **32**(5), 2737–2764 (2010)
20. Wirtz, D.: Model Reduction for Nonlinear Systems: Kernel Methods and Error Estimation. epubli GmbH (2014)
21. Heidlauf, T.: Chemo-electro-mechanical modelling of the neuromuscular system. Institut fur Mechanik (Bauwesen), Lehrstuhl fur Kontinuumsmechanik, Research Group on Continuum Biomechanics and Mechanobiology, Universität Stuttgart (2015)
22. Shorten, P.R., O'Callaghan, P., Davidson, J.B., Soboleva, T.K.: A mathematical model of fatigue in skeletal muscle force contraction. J. Muscle Res. Cell Motil. **28**(6), 293–313 (2007)

Index-Aware MOR for Gas Transport Networks

Nicodemus Banagaaya, Sara Grundel and Peter Benner

Abstract We extend the index-aware model order reduction method (IMOR) to differential-algebraic equations (DAEs) arising from gas transport networks, which have a tractability index of one or two. Applying model order reduction (MOR) techniques to DAEs has to be done carefully and the techniques, in particular for nonlinear systems, can not handle arbitrary models. In previous work, MOR for DAEs arising from gas transport networks was done by rewriting DAEs into ordinary differential equations (ODEs) by index reduction. Then, standard MOR techniques could be applied. We propose an approach to create an ODE system and algebraic equations from the original DAE, which is done automatically. That means we can get a new decoupled system easily, even if we change the discretization scheme or the coupling condition in the gas network. We explain the details of the automatic decoupling for the linearized gas transport equation and show its efficiency on several numerical examples.

Keywords Nonlinear differential-algebraic equations · Decoupling · Gas transportation networks · Model order reduction · Tractability index

1 Introduction

We consider a gas transportation network consisting of several pipes. It is represented via a directed graph. All the edges of the graph are pipes and the nodes are either just interior nodes or supply nodes or demand nodes. We assume that at supply nodes

N. Banagaaya (✉) · S. Grundel · P. Benner
Max Planck Institute for Dynamics of Complex Technical Systems, Sandtorstr. 1,
39106 Magdeburg, Germany
e-mail: banagaaya@mpi-magdeburg.mpg.de

S. Grundel
e-mail: grundel@mpi-magdeburg.mpg.de

P. Benner
e-mail: benner@mpi-magdeburg.mpg.de

© Springer Nature Switzerland AG 2020
J. Fehr and B. Haasdonk (eds.), *IUTAM Symposium on Model Order Reduction of Coupled Systems, Stuttgart, Germany, May 22–25, 2018*, IUTAM Bookseries 36,
https://doi.org/10.1007/978-3-030-21013-7_14

a pressure boundary condition is given and at demand nodes a mass flux boundary condition is imposed. The one-dimensional isothermal Euler equations are often used in modelling natural gas transport through a pipe [1, 2], and this is what we will do in the following as well. Spatial discretization of these equations together with algebraic conditions on the nodes leads to a nonlinear dynamical system of the form [1]

$$E\dot{x} = Hx + f(x) + Bu, \quad Ex(0) = Ex_0, \tag{1a}$$

$$y = Cx + h(x), \tag{1b}$$

where $E \in \mathbb{R}^{n \times n}$ is singular, indicating that (1a) is a system of nonlinear differential-algebraic equations (DAEs). This implies that x_0 must be a consistent initial condition. $H \in \mathbb{R}^{n \times n}$, $B \in \mathbb{R}^{n \times m}$, $C \in \mathbb{R}^{\ell \times n}$, $h(x) \in \mathbb{R}^{\ell}$, $f(x) \in \mathbb{R}^n$, and the state vector $x \in \mathbb{R}^n$ includes states representing the gas mass flow and states representing the gas pressure at discretization points within the pipe network. The input function $u(t)$ includes the vector of supply pressures, and the vector of demand mass flows. The desired output vector y could be any combination of pressure values and mass fluxes, but is often choosen to collect the pressure at the demand nodes and the mass fluxes at the supply nodes.

DAEs are known to be difficult to simulate and the level of difficulty is measured using index concepts such as differential index, tractability index, etc. The higher the index, the more difficult to simulate the DAE. Moreover, the system (1a) is a hyperbolic balance law including friction and gravity effects, increasing the computational complexity due to strong coupling and stiffness of the problem. In general, the solutions of hyperbolic balance laws can blow-up in finite time which can lead to numerical integration challenges. Despite the ever increasing computational power, dynamic pipeline network simulation using the system (1a) is costly, since it involves to solve of a hyperbolic partial differential equations (PDEs) for each pipe, see [2]. We are interested in a fast and stable prediction of the dynamics of natural gas transport in the pipe networks, and therefore the application of model order reduction (MOR) is vital. MOR aims to reduce the computational burden by generating reduced-order models (ROMs) that are faster and cheaper to simulate, yet accurately represent the original large-scale system behavior. MOR replaces (1) by a ROM

$$E_r\dot{x}_r = H_r x_r + f_r(x_r) + B_r u, \quad E_r x_r(0) = E_r x_{r_0}, \quad y_r = C_r x_r + h_r(x_r), \tag{2}$$

where E_r, $H_r \in \mathbb{R}^{r \times r}$, $f_r \in \mathbb{R}^r$, $B_r \in \mathbb{R}^{r \times m}$ and $y_r \in \mathbb{R}^{\ell}$, $h_r \in \mathbb{R}^{\ell}$, $C_r \in \mathbb{R}^{\ell \times r}$ such that the reduced order of the state vector $x_r \in \mathbb{R}^r$ is $r \ll n$. A good ROM should have a small approximation error $\|y - y_r\|$ in a suitable norm $\|.\|$ for a desired range of inputs u. There exist many MOR methods for nonlinear systems such as POD, POD-DEIM, etc. However, applying these MOR methods directly to DAEs typically leads to ROMs which are ODEs. These may be inaccurate or very difficult to simulate [1, 3]. This is due to the fact that they do not respect the hidden constraints, the consistent initial conditions and the smoothness of the input data. In [1, 4], using the state-of-the-art transformation, discretized gas transport problems of the form (1a) are transformed into a system of ODEs and then POD with the DEIM is used to reduce the system size. However, this approach leads to stiff ROMs, which affects the choice of suitable numerical solvers strongly. Moreover, this approach can not be automated and it depends on the spatial discretization method. It is in particular unclear how to extend it to gas networks with network control elements, such as compressors, valves, regulators, etc. We propose an index-aware MOR (IMOR) method for DAEs arising from gas transport networks. This approach involves first the automatic decoupling of the given discretized and linearized gas transport DAEs into differential and algebraic parts, then each part can be reduced separately leading to easier-to-simulate ROMs.

The paper is organized as follows. In Sect. 2, we present the discretized dynamic DAE model arising from gas transport networks and its transformed ODE proposed in [1, 4]. In Sect. 3, we discuss how a linear DAE system can be decoupled automatically into differential and algebraic parts using IMOR [3]. In Sect. 4, we show the details of the automatic decoupling applied to the linearized DAE introduced in Sect. 2. In Sect. 5, we briefly discuss how to do MOR within the IMOR method, in particular how to treat the algebraic part. In the final section, we present some numerical examples, divided into small, medium and large-scale examples illustrating the performance of the proposed method.

2   Discretized Gas Transport Network DAE Model

In this section, we consider the spatially discretized system of a gas transportation network proposed in [1, 4] leading to a nonlinear DAE. We then present the index reduction of the derived nonlinear DAE proposed in [4]. The nonlinear dynamic system of gas transport in a network is described by the pressure at the supply nodes $\mathbf{p}_s \in \mathbb{R}^{n_s}$, the pressure at all other nodes $\mathbf{p}_d \in \mathbb{R}^{n_d + n_0}$, the difference of flux over a pipe segment $\mathbf{q}_- \in \mathbb{R}^{n_E}$ and the average of the mass flux over a pipe segment $\mathbf{q}_+ \in \mathbb{R}^{n_E}$, modelled over a graph with n_E edge segments, that are the size of the discretization, n_s supply nodes, n_d demand nodes and n_0 interior nodes. The resulting structure of the equation is

$$
\begin{aligned}
|\mathcal{A}_S^{\mathrm{T}}|\partial_t \mathbf{p}_s + |\mathcal{A}_0^{\mathrm{T}}|\partial_t \mathbf{p}_d &= -\mathbf{M}_L^{-1}\mathbf{q}_-, \\
\partial_t \mathbf{q}_+ &= \mathbf{M}_A(\mathcal{A}_S^{\mathrm{T}}\mathbf{p}_s + \mathcal{A}_0^{\mathrm{T}}\mathbf{p}_d) + \mathbf{g}(\mathbf{q}_+, \mathbf{p}_s, \mathbf{p}_d), \\
0 &= \mathcal{A}_0\mathbf{q}_+ + |\mathcal{A}_0|\mathbf{q}_- - \mathbf{B}_d\mathbf{d}(t), \\
0 &= \mathbf{p}_s - \mathbf{s}(t),
\end{aligned}
\tag{3}
$$

where $\mathbf{M}_L \in \mathbb{R}^{n_E \times n_E}$ and $\mathbf{M}_A \in \mathbb{R}^{n_E \times n_E}$ are diagonal matrices encoding parameters such as length, radius of the pipe segments as well as constants coming from the gas equation. The matrix $\mathbf{B}_d \in \mathbb{R}^{(n_d + n_0) \times n_d}$ is a matrix of ones and zeros making sure that the demand of the demand node is put at the right place in the mass flux equation. The matrix $\mathcal{A}_0$ is created by taking the incidence matrix of the graph representing the refined gas transportation network and removing the rows corresponding to the supply nodes, where $\mathcal{A}_S$ is the matrix created from the incidence matrix by only taking the rows corresponding to the supply nodes. The matrices $|\mathcal{A}_0|$ and $|\mathcal{A}_S|$ are the incidence matrices of the undirected graph defined as the component-wise absolute values of the incidence matrices of the directed graph, see [1]. The vectors $\mathbf{d}(t) = (\ldots, d_i(t), \ldots)^{\mathrm{T}} \in \mathbb{R}^{n_d}$ and $\mathbf{s}(t) = (\ldots, s_i(t), \ldots)^{\mathrm{T}} \in \mathbb{R}^{n_s}$ are demand mass flow and supply pressure, respectively, which are considered as input functions. The vector $\mathbf{g}(\mathbf{q}_+, \mathbf{p}_s, \mathbf{p}_d) = (\ldots, g_k(\mathbf{q}_+, \mathbf{p}_s, \mathbf{p}_d), \ldots)^{\mathrm{T}} \in \mathbb{R}^{n_E}$ is the discretization of the gravity and friction term and therefore represents the nonlinear part of the equation with

$$g_k(\mathbf{q}_+, \mathbf{p}_s, \mathbf{p}_d) = -g_k \psi_k(\mathbf{p}_d, \mathbf{p}_s)\frac{\Delta h_k}{L_k} - \lambda_k \frac{\mathbf{q}_+^k |\mathbf{q}_+^k|}{\psi_k(\mathbf{p}_d, \mathbf{p}_s)},$$

where $\psi_k(\mathbf{p}_d, \mathbf{p}_s)$ is the k-th entry of the vector-valued function:

$$\psi(\mathbf{p}_d, \mathbf{p}_s) = |\mathcal{A}_S^{\mathrm{T}}|\mathbf{p}_s + |\mathcal{A}_0^{\mathrm{T}}|\mathbf{p}_d \in \mathbb{R}^{n_E}.$$

The constant λ_k encodes friction and other specifics of the pipe segment k, whereas g_k represents the gravity and pipe specific parameters and Δh_k denotes the height difference of the pipe segment. These scalar parameters in the system and those defined earlier are known at least within some range of uncertainty. System (3) can be rewritten in the form (1a) leading to a system of nonlinear DAEs with dimension $n = n_E + n_E + n_d + n_0 + n_s$. The desired outputs in $\mathbb{R}^{n_s+n_d}$ can be obtained using the output equation

$$\mathbf{y} = \begin{pmatrix} \mathbf{y}_q \\ \mathbf{y}_p \end{pmatrix} = \begin{pmatrix} 0 & |\mathcal{A}_S| & 0 & 0 \\ 0 & 0 & \mathbf{B}_d^T & 0 \end{pmatrix} \begin{pmatrix} \mathbf{q}_- \\ \mathbf{q}_+ \\ \mathbf{p}_d \\ \mathbf{p}_s \end{pmatrix},$$

where $\mathbf{y}_q = |\mathcal{A}_S|\mathbf{q}_+$ and $\mathbf{y}_p = \mathbf{B}_d^T\mathbf{p}_d$. If we let $\mathbf{x} = \left(\mathbf{q}_-^T \ \mathbf{q}_+^T \ \mathbf{p}_d^T \ \mathbf{p}_s^T\right)^{\mathrm{T}} \in \mathbb{R}^n$, then the discretized gas flow model can be written in the form (1), where

$$\mathbf{E} = \begin{pmatrix} 0 & 0 & |\mathcal{A}_0^T| & |\mathcal{A}_S^T| \\ 0 & I & 0 & 0 \\ 0 & 0 & 0 & 0 \\ 0 & 0 & 0 & 0 \end{pmatrix}, \quad \mathbf{H} = \begin{pmatrix} -\mathbf{M}_L^{-1} & 0 & 0 & 0 \\ 0 & 0 & \mathbf{M}_A \mathcal{A}_0^{\mathrm{T}} & \mathbf{M}_A \mathcal{A}_S^{\mathrm{T}} \\ |\mathcal{A}_0| & \mathcal{A}_0 & 0 & 0 \\ 0 & 0 & 0 & I \end{pmatrix}, \quad \mathbf{f}(\mathbf{x}) = \begin{pmatrix} 0 \\ g(x) \\ 0 \\ 0 \end{pmatrix},$$

$$\mathbf{B} = -\begin{pmatrix} 0 & 0 \\ 0 & 0 \\ 0 & \mathbf{B}_d \\ I & 0 \end{pmatrix}, \quad \mathbf{C} = \begin{pmatrix} 0 & |\mathcal{A}_S| & 0 & 0 \\ 0 & 0 & \mathbf{B}_d^T & 0 \end{pmatrix}, \quad \mathbf{u} = \begin{pmatrix} s(t) \\ d(t) \end{pmatrix}, \quad \mathbf{h}(\mathbf{x}) = 0.$$

with $n_s + n_d$ inputs. In [4], simulation and MOR of (3) were discussed and the tractability index concept was used to classify the DAE. It was shown that gas transportation networks are of tractability index 1 if and only if they have only one supply node, otherwise they are of index 2. However, the transformation techniques using projection matrices and back substitution were used to rewrite (3) into nonlinear ODEs given by

$$
\begin{pmatrix} |\mathcal{A}_0|\mathbf{M}_L|\mathcal{A}_0^{\mathrm{T}}| & 0 \\ 0 & I \end{pmatrix} \begin{pmatrix} \partial_t \mathbf{p}_d \\ \partial_t \mathbf{q}_+ \end{pmatrix} = \begin{pmatrix} 0 & \mathcal{A}_0 \\ \mathbf{M}_A \mathcal{A}_0^T & 0 \end{pmatrix} \begin{pmatrix} \mathbf{p}_d \\ \mathbf{q}_+ \end{pmatrix} + \begin{pmatrix} |\mathcal{A}_0|\mathbf{M}_L|\mathcal{A}_S^{\mathrm{T}}|\partial_t \mathbf{s}(t) \\ \mathbf{g}(\mathbf{q}_+, \mathbf{s}(t), \mathbf{p}_d) \end{pmatrix} \tag{4}
$$
$$
+ \begin{pmatrix} 0 & -\mathbf{B}_d \\ \mathbf{M}_A \mathcal{A}_S^T & 0 \end{pmatrix} \begin{pmatrix} \mathbf{s}(t) \\ \mathbf{d}(t) \end{pmatrix}.
$$

After simulating the above ODE, $\mathbf{q}_-$ and $\mathbf{p}_s$ can then be computed in a post processing step, however it is not necessary to compute the desired output. We can observe the dimension of the original DAE (3) has been reduced to $n_d + n_0 + n_E$. Then using standard MOR methods can be applied to the index-reduced ODE (4).

3 Automatic Decoupling of Linear DAEs

In this section, we recall the automatic decoupling process, which can be used for any linear DAE with certain properties [3]. In order to use this approach, we have to first linearize our nonlinear DAE, then use the automatic decoupling. A solution of the system is then computed from the decoupled system. This approach can be summarized in Fig. 1.

We first linearize the nonlinear DAE (1) by computing a stationary solution $\mathbf{x}_s$ for a given static input $\mathbf{u}_s$. This means we have

$$
0 = \mathbf{H}\mathbf{x}_s + \mathbf{f}(\mathbf{x}_s) + \mathbf{B}\mathbf{u}_s.
$$

Fig. 1 Graphical representation of automatic decoupling of DAEs

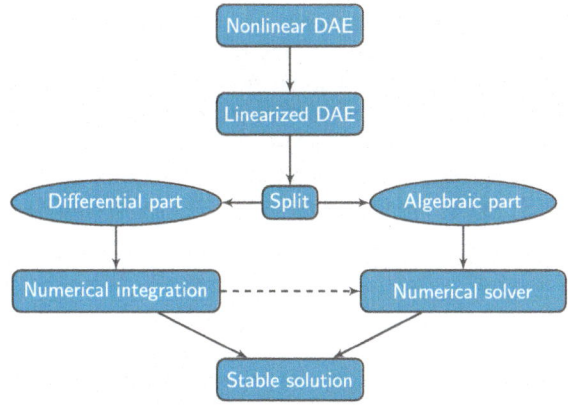

Then, using the Taylor series expansion around a steady-state operating point $(\mathbf{x}_s, \mathbf{u}_s)$, a linearized DAE system is obtained given by

$$\mathbf{E}\bar{\mathbf{x}}' = \mathbf{A}\bar{\mathbf{x}} + \mathbf{B}\bar{\mathbf{u}}, \quad \mathbf{E}\bar{\mathbf{x}}(0) = \mathbf{E}\bar{\mathbf{x}}_0, \tag{5a}$$

$$\bar{\mathbf{y}} = \mathbf{C}\bar{\mathbf{x}}, \tag{5b}$$

where $\mathbf{A} = \left.\dfrac{\partial \mathbf{F}}{\partial \mathbf{x}}\right|_{\mathbf{x}_s} = \mathbf{H} + \left.\dfrac{\partial \mathbf{f}}{\partial \mathbf{x}}\right|_{\mathbf{x}_s} \in \mathbb{R}^{n \times n}$, $\bar{\mathbf{x}} = \mathbf{x} - \mathbf{x}_s \in \mathbb{R}^n$ and $\bar{\mathbf{u}} = \mathbf{u} - \mathbf{u}_s \in \mathbb{R}^m$.
The linearized DAE system (5a) holds in a neighborhood of the stationary point $(\mathbf{x}_s, \mathbf{u}_s)$ for the nonlinear DAE (1).

Next, we split the linearized DAE (5) into differential and algebraic parts using special projectors and their respective bases as proposed in [3]. For convenience, we set $\bar{\mathbf{x}} = \mathbf{x}$ and $\bar{\mathbf{u}} = \mathbf{u}$. According to [3], in order to decouple linear constant coefficients DAEs into differential and algebraic parts, we can use the definition of the tractability index as the starting point. Assume (5) is solvable, i.e., $\det(\lambda \mathbf{E} - \mathbf{A}) \neq 0$, then we can define a matrix and projector chain by setting $\mathbf{E}_0 := \mathbf{E}$ and $\mathbf{A}_0 := \mathbf{A}$ given by

$$\mathbf{E}_{j+1} := \mathbf{E}_j - \mathbf{A}_j \mathbf{Q}_j, \quad \mathbf{A}_{j+1} := \mathbf{A}_j \mathbf{P}_j, \quad \text{for} \quad j \geq 0, \tag{6}$$

where $\mathbf{Q}_j \in \mathbb{R}^{n \times n}$ is a projector onto the null space of $\mathbf{E}_j$ and $\mathbf{P}_j = \mathbf{I} - \mathbf{Q}_j \in \mathbb{R}^{n \times n}$. Then if there exists an index γ such that $\mathbf{E}_\gamma$ is nonsingular and all $\mathbf{E}_j$ are singular for all $0 < j < \gamma - 1$ it is called the *tractability index*. In [4], the tractability index concept was used to classify DAEs arising from gas transportation networks. In this paper, we discus how the tractability index concept can be used to split the DAE into differential and algebraic parts. In order to obtain an appropriate tool for decoupling of the DAE (5) of index higher than one, an additional constraint $\mathbf{Q}_j \mathbf{Q}_i = 0$, $j > i$, is introduced. This class of projectors are sometimes called admissible projectors [5] or canonical projectors [6]. These projectors are numerically feasible and their construction is well discussed in [6]. A key step in forming the projectors in (6) is to find the initial projectors $\mathbf{Q}_j$ spanning the nullspaces of the usually sparse $\mathbf{E}_j$. Standard ways of identifying the nullspace include singular value decomposition (SVD) or alike, which do not utilize matrix patterns and can be expensive for large-size matrices. The most efficient way is to employ the sparse LU decomposition-based routine, called LUQ, see [7]. This same routine was also used to construct the projector bases introduced in [3]. According to [3], these projectors and projector bases can be used to split system (5) into an equivalent decoupled system given by

$$\mathbf{E}_p \partial_t \xi_p = \mathbf{A}_p \xi_p + \mathbf{B}_p \mathbf{u}, \quad \xi_p(0) = \xi_{p_0}, \tag{7a}$$

$$-\mathcal{L} \partial_t \xi_q = \mathbf{A}_q \xi_p - \mathcal{L}_q \xi_q + \mathbf{B}_q \mathbf{u}, \tag{7b}$$

$$\mathbf{y} = \mathbf{C}_p \xi_p + \mathbf{C}_q \xi_q, \tag{7c}$$

where $\mathcal{L} \in \mathbb{R}^{n_q \times n_q}$ is a nilpotent matrix with index γ and $\mathcal{L}_q \in \mathbb{R}^{n_q \times n_q}$, $\mathbf{E}_p \in \mathbb{R}^{n_p \times n_p}$ are always non-singular matrices. The subsystems (7a) and (7b) correspond to the differential and algebraic parts of system (5). $\xi_p \in \mathbb{R}^{n_p}$ and $\xi_q \in \mathbb{R}^{n_q}$ are the differential and algebraic variables, respectively. The value of the differential variable ξ_p is computed by applying any standard numerical integration scheme to (7a). After computing the value of the differential variable, the algebraic variables can be computed as follows. We can observe that (7b) can be rewritten as

$$\mathcal{L}_q \xi_q - \mathcal{L} \partial_t \xi_q = \mathbf{A}_q \xi_p + \mathbf{B}_q \mathbf{u} \Rightarrow (\mathrm{I} - \mathcal{L}\mathcal{L}_q^{-1} \partial_t) \mathcal{L}_q \xi_q = \mathbf{A}_q \xi_p + \mathbf{B}_q \mathbf{u}$$

$$\Rightarrow \mathcal{L}_q \xi_q = (\mathrm{I} - \mathbf{N}_q \partial_t)^{-1} \left(\mathbf{A}_q \xi_p + \mathbf{B}_q \mathbf{u} \right) \Rightarrow \mathcal{L}_q \xi_q = \sum_{j=0}^{\gamma-1} \mathbf{N}_q^j \partial_t^j \left(\mathbf{A}_q \xi_p + \mathbf{B}_q \mathbf{u} \right),$$

where $\mathbf{N}_q = \mathcal{L}\mathcal{L}_q^{-1}$ is also a nilpotent matrix with the same index γ as $\mathcal{L}$. Thus, (7b) can be rewritten as

$$\mathcal{L}_q \xi_q = \sum_{j=0}^{\gamma-1} \mathbf{N}_q^j \left(\mathbf{A}_q \partial_t^j \xi_p + \mathbf{B}_q \partial_t^j \mathbf{u} \right). \tag{8}$$

The above system can be simulated using numerical solvers. Thus, the algebraic variable ξ_q is computed by first applying numerical integration on (7a) to obtain ξ_p, and then apply numerical solvers for (8). Then the output solution can be obtained using (7c). Hence, instead of numerically integrating (7b) one has to numerically solve (8) which leads to stable solutions. For the two relevant values of tractability index in our situation, (5) becomes

- $\gamma = 1$:

$$\mathcal{L}_q \xi_q = \mathbf{A}_q \xi_p + \mathbf{B}_q \mathbf{u}. \tag{9}$$

- $\gamma = 2$:

$$\mathcal{L}_q \xi_q = \mathbf{A}_q \xi_p + \mathbf{B}_q \mathbf{u} + \mathbf{N}_q \left(\mathbf{A}_q \partial_t \xi_p + \mathbf{B}_q \partial_t \mathbf{u} \right). \tag{10}$$

In order to avoid numerical errors while solving the above system, $\partial_t \xi_p$ can be obtained by applying numerical solver for (7a) after obtaining the value of ξ_p, while $\partial_t \mathbf{u}$ can be computed explicitly or symbolically.

Finally, we discuss how the matrix coefficients of (7) can be computed as proposed in [3]. If (5) is of tractability index $\gamma = 1$, the matrix coefficients of (7) are given by

$$\mathbf{E}_p = \hat{\mathbf{p}}_0^\mathrm{T} \mathbf{E}_0 \mathbf{p}_0 \in \mathbb{R}^{n_p \times n_p}, \quad \mathbf{A}_p = \hat{\mathbf{p}}_0^\mathrm{T} \mathbf{A}_0 \mathbf{p}_0 \in \mathbb{R}^{n_p \times n_p}, \quad \mathbf{B}_p = \hat{\mathbf{p}}_0^\mathrm{T} \mathbf{B} \in \mathbb{R}^{n_p \times m},$$

$$\mathcal{L} = 0, \quad \mathcal{L}_q = \hat{\mathbf{q}}_0^\mathrm{T} \mathbf{A}_0 \mathbf{q}_0 \in \mathbb{R}^{n_q \times n_q}, \quad \mathbf{A}_q = \hat{\mathbf{q}}_0^\mathrm{T} \mathbf{A}_0 \mathbf{p}_0 \in \mathbb{R}^{n_p \times n_q},$$

$$\mathbf{B}_q = \hat{\mathbf{q}}_0^\mathrm{T} \mathbf{B} \in \mathbb{R}^{n_q \times m}, \quad \mathbf{C}_p = \mathbf{C} \mathbf{p}_0 \in \mathbb{R}^{\ell \times n_p}, \quad \mathbf{C}_q = \mathbf{C} \mathbf{q}_0 \in \mathbb{R}^{\ell \times n_q} \quad \text{and} \quad \xi_{p_0} = \mathbf{p}_0^{*\mathrm{T}} \mathbf{P}_0 \mathbf{x}(0).$$

The columns of the matrices

- $\mathbf{q}_0 \in \mathbb{R}^{n \times n_q}$ and $\mathbf{p}_0 \in \mathbb{R}^{n \times n_p}$ are linearly independent and span the column spaces of projectors $\mathbf{Q}_0 \in \mathbb{R}^{n \times n}$ and $\mathbf{P}_0 \in \mathbb{R}^{n \times n}$, respectively. $\mathbf{p}_0^{*T} \in \mathbb{R}^{n_p \times n}$ is the left inverse of basis $\mathbf{p}_0$ such that $\mathbf{p}_0^{*T} \mathbf{p}_0 = I_{n_p}$.
- $\hat{\mathbf{p}}_0 \in \mathbb{R}^{n \times n_p}$ and $\hat{\mathbf{q}}_0 \in \mathbb{R}^{n \times n_q}$ are linearly independent and span the null spaces of the matrices $\mathbf{q}_0^T \mathbf{A}_0^T \in \mathbb{R}^{n_q \times n}$ and $\mathbf{E}_0^T \in \mathbb{R}^{n \times n}$, respectively.

In this paper, we restrict ourselves to tractability index $\gamma = 1$, but in the case tractability index $\gamma = 2$, the matrix coefficients of (7) can be computed as discussed in [3].

The proposed automatic decoupling method allows efficient simulation of nonlinear DAEs arising from gas transportation networks without worrying about consistent initial conditions, hidden constraints and smoothness of the input data. Hence, the index problem is eliminated. Moreover, the decoupled system is well suited due to well-conditioned projectors and projector bases required for the decoupling strategy. The decoupling strategy allows automatic decoupling of nonlinear DAEs arising from gas transportation networks of any structure.

4   Decoupling Analysis of DAEs Arising From Gas Transport Networks

In this section, we discuss how the nonlinear DAE (3) arising from the spatial discretization of the gas transport network proposed in [4] can be explicitly decoupled using an automatic approach discussed in the previous section. Linearization of (3) leads to a linear DAE in the form (5). Following the discussion presented in the previous section, system (5) can be decoupled as follows. Setting

$$
\mathbf{E}_0 = \mathbf{E} = \begin{pmatrix} 0 & 0 & |\mathcal{A}_0^T| & |\mathcal{A}_S^T| \\ 0 & I & 0 & 0 \\ 0 & 0 & 0 & 0 \\ 0 & 0 & 0 & 0 \end{pmatrix} \quad \text{and} \quad \mathbf{A}_0 = \overline{\mathbf{A}} = \begin{pmatrix} -\mathbf{M}_L^{-1} & 0 & 0 & 0 \\ 0 & \mathbf{A}_{22} & \mathbf{A}_{23} & \mathbf{A}_{24} \\ |\mathcal{A}_0| & \mathcal{A}_0 & 0 & 0 \\ 0 & 0 & 0 & I \end{pmatrix},
$$

where $\mathbf{A}_{22} = \left. \frac{\partial \mathbf{g}}{\partial \mathbf{q}_+} \right|_{\mathbf{x}_s} \in \mathbb{R}^{n_E \times n_E}$, $\mathbf{A}_{23} = \mathbf{M}_A \mathcal{A}_0^T + \left. \frac{\partial \mathbf{g}}{\partial \mathbf{p}_q} \right|_{\mathbf{x}_s} \in \mathbb{R}^{n_E \times n_d}$,

$\mathbf{A}_{24} = \mathbf{M}_A \mathcal{A}_S^T + \left. \frac{\partial \mathbf{g}}{\partial \mathbf{p}_s} \right|_{\mathbf{x}_s} \in \mathbb{R}^{n_E \times n_s}$, we can now construct the projector and matrix sequence as follows. We first construct projectors

$$
\mathbf{Q}_0 = \left(\begin{array}{cc|cc} I & 0 & 0 & 0 \\ 0 & 0 & 0 & 0 \\ \hline 0 & 0 & & \\ 0 & 0 & & \mathbf{Q} \end{array} \right) \in \mathbb{R}^{n \times n} \quad \text{and} \quad \mathbf{P}_0 = I - \mathbf{Q}_0 = \left(\begin{array}{cc|cc} 0 & 0 & 0 & 0 \\ 0 & 0 & 0 & 0 \\ \hline 0 & 0 & & \\ 0 & 0 & & \mathbf{P} \end{array} \right) \in \mathbb{R}^{n \times n}, \quad (11)
$$

such that $\mathbf{Q}_0$ projects onto the nullspace of $\mathbf{E}_0$, where $\mathbf{Q} \in \mathbb{R}^{(n_d+n_s)}$ is the projector onto the nullspace of $\mathbf{E}_{13} = \left(|\mathcal{A}_0^{\mathsf{T}}| \, |\mathcal{A}_S^{\mathsf{T}}| \right)$ and $\mathbf{P} \in \mathbb{R}^{(n_d+n_s)}$ is its complementary projector. We note that projectors $\mathbf{Q}_0$ and $\mathbf{P}_0$ are not unique. Using the definition (6) of tractability index, we obtain $\mathbf{E}_1 = \mathbf{E}_0 - \mathbf{A}_0 \mathbf{Q}_0$. If we assume $\mathbf{E}_1$ to be nonsingular, then DAE (5) is of tractability index 1, since in [4], it was shown that gas networks are of tractability index 1 if and only if they have only one supply node. This implies that $\mathbf{E}_1$ is nonsingular if $n_s = 1$ and singular if $n_s > 1$. Since we are considering gas networks with one supply node, (5) is an index 1 system and its decoupled system can be derived from (7) leading to

$$\mathbf{E}_p \partial_t \xi_p = \mathbf{A}_p \xi_p + \mathbf{B}_p \mathbf{u}, \quad \xi_p(0) = \xi_{p_0}, \tag{12a}$$

$$\mathcal{L}_q \xi_q = \mathbf{A}_q \xi_p + \mathbf{B}_q \mathbf{u}, \tag{12b}$$

$$\mathbf{y} = \mathbf{C}_p \xi_p + \mathbf{C}_q \xi_q, \tag{12c}$$

where

$$\mathbf{E}_p = \hat{\mathbf{p}}_0^{\mathsf{T}} \mathbf{E}_0 \mathbf{p}_0 \in \mathbb{R}^{n_p \times n_p}, \quad \mathbf{A}_p = \hat{\mathbf{p}}_0^{\mathsf{T}} \mathbf{A}_0 \mathbf{p}_0 \in \mathbb{R}^{n_p \times n_p}, \quad \mathbf{B}_p = \hat{\mathbf{p}}_0^{\mathsf{T}} \mathbf{B} \in \mathbb{R}^{n_p \times m},$$

$$\mathcal{L}_q = \hat{\mathbf{q}}_0^{\mathsf{T}} \mathbf{A}_0 \mathbf{q}_0 \in \mathbb{R}^{n_q \times n_q}, \quad \mathbf{A}_q = \hat{\mathbf{q}}_0^{\mathsf{T}} \mathbf{A}_0 \mathbf{p}_0 \in \mathbb{R}^{n_q \times n_p}, \quad \mathbf{B}_q = \hat{\mathbf{q}}_0^{\mathsf{T}} \mathbf{B} \in \mathbb{R}^{n_q \times m},$$

$$\mathbf{C}_p = \mathbf{C} \mathbf{p}_0 \in \mathbb{R}^{\ell \times n_p}, \quad \mathbf{C}_q = \mathbf{C} \mathbf{q}_0 \in \mathbb{R}^{\ell \times n_q} \quad \text{and} \quad \xi_{p_0} = \mathbf{p}_0^{*\mathsf{T}} \mathbf{P}_0 \mathbf{x}(0) \in \mathbb{R}^{n_p}.$$

The columns of the matrices

$$\mathbf{q}_0 = \begin{pmatrix} I & 0 \\ 0 & 0 \\ \hline 0 & \\ 0 & \mathbf{q} \end{pmatrix} \in \mathbb{R}^{n \times n_q} \quad \text{and} \quad \mathbf{p}_0 = \begin{pmatrix} 0 & 0 \\ \hline I & 0 \\ 0 & \\ 0 & \mathbf{p} \end{pmatrix} \in \mathbb{R}^{n \times n_p}$$

are linearly independent and span the column spaces of $\mathbf{Q}_0$ and $\mathbf{P}_0$ in (11), respectively. Let k_q be the dimension of the nullspace of $\mathbf{E}_{13}$, and $k_p = (n_d + n_s) - k_q$. Then, $\mathbf{q} \in \mathbb{R}^{(n_d+n_s) \times k_q}$ and $\mathbf{p} \in \mathbb{R}^{(n_d+n_s) \times k_p}$ are matrices whose columns are linearly independent and span the column spaces of $\mathbf{Q}$ and $\mathbf{P}$ in (11), respectively. Hence, the left inverse $\mathbf{p}_0^{*\mathsf{T}} \in \mathbb{R}^{n_p \times n}$ of basis $\mathbf{p}_0$ is given by $\mathbf{p}_0^{*\mathsf{T}} = \begin{pmatrix} 0 \, I & 0 \, 0 \\ \hline 0 \, 0 & \mathbf{p}^{*\mathsf{T}} \end{pmatrix} \in \mathbb{R}^{n \times n_p}$. where $\mathbf{p}^{*\mathsf{T}}$ is the left inverse of $\mathbf{p}$. Finally, column matrices $\hat{\mathbf{p}}_0 \in \mathbb{R}^{n \times n_p}$ and $\hat{\mathbf{q}}_0 \in \mathbb{R}^{n \times n_q}$ are defined as $\hat{\mathbf{p}}_0 \in \mathrm{Ker}(\mathbf{q}_0^{\mathsf{T}} \mathbf{A}_0^{\mathsf{T}})$ and $\hat{\mathbf{q}}_0 \in \mathrm{Ker}(\mathbf{E}_0^{\mathsf{T}})$, respectively. We can observe that the linearized DAE of (3) has been decoupled into $n_p = n_E + k_p$ differential equations, and $n_q = n_E + k_q$ algebraic equations. The differential part has the same dimension as an implicit ODE (4) for gas transport networks with one supply node.

5 Index-Aware MOR for Gas Transport Networks

In this, section we discuss how the decoupled system (7) can be reduced using the index-aware MOR methods. IMOR replaces (7) by an IROM [3]

$$\mathbf{E}_{p_r}\partial_t\xi_{p_r} = \mathbf{A}_{p_r}\xi_{p_r} + \mathbf{B}_{p_r}\mathbf{u}, \quad \xi_{p_r}(0) = \xi_{p_{r_0}}, \tag{13a}$$

$$-\mathcal{L}_r\partial_t\xi_{q_r} = \mathbf{A}_{q_r}\xi_{p_r} - \mathcal{L}_{q_r}\xi_{q_r} + \mathbf{B}_{q_r}\mathbf{u}, \tag{13b}$$

$$\mathbf{y}_r = \mathbf{C}_{p_r}\xi_{p_r} + \mathbf{C}_{q_r}\xi_{q_r}, \tag{13c}$$

where

$$\mathbf{E}_{p_r} = \mathbf{V}_p^{\mathsf{T}}\mathbf{E}_p\mathbf{V}_p, \mathbf{A}_{p_r} = \mathbf{V}_p^{\mathsf{T}}\mathbf{A}_p\mathbf{V}_p \in \mathbb{R}^{r_p \times r_p}, \quad \mathbf{B}_{p_r} = \mathbf{V}_p^{\mathsf{T}}\mathbf{B}_p \in \mathbb{R}^{r_p \times m},$$

$$\xi_{p_{r_0}} = \mathbf{V}_p^{\mathsf{T}}\xi_{p_0} \in \mathbb{R}^{r_p \times n_p}, \quad \mathcal{L}_r = -\mathbf{V}_q^{\mathsf{T}}\mathcal{L}\mathbf{V}_q \in \mathbb{R}^{r_q \times r_q}, \quad \mathcal{L}_{q_r} = \mathbf{V}_q^{\mathsf{T}}\mathcal{L}_q\mathbf{V}_q \in \mathbb{R}^{r_q \times r_q},$$

$$\mathbf{A}_{q_r} = \mathbf{V}_q^{\mathsf{T}}\mathbf{A}_q\mathbf{V}_p \in \mathbb{R}^{r_q \times r_p}, \quad \mathbf{B}_{q_r} = \mathbf{V}_q^{\mathsf{T}}\mathbf{B}_q \in \mathbb{R}^{r_q \times m}, \quad \mathbf{C}_{p_r} = \mathbf{C}_p\mathbf{V}_p \in \mathbb{R}^{\ell \times n_p},$$

$$\mathbf{C}_{q_r} = \mathbf{C}_q\mathbf{V}_q \in \mathbb{R}^{\ell \times n_q}.$$

$\mathbf{V}_p \in \mathbb{R}^{n_p \times r_p}$ is constructed using any standard MOR method such as POD, empirical balanced truncation methods, etc., applied to the ODE subsystem

$$\mathbf{E}_p\partial_t\xi_p = \mathbf{A}_p\xi_p + \mathbf{B}_p\mathbf{u}, \quad \xi_p(0) = \xi_{p_0}, \tag{14a}$$

$$\mathbf{y}_p = \mathbf{C}_p\xi_p. \tag{14b}$$

After constructing $\mathbf{V}_p$ as above, the projection matrix $\mathbf{V}_q \in \mathbb{R}^{n_q \times r_q}$ can be constructed as follows. Substituting $\xi_p \approx \mathbf{V}_p\xi_{p_r}$ into (8) leads to

$$\mathcal{L}_q\xi_q \approx \sum_{j=0}^{\gamma-1}\mathbf{N}_q^j\left(\mathbf{A}_q\mathbf{V}_p\partial_t^j\xi_{p_r} + \mathbf{B}_q\partial_t^j\mathbf{u}\right). \tag{15}$$

We can observe that, for the algebraic variable ξ_q, we have the restriction

$$\mathcal{L}_q\xi_q \in \mathcal{W}_q = \mathcal{K}_\gamma(\mathbf{N}_q, \mathbf{R}_q) = \mathrm{Span}\left(\mathbf{R}_q, \mathbf{N}_q\mathbf{R}_q, \ldots, \mathbf{N}_q^{\gamma-1}\mathbf{R}_q\right),$$

where $\mathbf{R}_q = \left(\mathbf{B}_q, \mathbf{A}_q\mathbf{V}_p\right) \in \mathbb{R}^{n_q \times (r_p+m)}$. Then,

$$\xi_q \in \mathcal{V}_q = \mathcal{L}_q^{-1}\mathcal{W}_q = \mathcal{K}_\gamma(\mathcal{L}_q^{-1}\mathbf{N}_q, \mathcal{L}_q^{-1}\mathbf{R}_q).$$

We denote by $\mathbf{V}_q$ the orthonormal basis of $\mathcal{V}_q$ which can be computed using the singular value decomposition (SVD) and truncating the smallest singular values. For index $\gamma = 1$ gas transport network models: $\mathbf{V}_q = \mathrm{orth}(\mathcal{L}_q^{-1}\mathbf{R}_q)$ while for index $\gamma = 2$ gas transport network models: $\mathbf{V}_q = \mathrm{orth}([\mathcal{L}_q^{-1}\mathbf{R}_q, \mathcal{L}_q^{-1}\mathbf{R}_q\mathbf{N}_q])$. For $\gamma > 2$, one would in general use Arnoldi for this computation. As a result, both the differential and algebraic subsystems are reduced while preserving the index of the original DAE. An alternative way is to construct $\mathbf{V}_q$ using POD by generating snapshots using (8).

6 Numerical Examples

In this section, we illustrate the performance of the proposed automatic decoupling and IMOR methods for nonlinear DAEs arising from gas transportation networks. We compute the relative error in the format Re.error $= \|\mathbf{y} - \mathbf{y}_r\|_2 / \|\mathbf{y}\|_2$. The output error is defined as max(Re.error(pressure), Re.error(mass flow)). In all our experiments the speedup is a number that measures the relative performance of simulating the original and the reduced-order model. Simulations were done using MATLAB®Version R2012b on a Unix desktop.

Example 1 In this example, we compare different gas transport network models derived from the same nonlinear DAE (3). These are: the linearized DAE model (5) and the linear decoupled model (7). We consider a small size gas transport network obtained from [8]. It consists of 17 nodes, 16 pipes, 1 supply node and 8 demand nodes. Spatial discretization leads to a nonlinear DAE of the form (3) with $n = 55$, $m = \ell = 9, m_s = 1, m_d = 8$. We consider steady pressure at the supply node and mass flow as step function at demand nodes as input functions as shown in Fig. 2.

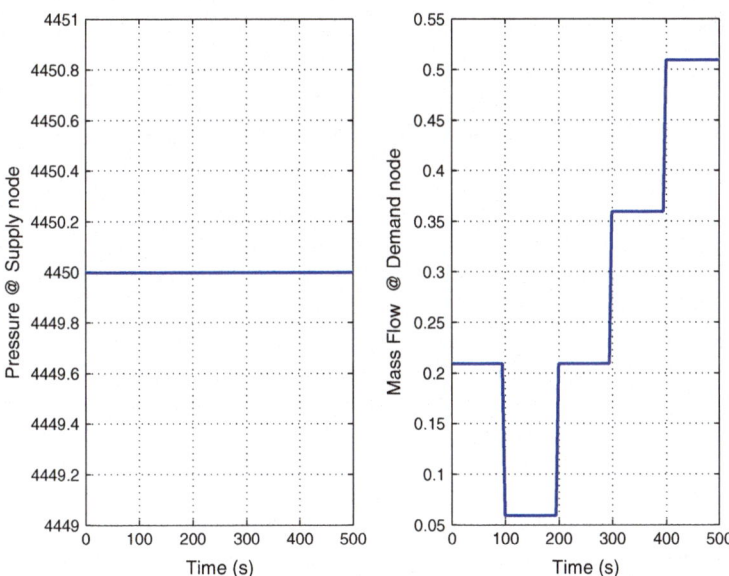

Fig. 2 Input functions in the time interval $t \in [0, 500s]$.

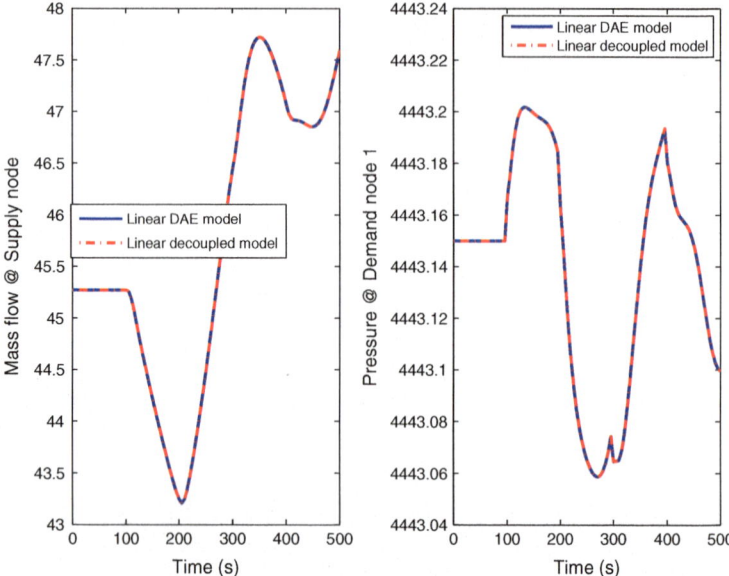

Fig. 3 Comparison of the output solutions in the time interval $t \in [0, 500s]$.

The linearized DAE model is of the form (5) with dimension 55, while the decoupled system (7) has $n_p = 36$ differential equations and $n_q = 19$ algebraic equations. We used the implicit Euler integration scheme to simulate the linear DAE. We also used the same method to simulate the ODE part and the LU method for solving the algebraic part of the decoupled system. Using the same time steps and time interval, we simulated all the models and some of the results are presented in Fig. 3. We can observe that the solutions of the linearized DAE and the linear decoupled models coincide as expected.

In Examples 2 and 3, we use gas transport network models from [9] with only one supply pressure node to illustrate the performance of the IMOR method proposed in Sect. 5. Both networks lead to index 1 DAEs. We apply the empirical balanced truncation (EBT) [10] and POD methods to reduce the differential part and by induction the algebraic part is also reduced to its minimum realization leading to ROMs which are also DAEs of the same index. We call the corresponding methods the index-aware EBT (I-EBT) and index-aware POD (I-POD). For comparison, we also applied EBT and POD to the decoupled original model.

Example 2 In this example, we are interested in comparing different standard MOR methods with that of the IMOR methods. We consider a medium size gas transport pipe network with 200 pipes generated using the following data. The length, diameter and average roughness of each pipe are chosen constant given by 100 m, 1 m and

Table 1 Comparison of the standard MOR and IMOR methods

Method	Decoupled model		IROM		Red. size	%Red.	Offline	Out. error	Speed-up
	n_p	n_q	r_p	r_q	r				
I-POD	400	201	2	4	6	99.0	0.07	1.2×10^{-5}	4.1
POD	–	–	–	–	2	99.7	0.07	3.7×10^{-5}	5.5
I-EBT	400	201	7	9	16	97.3	24.0	7.2×10^{-5}	4.4
EBT	–	–	–	–	16	97.3	37.9	7.3×10^{-5}	2.4

1.0×10^{-3} m, respectively. The gas composition through the network is methane with specific gas constant $518.26 \, \text{J/KgK}$ at supply inputs as shown in the first row of Fig. 6 in the time interval $t \in [0, 86400s]$. This leads to a nonlinear DAE (3) of dimension $n = 601$ which we linearized and decoupled into $n_p = 400$ differential equations and $n_q = 201$ algebraic equations. For comparison, the size of ROMs for different MOR methods is determined by making sure that the output error is below 10^{-4} and the results are presented in in Table 1.

We observed that direct reduction using EBT and POD methods lead to ODE ROMs which are very close to DAEs which affects the choice of numerical solvers, while the I-EBT and I-POD methods lead to DAE ROMs with the same index as the original system. We can also observe that the I-ROMs are computationally cheaper to construct compared to the ODE ROMs since they need lower offline costs especially with the EBT method. However, the standard MOR methods leads to slightly smaller ROMs compared to the IMOR methods. This is due to the fact the standard MOR methods eliminate the algebraic part and yield ODE ROMs while the IMOR methods preserves the algebraic part leading to a DAE ROM. For speed-up comparisons, we use the implicit Euler scheme for the linearized coupled DAE system. In Figs. 4 and 5, we compare the relative error of the pressure and mass flows for both the POD and EBT methods are varying sizes of the ROMs. Figure 6 shows the output solution of the ROMs.

Example 3 In this example, we are interested in comparing the speed-ups of the POD with that of the I-POD. We consider a large-scale gas transport pipeline network with 5,000 pipes. This model was generated numerically using the following data. The length, diameter and average roughness of each pipe are chosen constants given by $3,630$ m, 1.422 m and 1.0×10^{-6} m, respectively. The gas composition is methane with specific gas constant $518.26 \, \text{J/KgK}$ at supply inputs as shown in the first row of Fig. 7 in time interval $t \in [0, 2400]$. This leads to a nonlinear DAE system (3) of dimension $n = 15,001$ which we linearized and decoupled into $n_p = 10,000$ differential equations and $n_q = 5,001$ algebraic equations. Generating matrices of the decoupled system took 370.6 s. This implies that decoupling is computationally

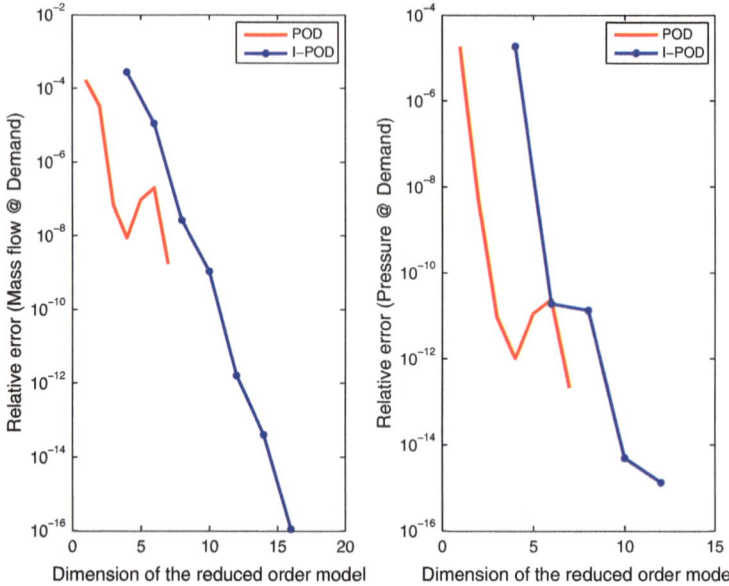

Fig. 4 Comparison of the relative errors of the POD models

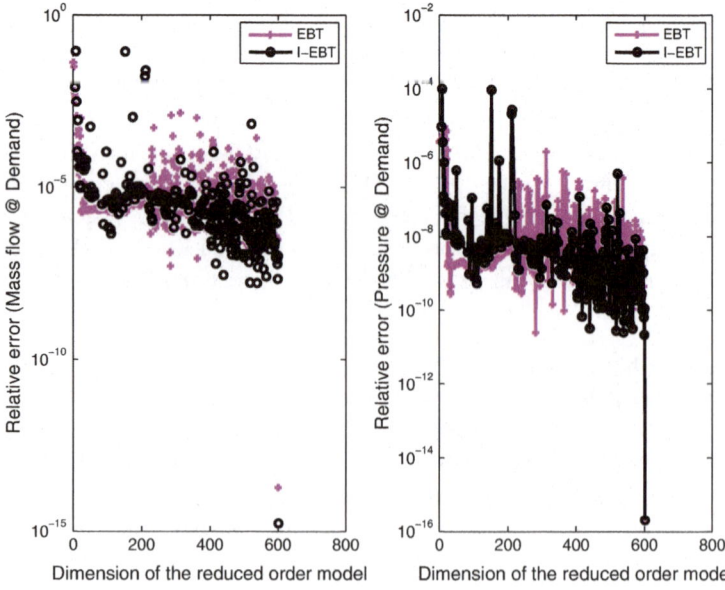

Fig. 5 Comparison of the relative errors of the EBT models

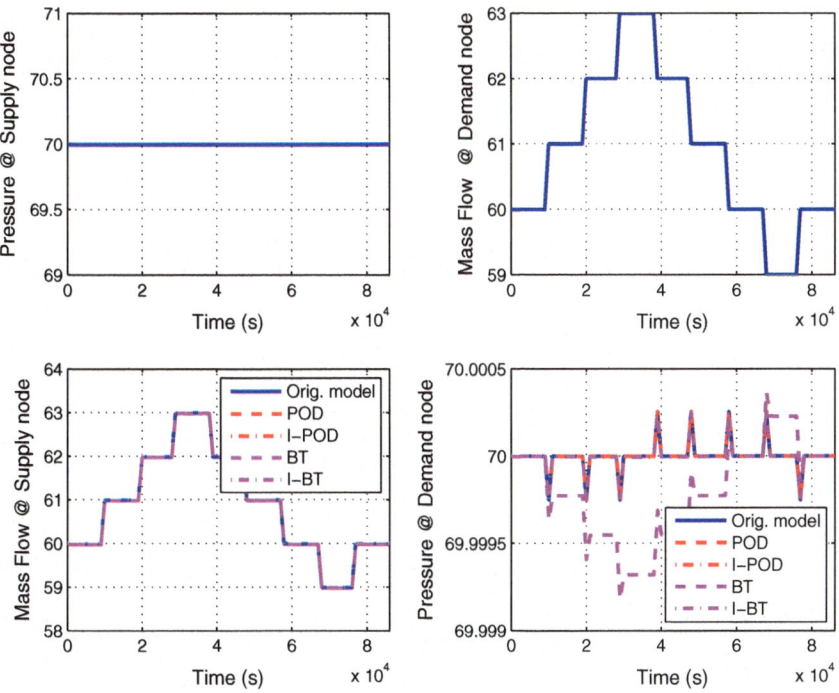

Fig. 6 Comparison of the supply mass flow and demand pressure of the ROMs

Table 2 Comparison of the POD and I-POD methods

Method	Decoupled model		IROM			%Red.	Out. error	Speed-up
	n_p	n_q	r_p	r_q	r			
I-POD	10, 000	5, 001	4	6	10	99.93	2.0×10^{-6}	644.6
POD	–	–	–	–	4	99.97	1.8×10^{-5}	575.8

efficient. For comparison the size of ROMs for different POD methods is determined by making sure that the output error is below 10^{-4} and the results are presented in Table 2, where, we can observe that I-POD is 1.12 times faster compared to the direct reduction using POD. In Fig. 7, we compare the output solutions of ROMs which coincide as expected.

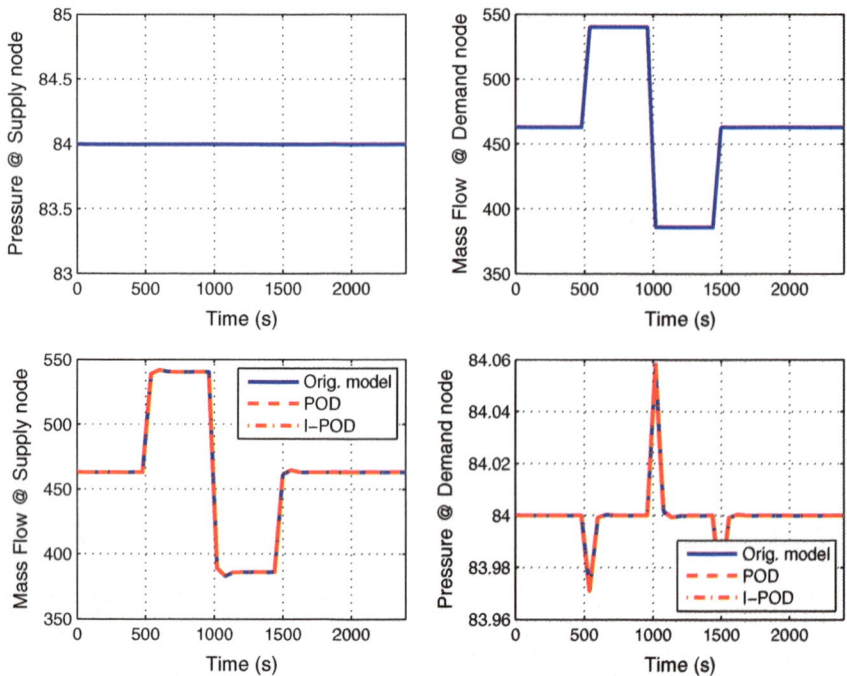

Fig. 7 Comparison of the supply mass flow and demand pressure of the ROMs

7 Conclusion

We have proposed an automatic decoupling strategy and an index-aware MOR method for linear DAEs arising from gas transportation networks. The automatic decoupling strategy is based on the foundations of tractability index and is independent of the spatial discretization method applied on the gas transportation networks. Moreover, the decoupling strategy can be applied on general structured gas transport networks with network control elements such as valves, compressors, regulators, etc. The main advantage of the decoupling strategy is that, it allows the use of standard numerical methods for simulation and model reduction of gas transport networks since it eliminates the index problem which normally causes numerical troubles. The IMOR method leads to ROMs which are also decoupled which makes them easier to simulate. Our decoupling strategy does not experience numerical troubles due to the well-conditioned projectors and projector bases. In cases of ill-conditioned matrices, incidence matrices can be used to construct these projectors and projector bases. This approach can also be applied directly to parametrized systems arising from gas transportation networks, if the projectors and the bases can be construct such that they are independent of the parameters. However, during the linearization process some information can be lost. Future research will deal with nonlinear DAEs without any kind of linearization.

References

1. Grundel, S.; Hornung, N.; Roggendorf. S.: Numerical aspects of model order reduction for gas transportation networks. In: Koziel, S., Leifsson, L., Yang, X.S., (eds.), Simulation-Driven Modeling and Optimization, pp. 1–28. Springer, Berlin (2016)
2. Herty, J., Mohring, J., Sachers, V.: A new model for gas flow in pipe networks. Math. Methods Appl. Sci. **33**(7), 845–855 (2010)
3. Banagaaya, N.: Index-aware model order reduction methods. PhD thesis, Eindhoven University of Technology, Eindhoven, Netherlands (2014)
4. Grundel, S., Jansen, L., Hornung, N., Clees, T., Tischendorf, C., Benner, P.: Model order reduction of differential algebraic equations arising from the simulation of gas transport networks. In: Schöps, S., Bartel, A., Günther, M., ter Maten, W.J.W., Müller, P.C. (eds.) Progress in Differential-Algebraic Equations, pp. 183–205. Springer, Berlin (2014)
5. Lamour, R., März, R., Tischendorf, C.: Differential Algebraic Equations: A Projector Based Analysis. Differential-Algebraic Equations Forum. Springer, Berlin (2013)
6. März, R.: Canonical projectors for linear differential algebraic equations. Comput. Math. Appl. **31**(4/5), 121–135 (1996)
7. Zhang, Z., Wong, N.: An efficient projector-based passivity test for descriptor systems. IEEE Trans. Comput.-Aided Des. Integr. Circuits Syst. **29**(8), 1203–1214 (2010)
8. Grundel, S., Hornung, N., Klaassen, B., Tischendorf, C., Benner, P., Clees, T.: Computing surrogates for gas network simulation using model order reduction. In: Koziel, S., Leifsson, L. (eds.), Surrogate-Based Modeling and Optimization, pp. 189–212. Springer, Berlin (2013)
9. Chaczykowski, M.: Sensitivity of pipeline gas flow model to the selection of the equation of state. Chem. Eng. Res. Des. **87**(12), 1596–1603 (2009)
10. Condon, M., Ivanov, R.: Empirical balanced truncation of nonlinear systems. Nonlinear Sci. **14**(5), 405–414 (2004)

Polynomial Tensor-Based Stability Identification of Milling Process: Application to Reduced Thin-Walled Workpiece

Chigbogu G. Ozoegwu

Abstract This work presents the stability analyses of milling process using a Full-discretization Method (FDM) which is constructed in the framework of second order polynomial tensor approximation of the cutting states. The proposed method is applied to the frequent milling model where the workpiece is considered rigid and the tool is considered compliant and, also, to the case where the thin-walled workpiece is considered flexible and the tool is considered rigid. The rigid tool is treated as a lumped parameter problem while the flexible thin-walled workpiece, being a continuum with very many degrees of freedom (DOF), is treated as a reduced order Finite Element problem. The computed numerical results agree with established results. The method is therefore applicable to the knowledge-based optimization of the milling of aero-structures. For future research, a foundation has been formed for the approach to be generalized for all orders of approximation for full computerization and accuracy optimization of the stability lobes of reduced order milling models using the FDM.

Keyword Model order reduction · Finite element · Full-discretization method · Thin-walled workpiece chatter stability lobe · Polynomial tensor

1 Introduction

Stability analysis is important for the optimization of surface quality and productivity of milling process. The out-of-process strategy, which simultaneously guarantees high productivity and reasonable surface integrity, is popular for the knowledge-based selection of cutting process conditions. The strategy is based on modelling and stability analysis of the regenerative dynamics of the cutting process utilizing the results of experimental or theoretical modal analysis. The regenerative delay differential equation models of milling process are periodic [1], and the computational

C. G. Ozoegwu (✉)
Department of Mechanical Engineering, University of Nigeria, Nsukka, Nigeria
e-mail: chigbogug@yahoo.com; chigbogu.ozoegwu@unn.edu.ng

© Springer Nature Switzerland AG 2020
J. Fehr and B. Haasdonk (eds.), *IUTAM Symposium on Model Order Reduction of Coupled Systems, Stuttgart, Germany, May 22–25, 2018*, IUTAM Bookseries 36,
https://doi.org/10.1007/978-3-030-21013-7_15

stability analyses of the models are mostly done in either the frequency domain or the time domain. The Frequency domain approaches are based on truncated Fourier series expansion of the periodic coefficients and the Nyquist stability criterion [2, 3] while the more recent time domain approaches are based on the time-domain discretization of the governing DDE and the construction of a finite-dimensional map [4–8].

Monolithic parts for the aerospace industry require high strength to weight ratio. As a result, the milling of thin-walled workpiece is a basic material processing requirement in the industry. During such a machining process, much of the blank workpiece is removed. The relatively high volume allowance has to be removed as fast as possible for the process to be economically viable, therefore, optimization of material removal rate (MRR) is key for the aero-machining industry. Since a thin-walled workpiece is a distributed system, the response is dependent on the location of the excitation (tool-workpiece interaction) and the extent of material removal. Stability analysis of milling of thin-walled workpiece has mostly been carried out in the frequency domain. A frequency domain approach was used for the stability analysis of the relative regenerative vibration between a flexible plate and a flexible milling tool [9, 10]. The three-dimensional stability boundaries of a thin-walled plate fixed at the base and at one side were identified in the frequency domain in the works [11, 12] and used to highlight the productive stable cutting depths as a function of spindle speed and tool location. A nonlinear dependence of cutting force coefficients on axial depth of cut was considered in a frequency domain construction of the stability boundaries of the three-dimensional dynamics of thin-wall workpiece milling in [13]. Three-dimensional stability lobes were identified for the optimization of stable milling of flexible workpiece in [14], considering the in-process structural variation due to material removal and tool location. Recently, methods based on thin plate theory and mode superposition principle were used to prepare the regenerative dynamics of thin-walled workpiece milling for stability analysis in the frequency domain [15, 16].

The regenerative chatter stability of thin-walled workpiece milling was studied with the semi-discretization method in the work [17]. A semi-discrete-time-domain method was established for the stability analysis of flexible workpiece milling considering process damping and non-uniform pitch effects [18], and this method was recently applied in identifying the enhanced stability boundaries of milling of thin-walled workpiece attached with appropriate additional masses [19]. Using the FDM, the stability boundaries of thin-walled workpiece milling were constructed in the works [20, 21]. The first order semi-discretization was recently used in the stability analysis of milling of a reduced-order model of thin-walled workpiece [22]. In the work, parametric model order reduction, which was based on modal truncation of the location-dependent modal matrices and cubic spline interpolation of the individual reduced modes, was introduced for reducing the high DOF delayed dynamics of a flexible workpiece. This work presents a new framework for the FDM, based on polynomial tensor approximation of milling states, for the stability analysis of the reduced-order model of thin-walled workpiece milling.

2 The Method

Different models of milling process can be represented in the delayed first order form

$$\dot{\mathbf{x}}(t) = \mathbf{A}\mathbf{x}(t) + \mathbf{B}(t)\mathbf{x}(t) - \mathbf{B}(t)\mathbf{x}(t - \tau), \tag{1}$$

where $t > 0$ is the time, $\mathbf{x}(t) \in \mathbb{R}^n$ is the milling state, $\tau \in \mathbb{R}_+$ is the discrete delay, $\mathbf{A} \in \mathbb{R}^{n \times n}$ is the coefficient matrix related to the transient response of the system and $\mathbf{B}(t) \in \mathbb{R}^{n \times n}$ is the coefficient matrix related the periodic excitation of the system. The discrete delay τ is also the period of the model for the case of uniform-pitch milling. The period of the system is divided into k equal discrete time intervals $[t_i, t_{i+1}]$ where $i = 0, 1, 2, \ldots (k - 1)$ and $t_i = i\tau/k = i\Delta t = i(t_{i+1} - t_i)$. Equation (1) is solved in each discrete interval to give

$$\mathbf{x}_{i+1} = e^{\mathbf{A}\Delta t}\mathbf{x}_i + \int_{t_i}^{t_{i+1}} e^{\mathbf{A}(t_{i+1}-t)}(\mathbf{B}(t)\mathbf{x}(t) - \mathbf{B}(t)\mathbf{x}(t - \tau))dt. \tag{2}$$

Since the milling states do not have exact analytical form, the integration problem can only be solved if the states $\mathbf{x}(t)$ and $\mathbf{x}(t - \tau)$ are interpolated/approximated accurately enough. This is the underlying problem the time domain methods sought to solve. In what follows, the proposed method for handling this problem is presented.

The general second order least squares approximation of $\mathbf{x}(t)$, as given in [23], is

$$\mathbf{x}(t) = \mathbf{a}^T(t) \left\{ \sum_{l=i-1}^{i+1} \mathbf{a}(t_l)\mathbf{a}^T(t_l) \right\}^{-1} \sum_{l=i-1}^{i+1} \mathbf{a}(t_l)\mathbf{x}_l, \tag{3}$$

where $\mathbf{a}(t) = \{1 \ t \ t^2\}^T$ is the vector of polynomial basis. The summations in Eq. (3) should be understood to apply to every element of the matrix $\mathbf{a}(t_l)\mathbf{a}^T(t_l)$ and the vector $\mathbf{a}(t_l)\mathbf{x}_l$. Equation (3) can therefore be written as

$$\mathbf{x}(t) = \mathbf{a}^T(t)\mathbf{T}^{-1}\mathbf{S}\mathbf{v}, \tag{4}$$

where $\mathbf{T}$ and $\mathbf{S}$ are 3 by 3 numerical matrices and $\mathbf{v}$ is a vector of discrete milling states. The matrix $\mathbf{T}$ is described as $\mathbf{T} = \sum_{l=i-1}^{i+1} \mathbf{a}(t_l)\mathbf{a}^T(t_l)$. The elements of $\mathbf{T}$ then become $\mathbf{T}_{mn} = \sum_{l=i-1}^{i+1} [\mathbf{a}(t_l) \odot \mathbf{a}(t_l)]_{mn}$ where the operation $\mathbf{a}(t_l) \odot \mathbf{a}(t_l)$ implies outer product of the numeric vector $\mathbf{a}(t_l)$ with itself. Since $\{\mathbf{a}(t_l)\}_m = t_l^{m-1}$ then $[\mathbf{a}(t_l) \odot \mathbf{a}(t_l)]_{mn} = t_l^{m+n-2}$, thus

$$\mathbf{T}_{mn} = \sum_{l=i-1}^{i+1} t_l^{m+n-2} = (\Delta t)^{m+n-2} \sum_{l=-1}^{1} l^{m+n-2}. \tag{5}$$

The matrix $\mathbf{S}$ and vector $\mathbf{v}$ are derived as follows; comparing Eqs. (3) and (4) gives $\mathbf{Sv} = \sum_{l=i-1}^{i+1} \mathbf{a}(t_l)\mathbf{x}_l = \left\{ \sum_{l=i-1}^{i+1} \mathbf{x}_l \sum_{i-1}^{i+1} t_l\mathbf{x}_l \sum_{i-1}^{i+1} t_l^2\mathbf{x}_l \right\}^{\mathrm{T}}$ $= \{\mathbf{Sv}\}_m = \sum_{l=i-1}^{i+1} [\mathbf{a}(t_l) \otimes \mathbf{x}_l]_m$ where $\otimes$ is the tensor or Kronecker product operator. On decomposing this into coefficient matrix $\mathbf{S}$ and a vector of states $\mathbf{v}$, it becomes obvious that $\mathbf{S}$ is a Vandermonde matrix with elements

$$S_{mn} = t_{i-2+n}^{m-1} = (n-2)^{m-1}(\Delta t)^{m-1},\qquad(6)$$

and the $\mathbf{v}$ is made of sub-vectors of states

$$\mathbf{v}_m = \mathbf{x}_{i+m-2}.\qquad(7)$$

Making use of Eqs. (5), (6) and (7), the approximation polynomial becomes

$$\mathbf{x}(t) = \frac{1}{2(\Delta t)^2}\left(t^2 - t\Delta t\right)\mathbf{x}_{i-1} + \frac{1}{(\Delta t)^2}\left((\Delta t)^2 - t^2\right)\mathbf{x}_i + \frac{1}{2(\Delta t)^2}\left(t^2 + t\Delta t\right)\mathbf{x}_{i+1}.$$
$$(8)$$

Following a similar logic, the delayed state $\mathbf{x}(t - \tau)$ is also approximated as

$$\mathbf{x}(t - \tau) = \mathbf{a}^{\mathrm{T}}(t)\mathbf{T}_\tau^{-1}\mathbf{S}_\tau\mathbf{v}_\tau,\qquad(9)$$

where

$$T_{\tau,mn} = \sum_{l=i}^{i+2} t_l^{m+n-2} = (\Delta t)^{m+n-2}\sum_{l=0}^{2} l^{m+n-2},\qquad(10)$$

$$B_{\tau,mn} = t_{i-1+n}^{m-1} = (n-1)^{m-1}(\Delta t)^{m-1},\qquad(11)$$

$$\mathbf{v}_{\tau,m} = \mathbf{x}_{i+m-1-k}.\qquad(12)$$

Making use of Eqs. (10)–(12), the delayed approximation polynomial becomes

$$\mathbf{x}(t - \tau) = \frac{1}{2(\Delta t)^2}\left(2(\Delta t)^2 - 3t\Delta t + s^2\right)\mathbf{x}_{i-k} + \frac{1}{(\Delta t)^2}\left(2t\Delta t - t^2\right)\mathbf{x}_{i+1-k}$$
$$+ \frac{1}{2(\Delta t)^2}\left(-t\Delta t + t^2\right)\mathbf{x}_{i+2-k}.\qquad(13)$$

Linear approximation of the periodic coefficient matrix $\mathbf{B}(t)$ is adopted, thus

$$\mathbf{B}(t) = \frac{1}{\Delta t}(\Delta t - t)\mathbf{B}_i + \frac{1}{\Delta t}t\mathbf{B}_{i+1}.\qquad(14)$$

In what follows in this section, the monodromy matrix is constructed. Equations (8), (13) and (14) are inserted in Eq. (2) and the integration executed and rearranged to give

$$\mathbf{x}_{i+1} = \mathbf{P}_i(\mathbf{F}_0 + \mathbf{G}_{22}\mathbf{B}_i + \mathbf{G}_{25}\mathbf{B}_{i+1})\mathbf{x}_i + \mathbf{P}_i(\mathbf{G}_{21}\mathbf{B}_i + \mathbf{G}_{24}\mathbf{B}_{i+1})\mathbf{x}_{i-1}$$
$$- \mathbf{P}_i(\mathbf{D}_{21}\mathbf{B}_i + \mathbf{D}_{24}\mathbf{B}_{i+1})\mathbf{x}_{i-k} - \mathbf{P}_i(\mathbf{D}_{22}\mathbf{B}_i + \mathbf{D}_{25}\mathbf{B}_{i+1})\mathbf{x}_{i+1-k}$$
$$- \mathbf{P}_i(\mathbf{D}_{23}\mathbf{B}_i + \mathbf{D}_{26}\mathbf{B}_{i+1})\mathbf{x}_{i+2-k}, \tag{15}$$

where $\mathbf{G}_{21} = \frac{1}{2(\Delta t)^3}\left(-\mathbf{F}_2(\Delta t)^2 + 2\mathbf{F}_3\Delta t - \mathbf{F}_4\right)$, $\mathbf{G}_{22} = \frac{1}{(\Delta t)^3}\left(\mathbf{F}_1(\Delta t)^3 - \mathbf{F}_2(\Delta t)^2 - \mathbf{F}_3\Delta t + \mathbf{F}_4\right)$, $\mathbf{G}_{23} = \frac{1}{2(\Delta t)^3}\left(\mathbf{F}_2(\Delta t)^2 - \mathbf{F}_4\right)$, $\mathbf{G}_{24} = \frac{1}{2(\Delta t)^3}\left(-\mathbf{F}_3\Delta t + \mathbf{F}_4\right)$, $\mathbf{G}_{25} = \frac{1}{(\Delta t)^3}\left(\mathbf{F}_2(\Delta t)^2 - \mathbf{F}_4\right)$, $\mathbf{G}_{26} = \frac{1}{2(\Delta t)^3}(\mathbf{F}_3\Delta t + \mathbf{F}_4)$, $\mathbf{D}_{21} = \frac{1}{2(\Delta t)^3}\left(2(\Delta t)^3\mathbf{F}_1 - 5(\Delta t)^2\mathbf{F}_2 + 4\Delta t\mathbf{F}_3 - \mathbf{F}_4\right)$, $\mathbf{D}_{24} = \frac{1}{2(\Delta t)^3}\left(2(\Delta t)^2\mathbf{F}_2 - 3\Delta t\mathbf{F}_3 + \mathbf{F}_4\right)$, $\mathbf{D}_{22} = \frac{1}{(\Delta t)^3}\left(2(\Delta t)^2\mathbf{F}_2 - 3\Delta t\mathbf{F}_3 + \mathbf{F}_4\right)$, $\mathbf{D}_{25} = \frac{1}{(\Delta t)^3}(2\Delta t\mathbf{F}_3 - \mathbf{F}_4)$, $\mathbf{D}_{23} = \frac{1}{2(\Delta t)^3}\left(-(\Delta t)^2\mathbf{F}_2 + 2\Delta t\mathbf{F}_3 - \mathbf{F}_4\right)$, $\mathbf{D}_{26} = \frac{1}{2(\Delta t)^3}(-\Delta t\mathbf{F}_3 + \mathbf{F}_4)$, $\mathbf{F}_0 = e^{\mathbf{A}\Delta t}$, $\mathbf{F}_1 = (\mathbf{F}_0 - \mathbf{I})\mathbf{A}^{-1}$, $\mathbf{F}_2 = (\mathbf{F}_1 - \Delta t\mathbf{I})\mathbf{A}^{-1}$, $\mathbf{F}_3 = \left(2\mathbf{F}_2 - (\Delta t)^2\mathbf{I}\right)\mathbf{A}^{-1}$, $\mathbf{F}_4 = \left(3\mathbf{F}_3 - (\Delta t)^3\mathbf{I}\right)\mathbf{A}^{-1}$ and $\mathbf{P}_i = (\mathbf{I} - \mathbf{G}_{23}\mathbf{B}_i - \mathbf{G}_{26}\mathbf{B}_{i+1})^{-1}$. The local discrete map for second order approximation becomes

$$
\begin{Bmatrix} \mathbf{x}_{i+1} \\ \mathbf{x}_i \\ \mathbf{x}_{i-1} \\ \vdots \\ \mathbf{x}_{i+3-k} \\ \mathbf{x}_{i+2-k} \\ \mathbf{x}_{i+1-k} \end{Bmatrix}
=
\begin{bmatrix}
\mathbf{M}_{11}^i & \mathbf{M}_{12}^i & \cdots & 0 & \mathbf{M}_{1,k-1}^i & \mathbf{M}_{1k}^i & \mathbf{M}_{1,k+1}^i \\
\mathbf{I} & 0 & \cdots & 0 & 0 & 0 & 0 \\
0 & \mathbf{I} & \cdots & 0 & 0 & 0 & 0 \\
\vdots & \vdots & \vdots & \vdots & \vdots & \vdots & \vdots \\
0 & 0 & 0 & 0 & \mathbf{I} & 0 & 0 \\
0 & 0 & 0 & 0 & 0 & \mathbf{I} & 0
\end{bmatrix}
\begin{Bmatrix} \mathbf{x}_i \\ \mathbf{x}_{i-1} \\ \mathbf{x}_{i-2} \\ \vdots \\ \mathbf{x}_{i+2-k} \\ \mathbf{x}_{i+1-k} \\ \mathbf{x}_{i-k} \end{Bmatrix}, \tag{16}
$$

where $\mathbf{M}_{11}^i = \mathbf{P}_i(\mathbf{F}_0 + \mathbf{G}_{22}\mathbf{B}_i + \mathbf{G}_{25}\mathbf{B}_{i+1})$, $\mathbf{M}_{12}^i = \mathbf{P}_i(\mathbf{G}_{21}\mathbf{B}_i + \mathbf{G}_{24}\mathbf{B}_{i+1})$, $\mathbf{M}_{1,k-1}^i = -\mathbf{P}_i(\mathbf{D}_{23}\mathbf{B}_i + \mathbf{D}_{26}\mathbf{B}_{i+1})$, $\mathbf{M}_{1k}^i = -\mathbf{P}_i(\mathbf{D}_{22}\mathbf{B}_i + \mathbf{D}_{25}\mathbf{B}_{i+1})$ and $\mathbf{M}_{1,k+1}^i = -\mathbf{P}_i(\mathbf{D}_{21}\mathbf{B}_i + \mathbf{D}_{24}\mathbf{B}_{i+1})$. The monodromy matrix for the system becomes

$$\boldsymbol{\psi} = \mathbf{M}_{k-1}\mathbf{M}_{k-2}\ldots\mathbf{M}_0. \tag{17}$$

3 Applications

The method is equally applicable to the frequently adopted model of milling process which assumes a flexible tool cutting a rigid workpiece and to the model of milling process which assumes a rigid tool cutting a flexible workpiece. While application of the established method is demonstrated for both cases, the latter case is of major interest here since that is the case for milling of thin-walled aero-structures.

3.1 Flexible Tool-Rigid Workpiece

For this milling case, a popular second order 2 DOF delayed model for symmetric tool is

$$\mathbf{M}\ddot{\mathbf{z}}(t) + \mathbf{D}\dot{\mathbf{z}}(t) + \mathbf{K}\mathbf{z}(t) = \mathbf{H}(t)(\mathbf{z}(t) - \mathbf{z}(t - \tau)), \tag{18}$$

where $\mathbf{z}(t) = \{z_x(t)\ z_y(t)\}^{\mathrm{T}}$, $\mathbf{M} = \begin{bmatrix} m_x & 0 \\ 0 & m_y \end{bmatrix}$, $\mathbf{D} = \begin{bmatrix} d_x & 0 \\ 0 & d_y \end{bmatrix}$, $\mathbf{K} = \begin{bmatrix} k_x & 0 \\ 0 & k_y \end{bmatrix}$,

$\mathbf{H}(t) = -w \begin{bmatrix} h_{xx}(t) & h_{xy}(t) \\ h_{yx}(t) & h_{yy}(t) \end{bmatrix}$. The regenerative vibrations in the feed and feed-normal directions are $z_x(t)$ and $z_y(t)$. Table 1 gives the values and the relationships between the directional modal parameters in terms of the elements of the matrices $\mathbf{M}$, $\mathbf{D}$ and $\mathbf{K}$. The quantities subscripted $"x"$ relate to the feed direction while those subscripted $"y"$ relate to the feed-normal direction. The specific force variations are given as

$$h_{xx}(t) = C_t \gamma (v\tau)^{\gamma-1} \sum_{j=1}^{N} g_j(t) \sin^{\gamma} \theta_j(t) \big(\chi \sin \theta_j(t) + \cos \theta_j(t) \big), \tag{19}$$

$$h_{xy}(t) = C_t \gamma (v\tau)^{\gamma-1} \sum_{i=1}^{N} g_j(t) \sin^{\gamma-1} \theta_j(t) \cos \theta_j(t) \big(\chi \sin \theta_j(t) + \cos \theta_j(t) \big), \tag{20}$$

$$h_{yx}(t) = C_t \gamma (v\tau)^{\gamma-1} \sum_{j=1}^{N} g_j(t) \sin^{\gamma} \theta_j(t) \big(\chi \cos \theta_j(t) - \sin \theta_j(t) \big), \tag{21}$$

$$h_{yy}(t) = C_t \gamma (v\tau)^{\gamma-1} \sum_{j=1}^{N} g_j(t) \sin^{\gamma-1} \theta_j(t) \cos \theta_j(t) \big(\chi \cos \theta_j(t) - \sin \theta_j(t) \big). \tag{22}$$

Table 1 Parameters of the system [5]

Mass	$m_x = m_y$	0.03993 kg
Natural frequency	$\omega_{nx} = \sqrt{k_x/m_x} = \omega_{ny} = \sqrt{k_y/m_y}$	5793 rads^{-1}
Tool damping ratio	$\zeta_x = d_x/(2\omega_{nx}m_x) = \zeta_y = d_y/(2\omega_{ny}m_y)$	0.011
Tangential cutting coefficient	C_t	6×10^8 Nm$^{-1-\gamma}$
Normal to tangential force ratio	χ	0.33333
Force law feed exponent	γ	1
Number of teeth	N	2

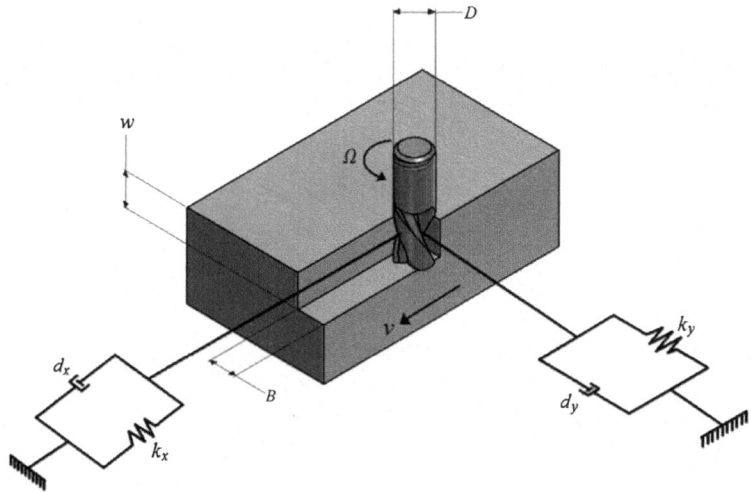

Fig. 1 The geometric and modal parameters of the flexible tool-rigid workpiece model

Equation (18) takes the first order form given in Eq. (1) where $\mathbf{A} = \begin{bmatrix} \mathbf{0} & \mathbf{I} \\ -\mathbf{M}^{-1}\mathbf{K} & -\mathbf{M}^{-1}\mathbf{D} \end{bmatrix}$ and $\mathbf{B}(t) = \begin{bmatrix} \mathbf{0} & \mathbf{0} \\ \mathbf{M}^{-1}\mathbf{H}(t) & \mathbf{0} \end{bmatrix}$. In the model, w is the depth of cut, $\theta_j(t) = \pi\Omega t/30 + 2\pi(j-1)/N$ is the angular displacement of the jth cutting edges (that is, $j = 1,2,...,N$), $g_j(t) = 0.5\big(1 + \mathrm{sgn}\big(\sin\big(\theta_j(t) - \tan^{-1}\mathcal{P}\big) - \sin\big(\theta_s - \tan^{-1}\mathcal{P}\big)\big)\big)$ is the screening function where $\mathcal{P} = (\sin\theta_s - \sin\theta_e)/(\cos\theta_s - \cos\theta_e)$, θ_s and θ_e are the start and end angles of the cutting interval, B is the radial depth of cut and D is the tool diameter. For up-milling, $\theta_s = 0$ and $\theta_e = \cos^{-1}(1 - 2\rho)$ while for down-milling, $\theta_s = \cos^{-1}(2\rho - 1)$ and $\theta_e = \pi$ where $\rho = B/D$ is the radial immersion. A illustration of this model showing the geometric and modal parameters is shown in Fig. 1.

3.1.1 Numerical Results and Discussions

On inserting the above model matrices in the constructed monodromy matrix and substituting the numerical values given in Table 1, the stability diagrams given in Fig. 2 are computed. The results agree with the known results in [4, 23].

3.2 Rigid Tool-Flexible Workpiece

This milling case is illustrated in Fig. 3 showing the primary motion (spindle rotation Ω) and the secondary motion (feed v) of the tool. The thin-walled workpiece is

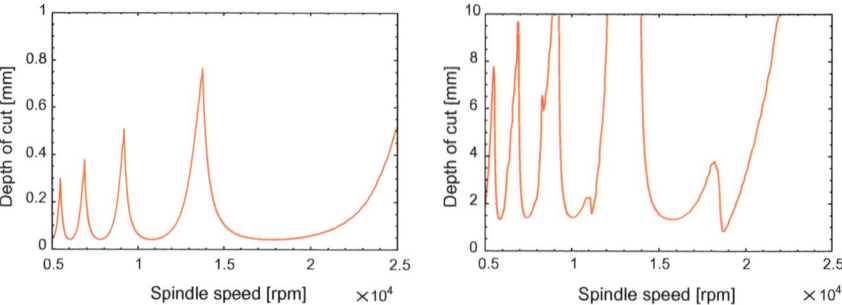

Fig. 2 The stability diagrams of the 2 DOF milling computed with the proposed method. (left) full radial immersion $\rho = 1$ and (right) low radial immersion $\rho = 0.05$

Fig. 3 An illustrative rigid tool-flexible workpiece milling

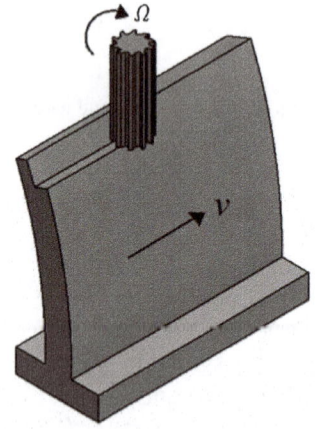

much more flexible than the tool and it is thus considered compliant while the tool is considered rigid.

3.2.1 Finite Element Model

Since the workpiece is a continuum, a large DOF second order delayed model is needed to capture the regenerative dynamics. The regenerative dynamics of the thin-walled workpiece is, therefore, subjected to Finite Element (FE) modelling to give a d_F-dimensional model for every location of the tool along the feed direction. The model for a tool-location, as specifically presented in [22], can be given as

$$\mathbf{M}\ddot{\mathbf{z}}(t) + \mathbf{D}\dot{\mathbf{z}}(t) + \mathbf{K}\mathbf{z}(t) = \mathbf{B}_H\mathbf{H}(t)\mathbf{C}_H(\mathbf{z}(t) - \mathbf{z}(t - \tau)), \qquad (23)$$

where $\mathbf{z}(t) \in \mathbb{R}^{d_F}$ is the state, $\mathbf{M} \in \mathbb{R}^{d_F \times d_F}$ is the mass matrix, $\mathbf{D} \in \mathbb{R}^{d_F \times d_F}$ damping matrix, $\mathbf{K} \in \mathbb{R}^{d_F \times d_F}$ is the stiffness matrix, $\mathbf{B}_H \in \mathbb{R}^{d_F \times n_d}$, $\mathbf{H}(t) \in \mathbb{R}^{n_d \times n_d}$, $\mathbf{C}_H \in$

$\mathbb{R}^{n_d \times d_F}$ and n_d is the number of directions of regenerative response. The matrices $\mathbf{B}_H$ and $\mathbf{C}_H$ serve to project the effect of the n_d-dimensional specific force variation matrix $\mathbf{H}(t)$ to the high DOF d_F of the thin-walled workpiece. If it is assumed that the axial component of cutting force is relatively low, which is a realistic assumption when helix angle is ignored, the regenerative responses in the feed and feed-normal directions are then considered so that $n_d = 2$. The FE model that is adopted from [22] has $d_F = 30888$ DOF.

3.2.2 Model Order Reduction

Since the degree of freedom d_F of the FE model is large, the computational process will be costly or impossible. Model order reduction is necessary to overcome this problem. Based on modal truncation, a parametric model order reduction of Eq. (23) gives [22]

$$\tilde{\mathbf{M}}(p)\ddot{\tilde{\mathbf{z}}}(t) + \tilde{\mathbf{D}}(p)\dot{\tilde{\mathbf{z}}}(t) + \tilde{\mathbf{K}}(p)\tilde{\mathbf{z}}(t) = \tilde{\mathbf{B}}_H(p)\mathbf{H}(t)\tilde{\mathbf{C}}_H(p)(\tilde{\mathbf{z}}(t) - \tilde{\mathbf{z}}(t - \tau)), \quad (24)$$

where $\tilde{\mathbf{z}}(t) \in \mathbb{R}^{d_R}$ is the reduced state, $\tilde{\mathbf{M}}(p) = \mathbf{V}^T\mathbf{M}\mathbf{V} \in \mathbb{R}^{d_R \times d_R}$ is the reduced mass matrix, $\tilde{\mathbf{D}}(p) = \mathbf{V}^T\mathbf{D}\mathbf{V} \in \mathbb{R}^{d_R \times d_R}$ is the reduced damping matrix, $\tilde{\mathbf{K}}(p) = \mathbf{V}^T\mathbf{K}\mathbf{V} \in \mathbb{R}^{d_R \times d_R}$ is the reduced stiffness matrix, $\tilde{\mathbf{B}}_H(p) = \mathbf{V}^T\mathbf{B}_H \in \mathbb{R}^{d_R \times n_d}$, $\tilde{\mathbf{C}}_H(p) = \mathbf{C}_H\mathbf{V} \in \mathbb{R}^{n_d \times d_R}$ and $\mathbf{V} \in \mathbb{R}^{d_F \times d_R}$ is the projection matrix. The matrix $\mathbf{V}$ is a concatenation of the first d_R eigenvectors where $d_R \ll d_F$. The parameter p here indicates the location-dependent dynamics along the feed direction where $p = 0$ at the beginning of tool pass and $p = 1$ at the end of the pass. Details can be found in [22] where each p-dependent matrix derived from cubic spline interpolation of the values at a finite number of location zones under a consistent system of coordinates $\tilde{\mathbf{z}}(t)$. The first order form of the reduced model is still Eq. (1) but with $\mathbf{A} = \begin{bmatrix} \mathbf{0} & \mathbf{I} \\ -\left(\tilde{\mathbf{M}}(p)\right)^{-1}\tilde{\mathbf{K}}(p) & -\left(\tilde{\mathbf{M}}(p)\right)^{-1}\tilde{\mathbf{D}}(p) \end{bmatrix}$, and $\mathbf{B}(t) = \begin{bmatrix} \mathbf{0} & \mathbf{0} \\ \left(\tilde{\mathbf{M}}(p)\right)^{-1}\tilde{\mathbf{B}}(p)\mathbf{H}(t)\tilde{\mathbf{C}}(p) & \mathbf{0} \end{bmatrix}$. This means that the proposed method is amenable to the reduced-order models of milling with varying workpiece dynamics.

3.2.3 Numerical Results and Discussions

The parameters of the milling process and the workpiece are adopted from [22]. The radial immersion of the up-milling process is 0.25. The milling tool has two teeth ($N = 2$), the feed per tooth is 0.1 mm, the tangential cutting coefficient is $C_t = 1.07 \times 10^8$ Nm$^{-1-\gamma}$, the force law feed exponent is $\gamma = 0.75$ and the normal to tangential force ratio is $\chi = 40/107$. The 0.17 m $\times$ 0.03 m $\times$ 0.15 m workpiece is a steel material with density $\rho = 7.8 \times 10^3$ kgm^{-3}, Young's modulus $E = 210 \times 10^9$ Nm^{-2}

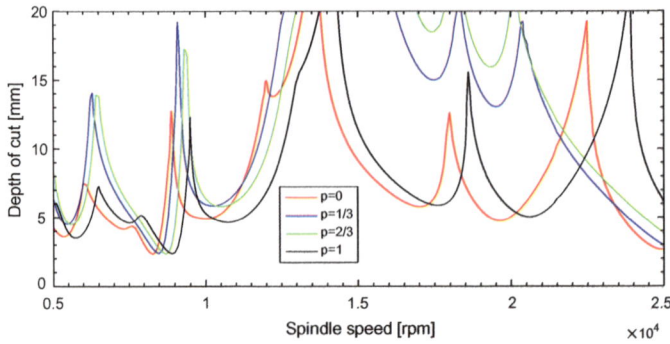

Fig. 4 The stability boundaries of milling the thin-walled workpiece at four locations

and Poisson ratio $v_{pr} = 0.3$. On inserting the reduced matrices in the constructed monodromy matrix and substituting the numerical values, the stability boundaries of the system are constructed using three eigenmodes, see Fig. 4. The results agree with the known results in [22]. The stability boundaries show the influence of both tool location and the extent of material removal on the stability lobes. The stability boundary at the location $p = 1$ differed from that at the location $p = 0$ because of material that had been removed. The effect of material removal on the stability boundary of cutting of thin-walled workpiece can be viewed as time-dependent. For the studied workpiece, the effect of location on system stability threshold is most felt in the higher speed domain. The implication of the position and time-dependence for the industrial application of using fixed spindle speed for milling is that any choice of cutting process parameters must satisfy any possible space-time existence of the stability boundary. More discussion of these stability boundaries can be found in [22].

3.2.4 Future Work

Since the algorithm works well, it would be germane to point out the possibility of generalizing the method with the aim of maximizing accuracy through identifying the most accurate approximation order p. This approach of accuracy maximization is especially important since it will serve as a way of compensating for any error introduced by modal truncation of the full order model. It is found that when polynomial tensor of degree p is used then $\mathbf{T} \in \mathbb{R}^{(p+1)\times(p+1)}$, $\mathbf{S} \in \mathbb{R}^{(p+1)\times(p+1)}$, $\mathbf{T}_\tau \in \mathbb{R}^{(p+1)\times(p+1)}$, $\mathbf{S}_\tau \in \mathbb{R}^{(p+1)\times(p+1)}$, $\mathbf{x}(t) \in \mathbb{R}^{d_R n_d}$, $\mathbf{v} \in \mathbb{R}^{d_R n_d(p+1)}$ and $\mathbf{v}_\tau \in \mathbb{R}^{d_R n_d(p+1)}$. The generalized tensor elements and sub-vectors found to be specific for the milling states are $\mathrm{T}_{mn} = (\Delta t)^{m+n-2} \sum_{l=-p+1}^{1} l^{m+n-2}$, $\mathrm{S}_{mn} = (n - p)^{m-1} (\Delta t)^{m-1}$,

$$\mathbf{v}_m = \mathbf{x}_{i+m-p}, \quad \mathrm{T}_{\tau,mn} = (\Delta t)^{m+n-2} \sum_{l=0}^{p} l^{m+n-2}, \quad \mathrm{S}_{\tau,mn} = (n - 1)^{m-1} (\Delta t)^{m-1},$$

$\mathbf{v}_{\tau,m} = \mathbf{x}_{i+m-1-k}$. The future direction of this research would be to construct the monodromy matrix of the reduced thin-walled workpiece with the generalization and, somewhat, compensate for any truncation inaccuracy by identifying and using the most accurate order of approximation.

4 Conclusions

A new FDM was proposed for analytical stability identification of the regenerative milling of thin-walled workpiece. The method is based on a second order polynomial tensor approximation of the milling states. The stability boundaries of a parametric FE model of the studied thin-walled workpiece, which has $d_F = 30888$ DOF but reduced to 3 DOF via modal truncation, were successfully constructed with the new method. The revealed effects of tool location and time (the extent of material removal) agrees qualitatively and quantitatively with the known results in [22]. A foundation has been formed in this work for future research into minimizing computational error, and thus compensating for the effects of model order reduction on computational precision, by generalizing the method so that identifying the most accurate approximation order becomes possible.

Acknowledgements The described research was partially done while I visited the ITM at the University of Stuttgart in the year 2018. This stay was funded by the Priority Program SPP 1897 'Calm, Smooth, Smart' of the DFG (German Research Foundation). This support is highly appreciated. I would like to acknowledge Dominik Hamann for making available data for testing the proposed algorithm and engaging in helpful discussions.

References

1. Sridhar, R., Hohn, R.E., Long, G.W.: A stability algorithm for the general milling process: Contribution to machine tool chatter research. Trans. ASME J. Eng. Ind. **90**(2), 330–334 (1968)
2. Minis, I., Yanushevsky, R., Tembo, A., Hocken, R.: Analysis of linear and nonlinear chatter in milling. CIRP Ann. - Manuf. Technol. **39**(1), 459–462 (1990)
3. Altintaş, Y., Budak, E.: Analytical prediction of stability lobes in milling. CIRP Ann. - Manuf. Technol. **44**(1), 357–362 (1995)
4. Insperger, T., Stépán, G.: Updated semi-discretization method for periodic delay-differential equations with discrete delay. Int. J. Numer. Methods Eng. **61**(1), 117–141 (2004)
5. Bayly, P.V., Mann, B.P., Peters, D.A., Schmitz, T.L., Stepan, G., Insperger, T.: Effects of radial immersion and cutting direction on chatter instability in end-milling. In: ASME International Mechanical Engineering Congress and Exposition, pp. 1–13 (2002)
6. Butcher, E.A., Nindujarla, P., Bueler, E.: Stability of up- and down-milling using chebyshev collocation method. In: 5th International Conference on Multibody Systems, Nonlinear Dynamics, and Control, Parts A, B, and C, vol. 6, pp. 841–850 (2005)
7. Ding, Y., Zhu, L.M., Zhang, X.J., Ding, H.: A full-discretization method for prediction of milling stability. Int. J. Mach. Tools Manuf **50**(5), 502–509 (2010)
8. Ozoegwu, C.G.: Least squares approximated stability boundaries of milling process. Int. J. Mach. Tools Manuf **79**, 24–30 (2014)
9. Bravo, U., Altuzarra, O., López De Lacalle, L.N., Sánchez, J.A., Campa, F.J.: Stability limits of milling considering the flexibility of the workpiece and the machine. Int. J. Mach. Tools Manuf **45**(15), 1669–1680 (2005)

10. Herranz, S. Campa, F.J., De Lacalle, L.N.L., Rivero, A., Lamikiz, A., Ukar, E., Sánchez, J.A., Bravo, U.: The milling of airframe components with low rigidity: A general approach to avoid static and dynamic problems. Proc. Inst. Mech. Eng. Part B J. Eng. Manuf. **219**(11), 789–801 (2005)

11. Thevenot, V., Arnaud, L., Dessein, G., Cazenave-Larroche, G.: Integration of dynamic behaviour variations in the stability lobes method: 3D lobes construction and application to thin-walled structure milling. Int. J. Adv. Manuf. Technol. **27**(7–8), 638–644 (2006)

12. Thevenot, V., Arnaud, L., Dessein, G., Cazenave-Larroche, G.: Influence of material removal on the dynamic behavior of thin-walled structures in peripheral milling. Mach. Sci. Technol. **10**(3), 275–287 (2006)

13. Adetoro, O.B., Sim, W.M., Wen, P.H.: An improved prediction of stability lobes using nonlinear thin wall dynamics. J. Mater. Process. Technol. **210**(6–7), 969–979 (2010)

14. Budak, E., Tunç, L.T., Alan, S., Özgüven, H.N.: Prediction of workpiece dynamics and its effects on chatter stability in milling. CIRP Ann. - Manuf. Technol. **61**(1), 339–342 (2012)

15. Song, Q., Liu, Z., Wan, Y., Ju, G., Shi, J.: Application of Sherman-Morrison-Woodbury formulas in instantaneous dynamic of peripheral milling for thin-walled component. Int. J. Mech. Sci. **96–97**, 79–90 (2015)

16. Shi, J., Song, Q., Liu, Z., Ai, X.: A novel stability prediction approach for thin-walled component milling considering material removing process. Chinese J. Aeronaut. **30**(5), 1789–1798 (2017)

17. Song, Q., Ai, X., Tang, W.: Prediction of simultaneous dynamic stability limit of time-variable parameters system in thin-walled workpiece high-speed milling processes. Int. J. Adv. Manuf. Technol. **55**(9–12), 883–889 (2011)

18. Eksioglu, C., Kilic, Z.M., Altintas, Y.: Discrete-time prediction of chatter stability, cutting forces, and surface location errors in flexible milling systems. J. Manuf. Sci. Eng. **134**(6), 1–13 (2012)

19. Wan, M., Bin Dang, X., Zhang, W.H., Yang, Y.: Optimization and improvement of stable processing condition by attaching additional masses for milling of thin-walled workpiece. Mech. Syst. Signal Process. **103**, 196–215 (2018)

20. Li, Z., Sun, Y., Guo, D.: Chatter prediction utilizing stability lobes with process damping in finish milling of titanium alloy thin-walled workpiece. Int. J. Adv. Manuf. Technol. **89**(9–12), 2663–2674 (2017)

21. Zhang, Z., Li, H., Liu, X., Zhang, W., Meng, G.: Chatter mitigation for the milling of thin-walled workpiece. Int. J. Mech. Sci. **138–139**, 262–271 (2018)

22. Hamann, D., Eberhard, P.: Stability analysis of milling processes with varying workpiece dynamics. Multibody Syst. Dyn. **42**(4), 383–396 (2018)

23. Ozoegwu, C.G., Omenyi, S.N., Ofochebe, S.M.: Hyper-third order full-discretization methods in milling stability prediction. Int. J. Mach. Tools Manuf **92**, 1–9 (2015)

Index

© Springer Nature Switzerland AG 2020
J. Fehr and B. Haasdonk (eds.), *IUTAM Symposium on Model Order Reduction
of Coupled Systems, Stuttgart, Germany, May 22–25, 2018*, IUTAM Bookseries 36,
https://doi.org/10.1007/978-3-030-21013-7

Printed by Printforce, the Netherlands